# THE
# GLOBAL
# BANKERS

# THE GLOBAL BANKERS

## ROY C. SMITH

T·T

TRUMAN TALLEY BOOKS

E. P. DUTTON

NEW YORK

Published in the United States by Truman Talley Books · E. P. Dutton,
a division of Penguin Books USA Inc.,
2 Park Avenue, New York, N.Y. 10016.

Published simultaneously in Canada
by Fitzhenry and Whiteside, Limited, Toronto.

Library of Congress Cataloging-in-Publication Data

Smith, Roy C., 1938–
The global bankers : a top investment banker explores the new
world of international deal-making and finance / Roy C.
Smith.—1st ed.
p. cm.
Bibliography: p.
Includes index.
ISBN 0-525-24797-1
1. Banks and banking, International.   I. Title.
HG3881.S543   1989
332.1'5—dc19                        89-4724
                                       CIP

Designed by REM Studio

1 3 5 7 9 10 8 6 4 2

First Edition

*To my loving wife, Marianne,
and to my colleagues at Goldman, Sachs & Co.,
without whom this book could not have been.*

# Contents

# Part III The Floating World of Japanese Finance

# Part IV Looking to the Millennium

# THE
# GLOBAL
# BANKERS

# Introduction

"**M**oney talks," they say. Accumulated money, or wealth, assures comfort, security, influence, and importance—all highly valued attainments in modern society. Throughout history the people of this planet have arranged their lives in accordance with the distribution of wealth and have sought opportunities to possess it. The pursuit of wealth, by nations, by corporations, and by individuals, has been one of life's few constants.

The many ways to accumulate wealth have been known for a long time. They have been applied in many different sets of circumstances, again and again over centuries. Princes have accumulated wealth by conquest and plunder, by exploration and discovery, by taxation and borrowing, and by expropriation, often without compensation. The aristocracy became wealthy by owning land, which often was received in the first place as a reward for some service to royalty. Merchants and manufacturers succeeded by enterprise and cunning,

1

professionals by the sale of their skills, peasants by escaping their lot and becoming something else. As the wheel of fortune turned, however, some came up while others went down. Mostly such fates are described in terms of the ups and downs of people's wealth, the most measurable, if not the most important, of human achievements.

Wealth, like a fluid, must be contained in a vessel in order to have the benefit of it. If not contained, it slips away and is no longer one's own. The principal vessels of wealth are banks. To be used, wealth must be transferable (or *liquid* in banking parlance). To be preserved, wealth must be protected, from theft and from mismanagement. All wealth, whether of governments, corporations, charitable institutions, or individuals, depends on banks to perform the various functions associated with its preservation and enhancement.

Banks descend to us from medieval times, when they functioned as pawnbrokers, goldsmiths, and dealers in foreign exchange and commodities. They became countinghouses. In time, people could borrow spendable money against possessions. Or they could acquire instruments of wealth, such as land or gold, by investing their excess cash savings. Or they could lend money to a bank in exchange for an agreed rate of interest; the bank would relend the money to others at a higher rate of interest, keeping the difference as its profit. Banks could also transfer money from city to city easily and safely and arrange for clients to make investments in other assets such as securities.

Such skills as banks possess and the many services they offer are essential to the wealthy and, of course, to those individuals too who also use banks but who can only be regarded as wealthy when grouped together with others like themselves, that is, the great retail public. Yet if a bank makes even one of the many serious mistakes that it is possible for it to make, or is dishonest, its clients' money can be lost. This used to happen all the time in the United States, with infuriating regularity. Confidence in banks has not always been high. Clients, having gone to great trouble to build up and preserve their fortunes, want to be sure, first of all, that the chosen vessel of containment is safe, well-managed, and leakproof. There are many different kinds of banks attempting to provide various financial services to their clients. How these services have changed and been

2

influenced by powerful international economic and financial forces (which is what this book is all about) can be introduced by the following story.

Some years ago a very canny owner of a successful company was being pressed by one of the company's advisers to accept what appeared to be a very generous offer to be acquired by another company.

"George," said the adviser, "you've really got to do this deal; it's like money in the bank!"

"Yeah," George replied after a few minutes, "but what bank?"

What bank has always made a difference, especially in the days before 1933 in the United States when there was no Federal Deposit Insurance Corporation to offer a U.S. government guaranty of one's deposit in a bank. Certainly before 1933, George would have had to think twice about selling a solid and profitable business and putting the money on deposit with a bank that might collapse in the next business cycle or be robbed by Bonnie and Clyde.

Even today, George's deposit is insured only up to $100,000; if he sold the company he put his whole life's work into, he would get a lot more than that for it. He could spread it around so no single bank would have more than $100,000, but that's a nuisance. Besides, banks don't always pay as much for deposits—even those that are pledged not to be withdrawn for a while—as George could earn by investing the money in something else. George wasn't so sure he needed a bank, except for little things.

In the end, George decided to sell his company and to invest $100,000 in a bank deposit, which he could draw on to pay his bills or take a trip to Europe, or whatever, and to have the bank invest the rest of the money for him in securities. The bank that attracted George's business in the first place, because of its size, reputation, and variety of financial services, was happy to oblige and, for a fee, agreed to manage a portfolio of investments for George. These included very secure but low-yielding U.S. Treasury bonds, some tax-free bonds issued by the State of New York, and some common stocks in companies George knew something about.

The bank was itself a dealer in Treasury bonds and the State of

New York tax-free bonds, but it was not permitted under the terms of the Banking Act of 1933 (the so-called Glass-Steagall Act) to underwrite stocks and bonds of corporations or to make trading markets in them. The bank had to purchase the stocks for George's portfolio from a stockbroker; those stockbrokers that also originate and underwrite new issues of corporate stocks and bonds are called investment bankers, that is, bankers who buy and sell investments.

George was familiar with investment bankers because he had used one to help him find the highest-price buyer for his company when he sold it. The investment banker had studied George's company and had estimated its value to other companies based on what the stock market would pay for shares of the company at the time plus a premium over that price that an acquirer would pay in order to have control of the company. The investment banker had also introduced a dozen or so prospective buyers to George, some of whom he had never heard of. When the deal was struck, George was satisfied that he had secured the best price possible for his company, largely through the process that the investment banker had introduced and managed.

After a while, the investment banker that George had used came to call on him, suggesting that George take some of the money he had received from the sale of his company and invest it in a portfolio of common stocks recommended by the firm's research department. One of the firm's experienced stockbrokers would be at George's disposal to make suggestions and to effect purchases and sales for him directly through the firm's membership on the New York Stock Exchange. George decided to give it a try and asked his bank to transfer some of the money in his investment account with the bank to a brokerage account with the investment bank.

Over the next few years, George bought and sold a number of securities through the firm, including some high-risk but, he hoped, high-reward venture-capital investments. He now had his money spread over a number of different risk categories, with the lowest risk, more secure investments paying the least, and the higher risk securities paying the most. George had decided to keep his eyes open for good investment opportunities wherever they came from and accordingly had opened an account with a Japanese investment banker who had access to some interesting investment ideas in Japan. If you can't beat 'em, join 'em, thought George.

Meanwhile, George was getting itchy in retirement. He liked being wealthy, but he didn't like being idle. He decided to buy another business, but to do so he would need to borrow a substantial sum and he would need additional bank facilities to increase the company's production and inventory. His investment banker helped him both to identify a company to buy and to determine and negotiate a fair price.

When he went to his commercial bank to arrange the financing for the transaction, he was surprised when his bank told him that, although they were prepared to be very competitive in offering financing for his purchase of the business, they were not likely to be the most competitive suppliers of low-cost working-capital funds for the company itself. The bank explained that a number of European and Japanese banks were now operating in the United States and were zeroing in on this type of business and offering lower interest rates than most American banks then were. George discovered this to be true. Indeed, after talking to some of the foreign bankers, he was surprised to find so many possible ways to obtain financing from international sources for the new company's business.

George was particularly interested in the services of a major Swiss bank, which explained to him that in Europe banks were allowed to be in both the commercial and the investment-banking businesses. Such banks were called *universal banks* because they offered both of these basic financial services under one roof. As the company continues to expand around the world, George was told, he would want to talk with this bank about lines of credit or bond issues in the Eurobond market to raise financing, or perhaps to open a private numbered account for himself and his family in Zürich or Geneva.

George's acquisition was a success and was growing nicely. It now needed additional financing for working capital and new equipment. He found himself being called almost daily by all sorts of different banks and investment banks who had either ideas or information for him. George thought they all marketed their services very effectively, but it was too difficult to keep all the market opportunities straight and to evaluate one idea against another. So George instructed the company's chief financial officer to set up a capital-markets desk staffed by the company's brightest young people to handle all of the calls from bankers and to "put us in a position to take advantage of all the attractive financing possibilities around the world."

George, himself, had to concentrate on running the business day to day and on getting its overall competitive strategy right. Competition in the United States from foreign companies was very tough, though it had improved a bit lately because the dollar had weakened so much against the currencies of the major industrial countries. George was tempted to make a big push into international business himself, but he wondered how he could protect himself against foreign currency losses that seemed unavoidable. He seemed to recall, however, one of his recent visitors mentioning new methods of hedging against foreign-exchange risk through the use of something called financial futures and options.

While considering this, he had a call from London, from a British merchant bank with a good idea about a company George could buy in Hong Kong that manufactured products similar to George's. This company needed a partner in order to expand into low-cost production in Taiwan and Korea. George agreed to meet with the firm's representative to discuss the idea further. In passing, he asked the caller why the firm was called a *merchant bank*?

The representative replied, "We are a firm originally made up of merchants. We needed to finance our customers and suppliers around the world and one day found that we were doing rather less commerce than finance, which in any event was less risky and more remunerative at the time. So the firm became bankers, specializing in being bankers to merchants. We financed the tobacco trade in the United States before your independence and the cotton trade up until the Civil War. We were everywhere the British empire was, and indeed still are for the most part.

"As our merchant clients grew, and became more active in America, so did we. We financed your railroads, canals, steel mills, and oil wells; now everybody does it. We still manage a great deal of money for British pension funds and for private clients that is invested in America, but also much that is invested in Asia, Australia, and the rest of Europe. We like to think we make our living with our wits and our connections, just as the Medicis did in the fifteenth century and many others have done since. In France our equivalent numbers are called *banques d'affaires,* and privately owned banks in Holland, Germany, Austria, and the rest of Europe do much the same as we, though perhaps on a smaller scale.

"Today, we are more similar in size and activity to your investment banks than we are to commercial banks, which, again by size and activity, are more closely analogous to *clearing banks* in the U.K.''

George thought all of this was very interesting. He had not known of the history. He began to realize how much financial matters had become globalized, even for those with relatively simple requirements like himself. There were all sorts of banks, from many different countries, that were offering services to him. Often, if not most often, these services overlapped. Not only had all of their services converged, but their marketing efforts had converged too, on him and his business. Globalized finance was apparently no longer just for the Fortune 500 largest companies.

For example, he thought, my old bank manages some of my money, as do now two investment banks, one of which has put me into a portfolio of Japanese stocks that has done very well; the other has provided me with venture-capital ideas. The Swiss bank has offered financing in the United States and in the Eurobond market and a private investment account for me in Zürich, and a British merchant bank can invest the company's pension-fund assets in growth companies overseas.

Our capital-markets group is now evaluating corporate-financing ideas in the United States, in the Eurobond market, and even in Japan, where they say money is flowing out like mad. Our working capital is provided by European banks, and now I'm being shown an acquisition idea in the Far East by the merchant bank that has been doing this kind of thing since before the Revolutionary War. I guess all of this has been going on for a while, but now it seems as if it's being all brought together so I can analyze what I like and don't like, as if it were all one big financial menu. That's a big change from the

7

time I was running my last company just a few years ago. The financial world must be getting pretty small if all this stuff is finding its way to me, but I'm mighty glad it has.

George's reaction is typical of those who rely on banks and financial markets for services that will preserve and enhance their wealth. Whether they are seekers of low-cost finance, such as corporations or governments, or investors looking for the good opportunities, or both, as in George's case, the globalization of banking has greatly extended the range of opportunities available to users of these services. This in turn has created many growth and profit opportunities for banks, thus releasing ferocious competitive energy and attracting many new resources to the business.

Looking back, it would be very difficult to think of anything that has grown as much over the past two decades as the number of individual transactions in financial assets and instruments that have occurred all over the world. Not semiconductors, not wonder drugs, not armaments—probably nothing has grown organically so fast. Twenty years ago, such transactions would have been limited to bank loans, the issuance of stocks and bonds, and brokerage transactions. Today the range of activities and services is vast.

This era of change and development in finance is the consequence of many remarkable events, beginning perhaps with the buildup of increasingly large balance-of-payments deficits by the United States as other countries recovered from the economic devastation of World War II. After a decade of fighting the problem, the United States realized it could not stem the tide of its payments deficits alone and unilaterally discontinued the sale of U.S. gold reserves to foreign holders of dollars. Soon afterward, the fixed-rate foreign-exchange system that was installed at the end of World War II was abandoned and a floating-rate system replaced it. The consequences of this change were enormous. They led directly to a period of deregulation of capital and other controls around the world, ultimately freeing up financial resources to be invested wherever the opportunities were attractive.

The oil-price shock in 1973 was the next great adventure to be experienced. This placed enormous strains on foreign-exchange rates, which fortunately had the floating-rate system to work with at the time. Great pressure was also placed on price levels in all oil-importing countries, and inflation became rampant. To cope with it, the U.S. Federal Reserve effected in 1979 a radical turnaround in the way it managed monetary policy in the United States. No longer would the supply of money be controlled indirectly, by the fixing of interest rates—now the reverse would apply. Money supply would be controlled directly, and interest rates could settle where they would. The result was a period of extremely high interest rates and exceptionally high volatility in the markets for financial instruments and foreign exchange.

For those whose business was the management of money, these conditions produced further changes. To survive in the increasingly performance-oriented business of managing investments for pension funds and other institutions, managers had to become traders, not just holders, of securities. Many new ideas, products, and approaches to the art of investment management appeared, along with many more competitors and more capital. Telecommunications technology was there when it was needed. The financial boom of the 1980s followed.

A huge increase in outstanding U.S. government securities also resulted from the continuing budget deficits, bringing with it an exponential increase in the trading of the securities in the market. Economic expansion and lowering inflation and interest rates and confidence drove stock markets to record levels, not just in the United States, but all over the world. Market participation was truly international. Funds from Europe, Japan, and the Middle East flowed into the United States; U.S. investors also sought out promising investments abroad.

The Japanese market, for example, fueled by extraordinary financial surpluses, exceeded both the volume and price levels of the New York Stock Exchange. In some other, more exotic markets such as Spain, Italy, Korea, and Mexico, transaction activity came to be enormous by the standards of those countries.

We also saw during the 1980s the most active period ever in mergers, acquisitions, and takeovers, and in methods of financing them involving very high-yield debt securities, called *junk bonds,* and the conversion of house mortgages into marketable securities (called *collateralized mortgage obligations,* or *CMOs*), which became a giant new financial industry almost overnight.

All in all, it was an incredible decade for the world of finance, a rocket ride into a new financial environment—a global environment.

It has been, and continues to be, an exciting time for bankers of all types. Their business takes place with counterparties from all over the world. It is very quick-changing and opportunistic, though the risks too have increased. It can be very profitable, when one is right, or first or lucky, and, on other occasions, it can be disastrous. The business has attracted some of the best and brightest talent ever deployed in finance, drawn from universities, trading rooms, and corporations all over the world.

Beginning in the early 1980s, bankers all over the world went into an extremely active period of international expansion. The world of finance was booming in America, in Europe, and in Asia, and one had to be represented significantly in these areas to take advantage of the new global opportunities. This frenetic expansion would be slowed, but not halted, by the crashing of the world's principal stock exchanges on October 19, 1987, and the damage-control efforts that followed. The crash was certainly severe. Markets collapsed all at once in New York, London, Frankfurt, Zürich, Sydney, Tokyo, and Hong Kong, and in all other countries where free markets existed. Caught in the crash was the largest international stock issue ever done in the Western world, a $12-billion distribution of British Petroleum shares being sold by the British government through investment bankers in the United States, Canada, Japan, and Europe.

Losses attributable to the crash were extremely heavy, particularly if one appraises them on an aggregate basis for all the world's markets. Yet no major bank or investment bank went out of business as a result of the crash, and no major contracts were defaulted on, including the British Petroleum underwriting agreement. No major stock exchange closed. The markets for new issues of common stock and for mergers and acquisitions were derailed for a time. Yet by the end of the year, all of the affected stock markets had recovered much of their losses—indeed, Tokyo had resumed its climb to record price levels—and the merger market was on the brink of its most active period ever, which occurred in the first quarter of 1988. By the end of June, foreigners had resumed buying high-priced Japanese shares, the Japanese were buying U.S. securities, the dollar had stabilized on a plateau well above its lows, and things were almost back to normal.

Almost maybe, but not quite back to normal. Fears continued that it could all happen again. The confidence of the retail investor was not back to normal; neither was that of many market observers or the congressional watchdogs who were still trying to figure out what to do about what had gone wrong. Certainly the stock prices of the major banks and investment banks had not returned to precrash levels, the former having continuing difficulties with Third World loans, the latter suffering postcrash earnings drag and a Typhoid Mary public image.

At the end of 1988, many banking organizations were still suffering from the rigorous changes they had begun before the crash to cut expenses and improve competitive performances. Many banking-industry employees were laid off, and senior management positions changed in New York, London, and Tokyo. Several top-level changes took place at Salomon Brothers, for example, and at Chase Manhattan Bank the commercial banking "faction" displaced several senior investment bankers in a move known at the bank as the "revenge of the nerds." All of these changes are typical of events that occur in a highly competitive marketplace that is undergoing adjustment. What goes up, comes down; though usually it goes back up again after a time.

The fundamental characteristics of the new global financial market that contributed so much to the expansion of activity—deregulation, institutionalization of investments, and heavy application of telecommunications technology—are not going to change because of the comparatively mild bottom-line effects of the crash of 1987, even if there is another crash. Today's markets are resilient, having pooled vast sums of money into a single global market that is accessible all over the world.

This book is about global bankers and global banking today as it is carried out from its three continental centers, New York, London, and Tokyo. It deals with each of these regions in a separate part, tracing the international development of the markets and highlighting the principal distinctions and most interesting and topical features of each.

The aim is to identify who all these busy people are who practice global banking today and what it is that they do. Focus is given to the internationalization of finance in the United States, which has been born again after almost fifty years of retrenchment following the discouragement of the Great Depression and World War II. The extraordinarily innovative developments in European finance are examined, beginning with the rise of the Eurobond market and including the breathtaking free-market reconstruction of the London Stock Exchange and the surprising resurgence of pragmatic capitalism in socialistic Europe. The mysterious financial ways of the mighty Japanese are also illuminated. The ending is reserved for some comments, observations, and predictions about global banking in the next century that are based on my own more than twenty-year continuing love affair with it.

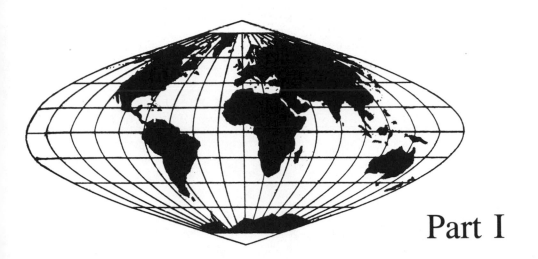

Part I

# The Internationalization of American Finance

# 1

# Born-Again
# Banking

International banking has been of vital importance to the United States throughout most of its history. The country was dependent on foreign capital raised by its bankers until about 1900, after which financial surpluses enabled it, again through its bankers, to become a major overseas lender until 1930. However, the Great Depression and World War II shattered investor confidence in foreign securities, and the markets were shut down for a generation. The skills of foreign finance, once so well developed, were allowed to decline. After the war, international banking activity resumed, cautiously at first but then, in keeping with the expansion of the world economy, enthusiastically. During the 1970s and 1980s, international banking was totally restructured. New opportunities and prospects were pursued vigorously. Overseas financial markets exploded with a vast array of new products, services, and techniques. But some mistakes of the past were repeated, and these have cast a heavy shadow over the future of some participants. The story of how we got here, from where we started, is a fascinating one. There have been many adventures along the way, and many more are continuing today.

American international bankers, both commercial bankers and investment bankers, were born again in the 1980s into an old but newly discovered belief in global finance, once again with a passionate and optimistic vision of its future. Some, however, have been carried away by opportunities of the moment that later went wrong; most have found the going more difficult than they orginally expected. But probably all have recognized that the role of global banking and finance in today's world of interdependent economies is here to stay and that to survive bankers must adapt.

The great American economic expansion of the nineteenth century was made possible by immigration of vast amounts of European capital and labor. Neither occurred as a result of any concerted government policies or foreign aid or resettlement programs. The capital and the labor were supplied voluntarily by free-market investors and by harassed or impoverished peasants and shopkeepers voting with their feet to find a better life. Except for the Civil War period, the migration of money and workers was continuous. Without these the United States today might be as underpopulated and as economically insignificant as Canada or Australia, both of which ended the nineteenth century as colonies of the British Empire.

The United States, of course, was not a colony. In 1900 it was a vast, richly endowed, self-governed nation of 75 million people whose enormous potential for development was to be sorted out by the trials and tribulations of free enterprise. Few regulations existed; the country was too immature politically to have them. Nothing politically or economically like the United States existed, or had ever existed. For many it was a godsend; anyone could come to live in the country and buy a stake in it or work for one. Foreign investment and migrants were welcomed, and of course both came and, for the most part, were well treated. More than 20 million immigrants settled in America between 1820 and 1900, and a lot of capital was attracted too.

European capital had been invested in American lands and enterprises from the earliest days of the Republic (and indeed, for many years before). The process of gathering the capital became organized in the early 1800s when prominent representatives of distinguished European banking houses established themselves in New York, Philadelphia, and Boston. A young Mr. David Parish had been sent to the United States in 1805 by Baring Brothers, distinguished London merchant bankers, whom he later surprised by underwriting a $16-million financing by the U.S. government to pay for the War of 1812 against England. In 1837, twenty-one-year-old, German-born August Belmont, later a very prominent New York banker, became agent in the United States for the Rothschilds. In 1860, J. P. Morgan left the English merchant bank that his American father ran to set up the House of Morgan in the United States. These and other firms like them provided the principal financial linkages between the Old World and the New.[1]

By the late 1800s, English and Yankee bankers were joined in the United States by German Jews, whose emigration had been effected under more humble circumstances but who, nonetheless, had prospered in trading businesses and emerged as prominent bankers after the Civil War. Firms such as J.&W. Seligman, Lehman Brothers, Kuhn Loeb, and Goldman Sachs appeared on the financial scene during this period. Like their Anglo-Saxon rivals they found the demand for finance for industrial growth to be substantial; there was plenty for all. Much of this finance had to be found in Europe, where well-organized channels and networks had been formed to tap excess savings and venture-capital resources in London, Paris, Berlin, and elsewhere. The colonial and industrial powers of Europe in the latter nineteenth century were the repositories of much of the world's wealth; it was the task of the American bankers to attract and deploy it into their country.

This they successfully did. A large business had developed in selling shares and bonds of American railroads and state and municipal securities to investors in Europe. There were many bankruptcies and defaults, but the funds continued to flow. The rewards matched the risks. America was the financial new world, primitive and savage, but it was where the action was; you had to be there. United

States banks, like Alex. Brown of Baltimore, set up their own representatives in London or arranged for European correspondents to distribute American securities to their clients.

Near the end of the century, however, American financial resources had grown enough to provide the majority of the capital needs for the growing economy. Financial markets were active in New York, Philadelphia, and other cities; a securities distribution system established to sell government bonds during the Civil War was revived. Individual entrepreneurs had become very wealthy. Banks and insurance companies had prospered and substantially increased the assets under their management. Wealth was no longer tied up in land and factories and equipment. Shares in these, or loans backed up by them, could be bought or sold on stock exchanges. As a consequence, liquid financial markets came into being for the first time on the American continent.

In the 1890s, some of the capital accumulating in the United States began to find its way across the ocean again, this time in search of comparatively high-grade, secure investments, which were hard to find in the United States at the time. Loans were organized for such blue-chip borrowers as the British Exchequer, the Bank of England, the German imperial government, the Kingdom of Sweden, the City of Paris, various German cities, and others. Twelve million dollars of British National War Loans were arranged in connection with the Boer War in South Africa. Nearly 250 foreign loans totaling over $1 billion were organized in the United States during the period from 1900 through 1913—mostly for first-rate, well-known borrowers.[2]

At the same time, many stock and bond issues of somewhat riskier U.S. corporations were arranged in New York, and a large proportion of these continued to be sold into Europe. Occasionally such issues would be denominated in European currencies, possibly for the protection of both European and U.S. investors whose experience with U.S. monetary stability had been disappointing over the last fifty years. By the turn of the century, financial traffic was moving both ways across the Atlantic.

And, apparently, across the Pacific too, much to everyone's surprise. Wall Street bankers were shocked by the announcement in March 1905 that the imperial Japanese government was attempting to raise

18

$75 million in the United States (together with an equal amount to be raised in London) to defray the considerable expenses of its war against Russia, then in its thirteenth month. It was fewer than forty years since the Japanese had emerged from the feudalism of the Tokugawa shoguns. They had been industrious and had developed their economy commendably, but Japan was still seen as an inferior, oriental Third World country with perhaps too great ambitions in China. The war had gone well for the Japanese so far against the limited number of Russian forces stationed in the Far East, but the bulk of informed opinion believed that the Japanese forces would be crushed by the superior Russian Grand Fleet that was then en route to the Far East from the Baltic.

This bold financial undertaking was led by the firm of Kuhn Loeb, whose head, Jacob Schiff, had emigrated from Frankfurt in 1865 and quite wisely married the daughter of Solomon Loeb and become a partner of the firm in 1875. Schiff, next to J. P. Morgan the best-known investment banker of his time, had built Kuhn Loeb into a leading syndicator and distributor of railroad securities. He maintained extensive contacts in Europe, in order to distribute there the United States securities that the firm underwrote. Active in Jewish affairs and philanthropies in New York, he was outraged by the brutal treatment of Jews in Russia since Nicholas II had become tsar in 1894. Faced with continuing domestic disorder, Nicholas had clamped down on the country and blamed many of its troubles on its Jewish community. A vicious massacre of Jews occurred at the village of Kishinev in 1903, which was followed by many savage pogroms elsewhere in Russia and in Poland. Schiff openly referred to Nicholas as a cruel and anti-Semitic tyrant, the "enemy of mankind."[3]

On the notion that the enemy of his enemy was his friend, Schiff sought out the Japanese envoy in New York, who was attempting to raise finance in the United States for the war, and more confidently than any other volunteered his services. Schiff and his firm were retained by the Japanese, whom they scarcely knew, and went all out to sell the $75 million of unsecured war bonds in the United States. This was a substantial challenge, for not only was the outcome of the war thought to be unpromising, and the Japanese completely unknown as creditors, but the issue was to be exceptionally large by

19

U.S. standards. The largest European bond issue ever offered in the United States at the time was a £30-million issue (about $130 million at the time) for the blue-chip Bank of England.[4]

Schiff's efforts, however, vastly exceeded all expectations. Subscriptions of over $500 million for the issue were secured from more than fifteen thousand investors, both incredible amounts for the time.[5] By no means were all of the subscriptions from Jewish investors. Americans of all types, many of them immigrants who had left their homelands to find a new life in a free country, supported the challenge of the feisty Japanese to the autocratic and unpopular Russian ruler. Two months after the loan was arranged, Schiff was vindicated. The Russian fleet was humiliated by the Japanese navy, which sank or captured fifty of the fifty-three warships dispatched from the Baltic nine months earlier. The war ended shortly thereafter, and a peace treaty brokered by Theodore Roosevelt was signed in Portsmouth, New Hampshire.[6] This was the first but certainly not the last time that Western investors rushed out to buy Japanese securities for the wrong reason and were well rewarded for their courage, if not their analytical powers.

Anyway, the event made Mr. Schiff look good, the Japanese went on to raise additional finance in New York under his leadership, the war bonds were repaid on time, and the Japanese were henceforth looked upon with far greater respect than before. In 1906 Schiff visited Japan, where he was introduced as the most distinguished visitor Japan had ever had from the United States and awarded the Second Order of the Sacred Treasure by the emperor.

Meanwhile, the Russians, seriously weakened at home by their shocking loss to the Japanese, went on to another decade of political disorder, revolution, and the rise of the communist party. Little credit is given in contemporary Russian history to the role of an American capitalist from New York whose sympathies for the Jewish proletariat of Russia led him to arrange financing for the war that helped to bring down the tsar. Schiff also subscribed 1 million rubles for a loan to the Kerensky government, which, along with all the other debts of the previous governments, was never repaid by the communists after they had swept into power.[7]

Many additional foreign offerings followed the Japanese issues over the next few years, and the U.S. capital market continued to function well in both directions, as an exporter of capital as well as an importer. Linkages with the still-dominant European financial centers were strong and efficient when World War I broke out in July 1914. The war, which was to result in America's emergence as the leading economic and financial power in the world, however, had unexpected and extraordinary effects on American financial markets.

Financial markets throughout Europe shuddered at the news of the assassination of Austrian archduke Franz Ferdinand on June 28, 1914. The stock exchanges tried to absorb the shock, but it was difficult to do so. Foreign-exchange and money markets became chaotic. A month later, when Russia announced general mobilization, the exchanges in Europe closed their doors. This left the New York Stock Exchange, for a few days, as the last major financial market doing unrestricted business. Investors from all over the world began cabling brokers in New York to dump their holdings of U.S. and other securities on the New York Exchange.

Bravely, the Exchange tried to stem the flow, but it was too much. On the morning of July 31, the New York Stock Exchange shut down, not to reopen for five months. Though some unlisted securities continued to trade and the outdoor New York Curb Exchange (the predecessor of the American Stock Exchange) condoned an unofficial "gutter" market in some securities, for all practical purposes there was no market for securities transactions or valuations in the United States until mid-December. The financial linkages with Europe had proved to be strong indeed; in this case they were overwhelming.

The next event was a foreign-exchange crisis. The dollar collapsed against sterling as Europeans sold their U.S. holdings and purchased what they thought to be more secure pound sterling assets with the proceeds. U.S. bonds held by foreigners were considered risky, as the United States might soon be cut off from foreign credit sources. Bankers in New York rallied and devised a plan whereby foreign obligations of U.S. entities would be repaid entirely, thereby signaling the market that it need not worry. The City of New York issued $100 million of gold-backed notes to bankers in the city; the proceeds, in gold, were deposited in Canada for the account of the Bank of England to be used to repay European holdings of New York City debt. Other issues followed, and the European cries for repayment subsided.[8]

Then came the tricky task of financing the war on behalf of the belligerents. The U.S. government was neutral, but the financial market was permitted to raise funds for both sides until U.S. participation in April 1917. The war was extremely expensive for all of its participants: Britain spent over $23 billion, France over $9 billion.[9] The money had to be raised where it could be found, and a lot could be found in the United States. Assets and other long-standing holdings of corporations were liquidated, and new loans were issued. However, much of what was raised in the United States was spent there on the purchase of food, clothing, and munitions. The early rounds of financing, which were unsecured, were justified on the grounds that the money was to be used for cash purchases of American goods, not for support of one side or the other. Still, ethnic ties in the German and Irish communities in particular were strong and created problems for the underwriters of the early issues sold for the Anglo-French effort. There were other complications as well; Jacob Schiff, despite strong family and business ties to Germany, would not contribute to the war effort of the Central Powers, nor would he support the Allied financings without an assurance from the British and the French governments that none of the money raised would go to support his old enemy the tsar, an assurance he was never given.[10]

All of the financings for the European war occurred in the private sector. Nearly $2 billion was arranged for the Anglo-French allies by J. P. Morgan and associates during 1914 to 1917, though the Austro-German alliance, the Central Powers, was only able to borrow about $35 million, a strong indication of where the market's sympathies lay.[11]

The war's end in November 1918 ushered in a new series of troubles. Germany was in chaos, the Russian civil war had begun, and the Spanish influenza epidemic that spread though Europe left 20 million dead, twice as many as had been killed in the war.[12] The U.S. and other economies went into a steep postwar recession, from which the much-weakened European economies never really recovered. The United States, however, enriched by the war and the prospects of continued domestic economic growth, snapped out of recession in 1921.

The years between 1921 and 1929 were extremely prosperous in the United States. Markets were buoyant, industry expansive, the government relatively passive. By the end of the decade, the United States accounted for 43 percent of the world's manufacturing output, a level it exceeded only in the years immediately following World War II.[13] The dollar was strong, and, as in the early years of the Reagan period sixty years later, it attracted substantial foreign investments to the United States. Many foreigners invested in the increasing supply of promising new issues of American securities. Others who held U.S. securities sold them and repatriated their profits. Still others, mainly foreign governments and corporations, sold new issues of stocks and bonds in the expansive New York market. The market offered an active volume of traffic in both directions across the Atlantic.

During the period from 1920 to 1929, an average of over $650 million of foreign bond and stock issues were brought to the market. In each year from 1921 through 1928, foreign issues accounted for 15 to 18 percent of all new issues floated on the New York market, a level of international involvement that has not been equaled since (in 1987, for example, a year of great emphasis on global finance, the comparable figure was less than 3 percent).

American investors, comprising small individual investors served by a growing community of brokers, welcomed foreign issues for the attractive interest rates they offered compared to U.S. bonds. On the whole, foreign bonds paid 1.5 to 2 percent more in interest than a "standard" U.S. bond (made up of a combination of various corporate and municipal bonds). European and Latin American bonds paid the highest rates; Asian (including Australian) and Canadian bonds paid the lowest. Issuers were willing to pay the rates in order to have access to the funds.

Forty-one percent of the foreign issues were European; 25 percent Canadian; 12 percent from the Far East (including Australia); and an astonishing 22 percent from Central and Latin America.[14] This last region had attracted large capital investments by U.S. corporations, mainly in the mining and petroleum sectors. Large American banks, following their customers abroad, set up shop in the area, and before long they were competing with one another for the business of the local governments, including the issuance of bonds in New York, for which the market was strong and the commissions high.

For the most part, U.S. banks stuck to lending local currencies, which they obtained through deposits in local branches, to U.S. and other respected corporations, and to financing shipments of goods to and from the United States. The banks were not then in the business of advancing long-term U.S.-dollar loans to foreign governments; such was the job of the bond market. The banks would participate in underwriting the bonds, which meant that they would occasionally be left with unsold portions of issues, and many banks made trading markets in the foreign bonds they underwrote, but the idea was not to accumulate a loan portfolio of these bonds. Bond-market investors, not the banks themselves, were the ones who carried the risk that the foreign government might not repay the loan.

Foreign bonds were controversial in the 1920s—they could hardly not be. They were promoted by those who argued (naïvely) that greater participation in international finance would better prepare the United States for the important world role that it must assume, and (pragmatically) that it was necessary, in the interests of all, for the financial markets to help European trading partners in particular to get back on their feet after the war so they could resume buying American goods. Both of these views were supported by Herbert Hoover, then secretary of commerce and widely recognized as one of the few American internationalists in high places. Hoover, however, had considerable business in developing regions, and he warned often of the need to protect investors against the promotions of speculators and unscrupulous bankers and brokers, particularly those involved with Latin American bonds. Some banks were later accused of throwing money at notoriously uncreditworthy foreign borrowers in order to get them to issue bonds so the banks could earn the large underwriting commissions involved.

Whereas Jacob Schiff, when deciding to arrange financing for a foreign client such as Japan, was mainly concerned with the cultural and social characteristics of the country and would make an assessment of the country's ability and willingness to repay what it had borrowed, most investors in the 1920s looked only at the interest rate.[15] Prosperity was in the air, optimism prevailed everywhere. Investors saw foreign bonds as an easy way to earn a much higher return on their investments than they could make on comparably rated U.S. bonds. Their behavior was similar to investors today purchasing low-quality "junk" bonds, or making high-yielding deposits in obscure savings-and-loan companies in order to get a higher return. Few had any idea of the resources or the capability of the borrowers to repay the loans or of the social and political conditions in the countries involved. The collapse of the Russian imperial government and the successor Kerensky government ended in default of all outstanding Russian bond issues. The prospects for repayment of outstanding German and Austrian bonds were highly uncertain in the 1920s. Those of a number of Latin American issuers were worse. Still, the parade of foreign bonds continued and included issues by German municipalities, many Central and South American republics, the new Republic of China, and others, government and corporate.

The parade continued, until it was rained on, torrentially, by the stock-market crash in October 1929.

The crash ravished the financial landscape in the United States and abroad. There were no safety nets to catch the plunging market, which panicked investors all over the world. The atmosphere changed overnight. The U.S. economy, already appearing sluggish in the second half of 1929, slowed further. People realized that the end of the great boom of the 1920s had come. Confidence sagged, commodity and real-estate prices dropped, the future was uncertain; but, nevertheless, the situation still looked salvageable.

In the early part of 1930, stock prices had recovered somewhat; money supply had been increased; and bond prices were strong, reflecting a movement of money out of the stock market into more secure investments. In addition, the Hoover government had made reassuring statements. New issues of foreign bonds resumed and were actually more active in 1930 than in the previous year, though the preference was for the better-quality names. Some weaker issuers, needing to refinance maturing dollar loans, began to find the cupboards bare. The assumption that they could always roll over maturing loans was proving to be false. The aggressive, solicitous bankers that had made the whole process so easy before for the issuers were nowhere to be found.

The United States enjoyed a large volume of exports during the 1920s, as industrial recovery progressed in Europe and because American businesses had taken over export customers of Europeans in Latin America and Asia during and after the war. The large volume of exports, together with the inflow of capital to the markets, created a balance-of-payments surplus. The surplus was balanced by U.S. purchases of gold from other countries and by the recycling of surplus capital in the United States to foreigners, albeit different ones from those who had invested in the United States. The system was working according to the laws of economics; the United States dominated world manufacturing and exports, which resulted in a large accumulation of funds in the United States, which had to be recirculated abroad in some form. This was accomplished through the accumulation of gold and foreign securities. In many ways, the United States was to the world then as Japan is to it today, except that the United States's relative position had been achieved over a much shorter period and was much more dominant.

The foreign debt was denominated in U.S. dollars, which meant that the countries who had to pay it back, with interest, had to earn dollars to do so. The best way for them to earn the dollars was to export commodities and other goods to the United States. Then U.S. imports would rise and its balance-of-payments surplus would decrease. The debt would be repaid, and the inequality in exports would be offset, leading toward the restoration of a healthier equilibrium between the countries.

But it was not to be. Everybody wanted to increase exports to the United States at the same time, which resulted in price wars and dumping charges. Farmers were upset, and factory workers were concerned. The protectionists of the day claimed all this involvement with foreigners, which had so seriously hurt our financial markets in 1914 and in 1929, was weakening the country in other, more insidious ways. Congress agreed and decided to retaliate against foreign countries that dumped their exports into the United States. In June

1930 the Smoot-Hawley Tariff Act was passed and President Hoover signed it. The bill raised tariffs by 20 percent; Canada and many European countries announced that they would introduce corresponding legislation in response. Trade with Latin America collapsed, along with its source of foreign-exchange earnings with which to service foreign debt. Political stability in Latin America could not be maintained: revolutions occurred in Brazil, Chile, and Argentina, the strongest economies in South America, and Mexico, which had already defaulted on its foreign debt, was on the ropes. It was not long before other countries followed suit; by the end of 1931, every Latin American borrower had defaulted except Argentina, and its bonds were trading at less than 25 percent of what investors had paid for them. Investors in these securities, mainly small individual investors and country banks, were wiped out. There was no protection in diversification; they all went.[16]

Meanwhile, German debt rescheduling and refinancing were not going well in Europe; the financial markets in Britain were in a stupor, and the French were selling their allies' currencies for gold. The only strong and rich economy in the world at the time was the United States, and it was turning inward behind an impregnable wall of tariffs and financial xenophobia.

Such, however, did not deter a Mr. Allin Dakin, the intrepid author of an informative article on foreign loans in a 1932 edition of the *Harvard Business Review*, then a student publication, who claimed that the only solution to the foreign-debt problem was for the U.S. government, that is, the State Department, to supervise the issuance of new refunding bonds and to become involved in negotiations with the obligors on behalf of the investors, or failing that, for the banks themselves to refinance all the bonds as they matured and to put bright young men (or women, presumably), who had been educated in sound business practices and were willing to live in the exotic regions in question, in charge of seeing to it that the countries sorted themselves out and paid what they owed on time. History does not record whether the author received any offers to put his ideas into practice.[17]

Despite the problems of 1930, by the beginning of 1931, many felt that the worst was over, that the main force of the storm had been withstood. They were wrong; in May 1931, the largest bank in

Austria, the Kreditanstalt, failed as a result of a run on deposits, bringing down with it several large banks in Germany. Before long the chain of losses had encircled the rest of Europe. A sterling crisis ensued, followed by Britain's departure from the gold standard. Europe was in financial chaos, and U.S banks were caught up in the spreading disasters too. The "concatenation of catastrophes from abroad," as President Herbert Hoover called these events, wiped out any progress the economy might have made after the first year of the Depression; indeed, it rapidly sank further as a result of efforts to protect the U.S. dollar and its gold reserves through restrictive monetary policies.[18]

The Great Depression followed, lasting ten dreary years and, at its worst, producing an unemployment rate in the United States of 25 percent. Banks, which failed in alarming numbers, were blamed for the excess of speculation, greed, and questionable practices that led to the crash, which many thought had started the whole thing. Other countries, suffering the Depression also, ran out of gold or foreign exchange with which to pay the interest and principal on the bonds they had floated in New York, and a great many others joined the Latin Americans in default.

After President Roosevelt's inauguration in 1933, the banks in the United States were closed to prevent panic withdrawals of deposits and an excess of gold shipments abroad. Many financial reforms took place that changed totally the character of banking and the government's role in it. Commercial banking and investment banking were separated by the Glass-Steagall Act of 1933. The Securities Act of 1933 and its companion, the Securities Exchange Act of 1934, were passed to protect investors against improper promotion and trading practices. The Securities and Exchange Commission (SEC) was created, headed initially by Joseph P. Kennedy, a notorious Wall Street speculator and the father of John F. Kennedy. The banking regulatory powers of the Federal Reserve were strengthened, and the Federal Deposit Insurance Corporation (FDIC), to insure bank deposits, was created.

The banking industry, guilty certainly of many market-rigging and other abuses during the 1920s, was made the principal culprit for the entire Depression, and for its punishment was to spend the next generation incarcerated in a fortress of governmental regulation. Banks were to accept deposits and make loans only under strict provisions designed to ensure their solvency. They were also subject to laws governing where they could and, more important, could not operate. They were protected from competition, but in exchange had to meet the public trust. They became, for all practical purposes, public utilities, like electric power and telephone companies.

Investment banks emerged as relatively small brokerage firms that could underwrite new issues of stocks and bonds. Banks could no longer participate in these activities and had to divest themselves of them. Brokers were regulated by the SEC and were not permitted to accept deposits. Margin loans were subject to Federal Reserve control.

There was very little financial business done on Wall Street during the 1930s. A former senior partner of Goldman Sachs, Edward Schrader, had been recruited to the firm from the Harvard Business School in 1936. Looking forward to his career at the firm and grateful for a job of any kind (despite his high academic qualifications), he showed up for work on his first day only to encounter one of the Sachs's in the elevator. Ed was from Indiana, and was by no means a flashy dresser, but he was wearing a pair of white buck shoes that he had purchased when he had come East so as to appear to be up to date. Mr. Sachs looked at the shoes.

"Are you an employee, sir?" he inquired of Ed, who said he was, shyly about to introduce himself.

"Then go home immediately and change those shoes. These are not happy times."

When Ed returned, he was sent to the cashier's cage for a few days' orientation. Unfortunately, the firm was struggling with the many problems of those unhappy times, and Ed was forgotten. He remained in the cage for eighteen months or so before he built up the courage to tell anyone about his problem. Toward the end of 1938, he was remembered and set to work in the underwriting department, where, because there was no underwriting business to do, he received a 20 percent pay cut at the beginning of 1939.[19] Wall Street looked very different to the yuppies of the Depression era than it did to those of the mid-1980s.

29

In any case, the situation was a long way from recovery even by the late 1930s. The foreign political environment was deteriorating fast in Europe, where Hitler had been elected chancellor of Germany in 1933, and in the Far East, where Japan had invaded China. The world economy, trade, and finance were in shambles. There was no coordination of policies among the leading powers. Each country did what it could to save its own skin and to advance its own interests. The situation would not be resolved until after World War II.

Banks had to salvage what they could under the circumstances. In Germany, repayment of dollar loans from foreign banks were frozen in 1934, subject to partial payment over the years, if dollars were available. Germany suffered a shortage of foreign exchange with which to repay the loans. There were many foreign-exchange controls in Germany at the time, including the requirement that those seeking to visit the country would have to purchase Reichmarks once they got there. According to the recollections of George Moore, who retired as chairman of Citicorp in 1970 after a forty-three-year career with the bank, Citibank managed to secure permission from the head of the German Central Bank, Hjalmar Schacht, to sell "travel marks" to foreign visitors. Citibank would collect payments on loans to its German borrowers in marks and sell these marks (at a discount) to tourists for dollars or other convertible currencies, which they could remit to New York. By the beginning of World War II, Moore explains, Citibank had managed to extract sufficient dollars from the sale of travel marks to redeem all of its loans to German borrowers, which totaled about $95 million at their peak in 1933.[20]

The war that followed burned away the economic conditions of the 1930s, but left devastation in its wake that had not been seen since the Black Plague in the fourteenth century. Winning the war, of course, required that substantial coordination of effort among the Allied nations take place. This turned out to be good practice for the economic planning that was to follow the war. Economists and historians knew that the seeds of World War II were sown during the peace negotiations following World War I, and had persuaded the politicians that a more compassionate settlement should follow the Japanese and German surrenders. The Allies met to plan the international financial system that was to go into effect at the war's end. In 1944, at Bretton Woods on Mount Washington, New Hampshire, the Allies agreed on the formation of the International Monetary Fund (IMF), to assist member countries in addressing short-term balance-of-payments problems to provide for more stable exchange rates, and a sister organization, the International Bank for Reconstruction and Development (the World Bank), which would offer long-term loans for the construction of basic economic infrastructure and other worthy development projects. These institutions would be owned by all of the world's countries that wished to join and to abide by their rules. The communist countries and ever-neutral Switzerland were the only nonjoiners.

Subsequently, the General Agreement on Tariffs and Trade (GATT) was similarly adopted, to provide a framework for fair-trade practices so the cutthroat practices of the 1930s might be avoided in the future.

Finally, there was the Marshall Plan, an exclusively American initiative aimed at helping former enemies and weakened allies regain their economic strength. The Marshall Plan and the Truman Doctrine, of which it was a part, were designed to deter the development of communism in Europe and to offer a unified "free-world" economy as a bulwark against its further spread.

Our postwar economic forefathers had done well. All of these endeavors were successful, and the world economy came back to life, stronger and better fortified and protected against the calamities of the 1930s than ever before.

All of this had been the work of the government, however, not the bankers operating in a free marketplace, although bankers were well represented among the government negotiators and appointees to senior positions in the new organizations. Still, the role of providing finance to foreign governments, especially long-term bond issues, was no longer left to the banks and investment banks, but had been assumed by governments themselves and their new supranational agencies.

After the war, the handful of American banks with extensive prewar international-branch systems began the process of picking through the rubble to see what might be salvaged. By the 1950s and 1960s, U.S.

31

corporations had begun to expand overseas again, enjoying an un-precedented lack of serious competition from international firms. Mo-tivated by the formation of the European Economic Community (EEC), in which they wished to participate, and the high value of the U.S. dollar, American multinational firms moved quickly to acquire com-panies or set up factories in Europe. U.S. banks followed them there and began enjoying a substantial deposit business as shell-shocked Europeans looked for a safe, reliable place to put their money.

George Moore recalls that he had been asked by Citibank's pres-ident, Stillman Rockefeller (whose cousin David Rockefeller was soon to become head of Citibank's chief New York rival, Chase Manhat-tan Bank), to look into the affairs of the international division of the bank in the mid-1950s to see what should be done with it. Moore had had little to do with the international business during his career, and apparently he was known for his disdain for it. After reviewing the division, Moore concluded that "the results of the past fifty years of foreign operations, net, net, net were red, red, red." The bank had acquired several others during its history, including one with sub-stantial overseas activities, and was in 1955 one of the small number of U.S. banks that endeavored to operate an extensive network of international branches. Citibank had eighty-three foreign branches in 1930, accounting for 29 percent of its loans, but these had shrunk to sixty-one branches, and 14 percent of the loans, by 1955.

Rockefeller also asked Moore to pick the best man in the do-mestic part of the bank and put him in charge of the London opera-tions. Moore nominated Walter Wriston for the job and finally suc-ceeded in persuading him to take it. Moore and Wriston then began a partnership that lasted until Moore's retirement in 1970 and his succession by Wriston.

The two of them looked in detail into what the international operations ought to be. They concluded that "immense business op-portunities existed in Europe, which so far they had failed to see." Fortunately all their American competitors had been blind also. They came up with an expensive but daring strategy. They would replace just about everybody. The old-timers were too cautious and too beaten down by the discouragements of the 1930s to rebuild the network on an aggressive, optimistic basis. They would move quickly, while their competitors were still asleep, and before the European banks became strong enough to keep them out of their markets. They would aim at basic commercial-banking services but make their mark by innovative, hard-driving, smart-thinking American business practices. Wriston replaced Moore as head of international when Moore became president of the bank in 1959. Under Wriston's leadership, the international business blossomed as never before, becoming by 1967, when Moore became chairman and Wriston president and chief executive, an important contributor to the bank's earnings with a network of 148 branches and 93 affiliates overseas (much larger than any other U.S. bank's foreign network) that accounted for about 30 percent of all loans. Soon foreign loans would exceed 50 percent.[21]

Moore and Wriston's international strategy was as original as it was opportunistic. To take the best people away from the traditional, client-oriented domestic business and have them running all over the world stirring up risky foreign business was not the conventional banker's strategy in the 1950s. But to the Citibank people, this was the best and only way to escape the considerable restraints on growth in the United States that were still imposed on banks as a result of the regulatory legacy of the 1930s. The regulatory burden did not apply overseas, however; therefore, overseas business was the only unrestricted activity into which the bank could enter. If the bank did not want to remain a public utility forever, with only commensurate rewards, then it would have to make the most of its international opportunities.

And there were many such opportunities, apart from following the clients abroad. There were opportunities in the Eurodollar market, then just coming together, and in accelerating growth by "purchasing" deposits in the marketplace and relending the money to various U.S. and non-U.S. clients. There were also numerous case-by-case opportunities in indigenous markets such as Japan, Brazil, the Philippines, Indonesia, and others. And there were numerous business possibilities in the fast-growing field of foreign-exchange and money-market trading. They went after everything.

Wriston, the son of the late Henry Wriston, a former president of Brown University and one of the best-known and most highly regarded university educators of his time, became the most influential banker of his. At a time when other bank heads seemed like bureaucrats, Wriston stood out as his own man. He did his own thinking, formed his own conclusions, and let the world know it. He fought the regulatory status quo as long as he was at the bank and was responsible for much of its wearing away through legal and other victories. He sponsored innovation, technology, and retail banking when everyone else wanted to give up on it, meritocratic appointments of very young executives, and a Darwinian survival-of-the-fittest working environment for executives. Citibank climbed from a weak third place in terms of U.S. banks, ranked by assets and profits at the time of his appointment, to top position. Today, with over $200 billion in assets, Citibank towers over all other U.S. banks and continues to be the one bank that all others copy shamelessly.

In the early 1970s, Wriston began claiming that Citicorp was a growth company, like many high-priced technology companies, and that therefore its stock price should be higher. Citicorp was growing at 15 percent per year, he would say, and could keep it up indefinitely. This being the case, he added, it ought to have a price-to-earnings (P/E) ratio of 15 to 20, typical of growth stocks, rather than a ratio of 5 to 6, which was typical of public utility companies.[22]

This was a radical thought. Why should a highly regulated, geographically hemmed-in banking business be confused with a growth stock? Because, Wriston would argue, of the capabilities of Citibank management to improve the profitability of permitted activities and to find new ones to add to the portfolio, but most of all because of the totally unconstrained possibilities for growth in financial services throughout the whole world. International banking opportunities would proliferate, he noted, in a world shorn of fixed exchange rates where capital transfers could be made free of controls and profitable markets in new floating-rate foreign exchange would soon develop.

International earnings would become the dominant source of profits for the bank, exceeding 80 percent of total profits in the late 1970s. By 1973 the market had bought Wriston's line. Citicorp stock reached its all-time high P/E ratio of 25. But Wriston hadn't counted on the quintupling of oil prices in 1973 and the world financial instability that followed, just as his predecessors had not seen the collapse of foreign trade and finance in the 1930s. The 15 percent growth became hard to hold on to, and the bank fell off the track in 1975, reporting erratic growth for several years afterward. By 1982 the stock had fallen back to an inglorious 5 times earnings, lower than many public utilities.

However, most other major banks, like the stock market, had bought Wriston's reasoning on foreign growth opportunities and were vigorously trying to catch up to Citibank. Few individuals have influenced a whole industry as much as Wriston in getting all the others to follow him and Moore in trying to escape the fate of being permanently just another dull company from a highly regulated business.

Part of the reason for the banks' low stock prices in the 1980s was their growing portfolio of long-term dollar loans to the governments and corporations of developing countries. Wriston's view, often and loudly trumpeted, was that these loans did not constitute a serious problem. "Countries don't go broke," he would say, "they reschedule." Sooner or later they pay up all that they owe, and just like the U.S. government, they will be fine in the long run if they are simply allowed to sell new debt to pay off maturing debt. If they default entirely, then they will be cut off from all credit sources, which they know would be disastrous for them.

Perhaps Wriston had read Allin Dakin's article in the 1932 *Harvard Business Review* and had lots of confidence in the bright young men and women he had sent down to Zaire and other such locations to straighten things out. Or perhaps he figured that given enough time, the problem would diminish in relative importance. Or maybe he was jawboning publicly to persuade all of those who were getting panicky about the Third World debt not to do so, for fear of making the situation worse. In any case his line has never varied, and though Third World debt is a big problem for all major banks today, Wriston must believe that in the long run the problem will be resolved. But, as John Maynard Keynes reminded us, "In the long run we are all dead."

Wriston also vigorously points out that the banks saved the world from financial collapse in the mid-1970s by recycling oil dollars from Middle Eastern producers back to the developing-country oil import-

35

ers, who would otherwise have been unable to meet their foreign-exchange obligations. If the banks were heroes in the 1970s, he would argue, how can they be the goats in the 1980s?

It is true that the action of the Organization of Petroleum Exporting Countries (OPEC) did create a financial crisis for all oil importers, especially those from the Third World. It is also true that none of the world's governments or any of the supranational monetary or finance agencies could do anything about it. The Arabs did not know much about finance, and they did not want their money invested in some risky foreign credit, or in overly long term securities. It was to be deposited in banks big enough to handle the size of transactions the Arabs required, and secure enough not to make them uneasy. These requirements eliminated all but about fifty banks in the world, all of whom were eager to get their share or more of this lucrative deposit business. Bankers were three to a bed in Riyadh, Kuwait, and Teheran during the mid-1970s because there weren't enough hotel rooms to take care of all of them. Nobody twisted the banks' arms to take the deposits, which often were at relatively low interest rates. These deposits would be used to fund profitable new loans, provided that the banks could find enough borrowers to take up all the money that was coming in.

The one place the banks didn't have trouble lending the funds was to large Third World debtors, who would pay relatively high rates to get the money. They were voracious and took all they could get, later claiming that the banks forced the money on them. The banks made a lot of money during this period, but it all started to come apart in 1979, when oil prices shot up again, and borrowers were having trouble meeting debt-service payments.

At this point, the Federal Reserve's watershed change in how it managed monetary policy occurred. Money supply would be squeezed tightly to choke off inflation. Interest rates would be left to settle where they might, and they settled high—so high that the debt-service burden of the Third World became too great for many countries to handle. The heavy rounds of rescheduling began with the Mexicans in 1982 and have continued ever since.

Of course lending to less-developed countries was not the only business of the banks. There was substantial activity in lending and transferring funds for large corporate clients. Initially these were, in the case of U.S. banks, their large domestic clients. Many of these firms had been expanding in Europe and Asia from the 1960s onward and asked banks for information and assistance in understanding different local environments and regulations and in providing U.S.-quality banking services locally. Naturally this type of assistance is linked to the total worldwide business flow that a particular client generates with a bank. Morgan Guaranty maintained a very effective special department that provided baseline coaching on all sorts of international topics for their major U.S. clients. The bank certainly hoped to generate specific pieces of business from this effort, but its main purpose was to help the financial officers of the client understand the international dimensions of their jobs. Likewise, in the early 1970s, the Tokyo representative of Chemical Bank was active in organizing through the American Chamber of Commerce "briefing breakfasts" for virtually all visiting top management of all major U.S. companies, most of whom, at that time anyway, knew very little about Japan. Prominent in the briefings, the Chemical man was remembered.

The overseas business of the corporate customer in the 1970s grew very fast, in both volume and complexity. Many corporate treasurers found themselves getting involved with international finance, almost against their will, and they needed their reliable service-oriented, English-speaking banks to provide it.

In the beginning of the postwar expansion abroad, companies had two types of foreign business, plants and joint ventures located abroad, and exports from the United States. As I have mentioned, manufacturing overseas was opportune in the 1950s and 1960s. Each unit had a financing officer in it whose main job was to handle the foreign-exchange requirements of the imports of raw materials of the plant and the exports of its finished goods, and the financing of working-capital requirements from local currency borrowings. Banks could be helpful in both of these areas, but to be so they had to be on top of the local corporate unit, which required many branches all over the world.

Export finance was handled at the head office in the United States, in conjunction with the banks who provided it. Some projects were more complex and involved substantial participation by the U.S. Export-Import Bank. The banks looked on such financings as very profitable forms of foreign lending, with most of it guaranteed by the U.S. government. The other export that the corporate head office handled was the export of capital to finance overseas investments. Normally, if a company wanted to build a plant in Brazil, it would estimate the minimum amount of U.S. dollars that it had to bring in to Brazil for the project and borrow this amount cost effectively in the U.S. capital market.

In the late 1950s, the United States began to encounter its decade-long problem of defending the value of the dollar set at the 1944 Bretton Woods conference. The postwar recovery of Western Europe and Japan had caused the United States to experience a balance-of-payments problem: more money was flowing out than was flowing in, and much of the money that accumulated abroad was being used to buy gold from the United States. U.S. reserves were being diminished, and this made people nervous. So the U.S. government, among other measures, imposed restrictions on foreign borrowing in the U.S. capital market so the amount of funds leaving the country in this manner would be checked, and it requested U.S. corporations voluntarily to restrain their exports of capital for direct investments abroad. Later this program was made mandatory. Its effect was that U.S. corporations would shift from borrowing the capital needed for the plant in Brazil in the U.S. market to borrowing it abroad. This simple shift had an enormous effect on the development of overseas capital markets and medium-term bank lending.

Europe actually was very liquid at this period. There were lots of dollars on deposit in U.S. and international banks in London and other cities in Europe. These expatriate dollars, called *Eurodollars,* were the dollars that accumulated abroad as a result of the U.S. balance-of-payments deficit, those not used to buy gold. Eurodollar deposits were not regulated by banking authorities either from the United States or from the country in which the deposit was made. Citibank invented the negotiable *certificate of deposit* in the early 1960s, which turned the deposit into a security that could be traded in the market. This was a major innovation that resulted in the creation in Europe of an unregulated money market for dollar-denominated instruments. It was not long afterward that the Eurobond was invented to take advantage of the network to create a market for corporate securities denominated in dollars.

The first Eurobond was a $15-million issue for Autostrade, the Italian state highway system, in June 1963, managed by the London merchant bankers S. G. Warburg. Like all European financial innovations, the issue was copied almost before the ink was dry. The big users of Eurobonds came to be U.S. companies complying voluntarily or otherwise with U.S. restraints on capital exports.

Companies had to pay somewhat more for this money than the cost of borrowing in the U.S. bond market, but they accepted it, grudgingly, as part of the cost of doing business abroad. From 1965 until 1974, when the regulations were rescinded, the vast majority of Eurobond issues were by U.S. companies. Their activity launched the Euromarket into a twenty-year period of continuing expansion and increasing sophistication. It is today comparable in size to the U.S. corporate bond market and serves as a major source of finance for corporations and governments all over the world. It is a preferred market for international finance because it is easy to use, avoids the strict securities regulations of the United States, and is extremely innovative and cost efficient for major users. The volume of foreign bonds experienced in the 1930s in the United States never returned. It reappeared instead in the Eurobond market.

Whereas U.S. banks were busy with the overseas corporate business of their clients, the investment banks had little to do internationally until the Eurobond market developed. Then, many felt that if they did not become involved, they would lose clients to Warburg or other European banks, or to those of their U.S. competitors who had long-standing historical ties to Europe. Possibly worse, they could lose underwriting business to European affiliates of U.S. commercial banks.

So the U.S. investment banks brought themselves up to speed in Eurobonds, but mostly only to the extent necessary to hold on to their own clients and to be able to demonstrate competence in this new field. There was just too much going on in the domestic markets for investment banks to spare much in the way of resources for Euro-

39

bonds, which even in the best of years would only amount to a dozen or so of the total number of issues that each of the major firms would lead.

I was a student at Harvard Business School in the mid-1960s, looking for a subject for a thesis. Motivated mainly by my wish to pick an interesting and noteworthy topic, I decided to do an analysis of the Eurobond market. This proved to be a mistake. There was very little information available and nothing much to get one's teeth into. It was very difficult to form any conclusions about the future of the market, which seemed to depend so much on whether the particular combination of U.S. balance-of-payments regulations and the absence of supervisory regulation of the market itself would continue. "Flash in the pan," some thought; "curious little sideshow, at best," said others. The thesis, for which at first I had such high hopes, came back from Harvard's demon professor of international economics, Raymond Vernon (on loan to the otherwise more practical Business School), marked "painfully superficial."

The work, however, aroused my interest in the possibility that investment banks might be developing an international side to their businesses. So I raised the question with those with whom I was interviewing for a job. Most firms had one or two senior people who traveled extensively abroad and handled international transactions when they came up, but rarely more. Several firms, especially First Boston, Kuhn Loeb, Smith Barney, and Dillon Read were trying to rebuild the market share in foreign issues that they had enjoyed in earlier days. There was not all that much business, however, once the Interest Equalization Tax, which effectively closed the market to foreigners, was enacted in 1963.

During interviews I spoke almost exclusively with bankers who were buoyant about the prospects for domestic expansion and whose only interest in the international side (like Walter Wriston before George Moore co-opted him) was to be sure they managed to avoid getting assigned to it. When you finally got to meet the guys who did the international work, they usually looked at you with disbelief when you expressed some knowledge of and an interest in international banking.

"You've got to learn the domestic business first," they all said (even the internationalists). "If there still is any of this Eurobond business going around when you do, you can get involved then. Of course, if you are any good, you probably won't be spared from the much more important domestic-client business you will then be handling."

I decided that I had made enough international mistakes for one year in selecting the Eurobond market as my thesis topic and went to work in the corporate-finance department of Goldman Sachs, where I was quite happily involved only in domestic deals.

The international department at Goldman Sachs wasn't much in the middle 1960s. Headed by a distinguished and urbane senior partner, Stanley Miller, who was the principal link to the firm's prewar clients, it consisted of a few elderly European gentlemen stockbrokers, a Belgian arbitrageur who later ran our Eurobond syndicates, and a couple of American salesmen who covered European institutional accounts. At the time, the business was strictly controlled by Mr. Miller, a strict, hands-on manager, whose permission had to be obtained in order to make an overseas telephone call.

Such calls were then expensive and always carried the danger that a young, overeager employee would insist on speaking to a very senior European customer in the middle of the night, thus alienating him for life. Stan knew these things. We didn't.

In order to participate in Eurobond financing, however, the international department had to get help from the corporate-finance department, whose designated international hitter was Michael Coles. Mike was picked for this specialty because he is an Englishman and therefore was thought to have some intuitive capacity for Eurothink, which only purebred foreigners can possess. This qualification, together with his knowledge of European geography (he knew where Switzerland was) and the fact that he had a passport, made him an ideal candidate for the job. Mike was one of the firm's more unusual people. He had graduated from high school in Britain and immediately joined the Royal Navy, where he was trained as a pilot, serving later as a teenage combat flyer in the Korean War. After ten years of navy life, Mike met an American girl he wanted to marry, resigned from the service, and applied to Harvard Business School. Never having attended a university was an obvious obstacle to his being accepted, but somehow he talked himself into the place, where he compiled a very distinguished record.

Mike saw the opportunity in being asked to handle the only financings the firm's largest clients (for example, Ford, General Electric, B. F. Goodrich) were then doing, but he didn't want to give up his spot in domestic corporate finance. He had come to America to live and to build a career in American finance. He, perhaps like Walter Wriston, did not want to get stuck in an obscure, low-volume, hard-traveling part of the firm, such as international banking. Naturally, this is what happened to him, but like Wriston, he did a great deal to improve the international business once he got into it.

In the late 1960s, the Goldman Sachs management committee decided that it needed to formalize its international operations and should hire a "real" international player to head them up. Stan Miller was less convinced, so one of the management partners took it upon himself to go over to Europe and stay there until he found the right man, rather like a frontier farmer's approach to finding a wife. The effort resulted in the hiring of a middle-aged Dutchman who was then an executive director at Warburg. His appointment was something of a disaster. He had no idea just how un-international the firm was at the time (the two most senior partners were untraveled and uncomfortable with foreigners), nor was he able to avoid inevitable clashes with Stanley Miller, who had not been eager to hire him in the first place. Further, he had no real idea of the sort of business that the firm ought to be trying to develop, as evidenced by his highly enthusiastic if unsuccessful efforts to finance a Peruvian toll road and an Indonesian salt mine.

By the end of 1968, the Dutchman had left the firm, but the management committee still wanted someone in charge. They picked Mike Coles, who had just become a partner of the firm as a result of his international contribution to the firm's large client business. Mike, more reluctantly than he could express, accepted the job and set to making a real business out of the international department. He hired new people, started to pursue business in Japan, made many visits to enhance the European awareness of the firm, and opened an office in London in 1970, which he went over to head himself in 1971. When

42

he did, I was recruited from the corporate finance department to sit in his chair in New York and to look after the international requirements of our major clients.

At the time, arranging a Eurobond issue was a fairly routine if lengthy process. It would begin with a meeting at the client's office at which the chief financial officer would present a tirade against the government's policies that made this unnecessary financing necessary. This completed, discussion would turn next to the formation of a Netherlands Antilles subsidiary that would actually issue the bonds, guaranteed by the parent company. This step was necessary because another set of U.S. regulations required that U.S. tax be withheld from payments of interest to foreigners. Naturally, the foreigners didn't care for this arrangement, and they would not buy bonds if the withholding tax applied. Borrowing through a Netherlands Antilles subsidiary was the way to get around this rule without attracting some other tax in the process. The subsidiary was strictly a paper company, represented by a plaque on the wall of a lawyer's office in Willemstad that no one ever visited. The government was in the contradictory position of having one set of regulations requiring overseas financing and another attempting to prevent it. Fortunately for everyone, there was a loophole.

With the issuing vehicle decided, and agreement reached on the size and maturity of the issue (never longer than fifteen years, usually less), conversation then turned to the marketing-and-syndication plan. Marketing was usually done by telexing a group of prospective European underwriters announcing that the issue was coming and providing a summary of the expected terms and the contents of the *offering circular* (the name given to prospectuses in Europe). There was also a time schedule, which noted that the underwriters were to accept the invitation to underwrite within a week or so, after which the issue would be priced based on market conditions prevailing at the time. There would also be a *road show*, which consisted of one or more teams of company executives and the managing underwriters traveling to various cities in Europe to make a presentation over a sumptuous lunch at a fine hotel, so all the prospective underwriters from that city could meet management and get psyched up for the deal. Depending on the deal, visits would usually start in London and

43

stop at Edinburgh, Paris, Geneva, Zürich, Frankfurt, Amsterdam, and sometimes Brussels. Quite often this fast trip through the financial capitals of Europe was the only part the U.S. clients liked, but it was logistically a madhouse for the lead managers. Not only did everything have to go right for the traveling group, frequently accompanied by company wives, but the meetings in the various cities had to be well attended, preferably by the right people. In our case, after the company folks had retired for the night, we had to telephone Stan Miller and tell him everything that had happened during the day and estimate how many bonds each firm we met would take.

Mike Coles recalls that "after a long day visiting banks and investors we would check in to some place like the Frankfurterhof and immediately book an overseas call through the hotel operator to Stan Miller. It would take about two hours for the call to come through, by which time we had relaxed, had a nice dinner, a bottle of wine or two, and were ready for bed. When Stan came on the line, about six or seven o'clock his time, he had just finished his day and was more inclined than not to be a bit testy. He insisted on a complete report, which often took an hour or so to deliver.

"Neither Stan nor we had any other way to know what the market was in those days, or where the deal should be priced. We were not involved in the secondary markets, such as they were, so Stan had us get on a plane and go ask the investors directly. It was all quite crude, but it seemed to work."

Putting together the syndicate was also complex. We always asked the company to review a proposed list of underwriters before finalizing it. The company always had numerous suggestions and requests to honor particular relationships, often quite remote ones. They also questioned all of our recommendations: Why were we recommending inclusion of one particular French bank, for example, over another? Naturally, we wanted some banks included to help our relationships with them, but we were willing to bend over backward to accommodate the company's requests, which could sometimes be used later in the process to give us some leverage with those banks so honored. Many of the special relationships would justify a special visit during the road show.

Finally, the time for pricing the issue would come, and after consulting with some of the banks, we would recommend to the company what the interest rate should be, the key point of the exercise, which had taken a couple of months to get to. We would discuss our recommendation for a couple of hours, sometimes changing it slightly, then announce the price to the syndicate and the market. A closing would occur a few weeks later, usually in London, with less for the bankers to do than for the lawyers. The junior corporate-finance people on the team in New York, who had put the paperwork together but not had much to do with the marketing of the issue, would be sent over to the closing as a bonus.

Much has changed since those days. Eurobonds today are put together by telephone, mainly as competitive bids, over a matter of a few hours. Marketing and syndication are entirely up to the bank that buys the deal. There is no longer a need for bond-marketing road shows or calls to Mr. Miller.

In 1980 I moved to London as partner in charge of the office. We then had 60 people in London, about 5 or 6 in Tokyo, and about the same in Zürich. But we had begun to trade Eurobonds in London, had developed a potent sales force to handle U.S. stocks, and were covering many European companies in an effort to handle their business. By 1984 we had 180 people in the London office, including a large group engaged in trading and sales of Eurobonds and Euromoney market securities, a group trading and selling Japanese and other non-U.S. stocks, a group dealing in foreign exchange, gold, and coffee, and we had bought a merchant-banking subsidiary from a major U.S. commercial bank and therefore had a license to accept deposits and make bank loans. We had managed Euromarket financings for various British, German, French, Swedish, Norwegian, Finnish, Danish, Dutch, Belgian, Italian, and other European issuers, and for issuers from Japan, Korea, Hong Kong, Malaysia, Thailand, Singapore, Australia, and New Zealand. We had also advised U.S., European, and Japanese clients in merger-and-acquisition transactions, including transactions in which no U.S. company was involved.

A lot was going on, and our New York colleagues began to notice. Perhaps what captured their attention most was the riveting fact that in 1982 the London office of the firm had sold more corpo-

45

rate bonds through new issues than the New York office had. The Eurobond market was hot; good-quality U.S. corporations could often finance in it at lower interest rates than they could in the United States. The reason for this was that the dollar was rising while interest rates were dropping. Nondollar investors in dollar bonds were profiting from both movements, whereas U.S. investors only benefited from the improvement in interest rates. Consequently, there was excess demand for Eurobonds, which sucked up issues that might have otherwise been done in New York and offered them instead in Europe, and by that time in Japan too, where the demand for U.S.-dollar bond issues was also very strong.

The volume and importance of international transactions to investment bankers had reached a point it had not experienced since the long-forgotten 1920s. Something like 20 to 25 percent of the Goldman Sachs bond business was being done with foreigners; the same for mergers and acquisitions; and something less than that, but only a little less, of the stock business was international. Real estate, leveraged buyouts, and venture capital all had international potential. Huge pockets of money were being formed in Japan, in Switzerland, in the United Kingdom, and in Germany. Investors from the Middle East were still active, and U.S. clients were starting to want to invest in international securities in a big way. Everything investment banks did now had an international component. The international business had been born again in investment banks.

New York–based department heads began to want some of their personnel in London, or Tokyo. Each leader of a functional area of the firm began to feel the importance of absorbing the international portion of the activity under his control under a more unified arrangement. Back-office and administrative people had to be there too in order to maintain appropriate controls. By 1988, even after some postcrash pruning, most major investment-banking firms had seven hundred to eight hundred people in London and half that number in Tokyo.

One consequence of the rising importance of international activities in investment banks has been a swamping of foreign offices by expatriate Americans who have been sent over to "upgrade things." There has always been a narcissism in New York firms that has led them to believe that nothing in finance is done as well outside of New York.

There are times when this proves to be true, and there are many times when it does not. Technical banking expertise is usually not enough on international beats, where so many distinctly different and constantly changing business practices prevail. It can take years to globalize a banker suitably, by which time he himself may be looked on from headquarters as being in need of upgrading. It will be a while yet before investment banks get this straightened out, and until then much of the considerable turmoil and turnover of personnel that have characterized their business abroad in recent years will continue. Thanks to Moore and Wriston, commercial banks were reborn a bit earlier than investment banks. They have enjoyed a worldwide expansion of their franchises, seen their products and ideas come into acceptance, and found a way to change the ''public-utility'' character, if not the soul, of their institutions. However, they must endure the debilitating consequences of a different kind of post-rebirth overindulgence: they have to work out from under the Damoclean sword of Third World debt. They also have to learn how to manage two sometimes incompatible businesses at once, the deposit-taking and lending businesses—banking—and the securities-trading and financial-advisory businesses—investment banking. Most of the banks have centered on the former business, which many now feel is in decline. Some have tried their hands at investment banking, learning in the process that the skills and management practices learned in one business do not necessarily work in the other, especially under highly competitive conditions and in fast-changing markets.

# 2

# Global
# Banking
# Today

During periods of great change, growth, or turmoil, it seems to be common for single words or phrases to emerge as symbolic representations of what is going on. For example, there have been many *revolutions*: the industrial revolution, the cultural revolution, the green revolution, and so on. There have been many *gaps*: the missile gap, the generation gap, and so on; many *crises*; many *booms*; *bubbles*; and *panics*. Colorful imagery all.

The phrasemakers of international finance, however, are grayer men. They appear to prefer four- and five-syllable mouthfuls ending in *tion*. They have not been idle. In the past few years, we financial people have been showered with new representational word images that have become stand-ins in our conversation for complex developments that are themselves still evolving. There is, however, no standard of exactness—if we are off a bit, it usually does not matter. Clunkers like *institutionalization, deregulation, securitization, inno-*

*vation, recapitalization,* and *internationalization* have become part of our financial vocabulary, without our always knowing what the words are supposed to mean. We are more comfortable with the simpler *tion* words, like *transaction, communication,* and *position,* but these words too have taken on new meanings in the present financial environment.

The prince among words blessed with our suffix is *globalization,* it being now in almost constant use in the context of capital markets. Clearly, what is intended to be represented by the word is the involvement of capital markets in different parts of the world with each other. By itself, such a concept is not remarkable, because it is not new. Both money and those requiring it have crossed borders in search of one another for centuries without having been globalized. To become so requires two modern ingredients that have been missing until recently.

The first of these must be that the border crossings not be just opportunistic, smash-and-grab operations, but involve a permanent, institutionalized nature that is based on linkages between markets. The second is that the linkages result in a significant increase in the number of transactions between markets so that they begin to have some influence on the rate-setting, or pricing, mechanisms. As the process occurs, those markets so linked become integrated, initially to a limited extent but more so as volume increases.

If the only place where such integration existed was between the United States and, say, Canada, then the whole subject might be represented by a different, more narrow expression such as *North Americanization.* But it is not—integration is in evidence currently among the capital markets of North America and those of Europe and Japan. So we refer to the *globalization* of capital markets, and the term seems to have become accepted as a new buzzword in the lingua franca of finance. Apparently it was preferred to *worldwidization* or even to *internationalization,* which after all has eight syllables and twenty letters and tends to mean something else anyway, namely, the process that individual firms must go through in order to become effective competitors in the globalized marketplace.

In 1986 the Bank for International Settlements published a volume titled *Recent Innovations in International Banking,* which had been prepared by a study group of representatives from the central bank of each of its ten member countries. The Bank for International Settlements (hereafter BIS) is something of an international financial anachronism. Its acts as a sort of central bank for central banks. It is no longer involved with settlements, though it was founded to coordinate the collection and rescheduling of German financial reparations following World War I. It has survived because it has been able to function objectively and usefully as a sort of international financial ombudsman. It has an excellent, multinational staff and is respected by the central banks of its member countries.

The BIS study group's comments about global integration of financial markets goes a long way to define what we *do* mean by globalization:

> The roots of the present trend towards a global integration of financial markets go back to the 1960's when the development of the Eurocurrency and Eurobond markets heralded the advent of truly international financial markets. However, owing to various regulations and exchange controls, the links between these international markets and the individual domestic markets remained in most cases rather loose or partial. It was only in the course of the 1970's, and particularly during the past five years [1981–1986], that international and domestic markets have become increasingly integrated. This has occurred as a result of macroeconomic developments, deregulatory measures, technological changes and financial innovations. Since these changes have been neither smooth nor uniform, the outlines of what could be called truly global financial markets often appear as a patchwork of individually integrated financial instruments and channels of intermediation.

Globalization is what is happening to all international financial markets. Foreign-currency and commodities markets have been efficiently integrated for a long time. Markets for short-term credit have been integrated to a large extent through the banking system, and bond markets have become much more integrated since the removal of hindrances to cross-border capital flows. The spread between interest rates in the London interbank market and the U.S. market for certificates of deposit or between the new-issue yields of Eurodollar and U.S. domestic bonds has narrowed as never before. The process is working too between financial instruments denominated in other

50

currencies, though perhaps less completely. Stock markets lag in the integration process, but even in this area, where each stock is different from all others, there is a rapidly growing traffic in border-crossing transactions.

During most of world financial history until 1971, foreign-exchange rates were fixed against some standard, usually gold. Currencies under pressure were defended in various ways, the most popular of which was through controls on the import and export of money. Such controls took various forms, from the total prohibition of certain transactions, such as the purchase of foreign currencies with domestic currency for the purpose of making investments abroad, to more complex tax and other regulations designed to restrain but not totally prevent international capital flows.

After the collapse of the original 1944 Bretton Woods agreement on fixed, postwar exchange rates, which occurred when the United States abandoned the gold standard and accepted a floating exchange-rate system in 1973, foreign-exchange controls were no longer necessary for the large industrial countries. There were no longer official exchange rates to be protected. The market would reprice currencies every day depending on different economic and other factors. The continual resetting of exchange rates (that is, their *floating*) by the market would set in motion forces of supply and demand that would ultimately cause adjustments that would correct balance-of-payments imbalances.

Deregulation of foreign-exchange controls began in 1974, when the United States withdrew the ones that it had imposed during the 1960s to defend the dollar, the Interest Equalization Tax and the Commerce Department's Foreign Direct Investment Controls. The dollar would now come and go as market forces dictated.

The movement to flexible exchange rates came none too soon. In 1973, the first oil-price explosion occurred, which resulted in a huge, unprecedented transfer of funds from the oil-consuming countries to the Middle East, requiring exceptional flexibility in the international transferability of capital, which the new system provided.

Beginning in 1979, further deregulation occurred with the scrapping of most remaining foreign-exchange controls by the British and

Japanese, actions that came to be emulated by other countries, leaving the principal industrial countries by the mid-1980s with virtually no regulatory barriers affecting capital flows among them.

The result was that large sums became free to move into foreign investments of one kind or another. British pension funds could freely invest in financial and real-estate assets outside the United Kingdom. Japanese insurance funds could invest in higher yielding securities outside Japan, in the United States or Germany or Hong Kong. Though such conservatively managed institutions partake of new opportunities only cautiously, between them they represented many hundreds of billions of dollars of potential investment, even a small percentage of which would make a great difference to international financial markets. The flowing of funds from domestic investments into international ones accelerated as more managers became familiar with, and confident in, their abilities to invest funds in the new environment. As their commitments increased, so did the liquidity and other dynamic qualities of the markets affected.

Deregulation, however, did not stop with cross-border foreign-exchange controls. It went further. Internal financial markets were to be made freer, themselves deregulated. This process began in the United States on May 1, 1975, with the so-called Mayday reforms of the New York Stock Exchange, which principally required negotiated as opposed to fixed commission rates on brokerage transactions. Other reforms and liberalizations also occurred in the United States—of laws governing pension funds, of antitrust restrictions on mergers and acquisitions, and of the competitive activities of banks. The "Big Bang," which totally reformed the British bond and equities markets, was announced in 1983 and completed in 1986. It was widely influential and set off a chain of other bangs around the world. Enhanced competition leading to enhanced performance in financial markets became the accepted standard in Australia, Canada, France, Germany, and to a lesser extent in the other countries of the Organization for Economic Cooperation and Development (OECD) and in Switzerland. As markets became more performance-oriented, turnover and innovation increased as did the exploratory migration of hitherto totally domestic funds into the terra incognita of international finance, where both exceptional rewards and excitement were expected.

A further aspect of deregulation was the slow dismantling of restrictions on the participation in domestic-securities markets by foreign financial firms. Banks had enjoyed comparatively liberal reciprocal branching privileges within the industrial countries for several years. By the end of the 1970s, the New York Stock Exchange had permitted foreigners to join, London had become the home of the Eurobond market and the principal location of U.S. securities brokers in Europe, and Tokyo had allowed foreign branches of securities firms to be opened. After the Big Bang, the U.K. markets were completely opened to foreign competition, even at the cost of seeing British firms lose market share or go out of business. The Tokyo Stock Exchange allowed foreign members to join in 1986. Foreigners suddenly developed real competitive status in Toronto, Sydney, Frankfurt, and Zürich. Although international financial houses had been active in these places for some time, they were now allowed to conduct domestic business there. This resulted in an increase in the presence of the firms abroad and of course in the amount of international transactions that they could precipitate.

The march of deregulation has created two additional effects. First, it has developed a momentum of its own that appears to be accepted by the member countries of the European Economic Community (EEC) as Europeans contemplate the 1992 target deadline for complete harmonization of financial regulation. If the EEC stays on course until then, the restrictions on financial transactions between member countries may actually be fewer than those between the various U.S. states, an event with important significance to the future financial- and capital-market activity level within Europe.

The second effect of these developments is the increasing requirement for various forms of reregulation of previously deregulated areas to provide common and prudent standards by which all participants must operate. For the EEC to harmonize its financial regulations, for example, it must deregulate access to individual markets by participants from others, but at the same time it must come up with new standards for regulations in those many areas where no regulations exist or where old ones need to be revised. There are also efforts under way at present to standardize regulations regarding bank-capital adequacy and to provide for standardized regulations regarding securities markets and dealers. These matters, which greatly affect the future of global banking, get further airing in Chapter 11.

In addition to the effects on the markets of deregulation, globalization has also been enhanced by changes in the institutional nature of the investor base that participates in international markets. In the period prior to 1980, the most active international investors were the Swiss, British, and Dutch managers of funds owned by individuals and a few internationally sophisticated institutions. The funds managers had developed their skills as a result of being required to find suitable investments for their clients outside of their home countries, in which investment opportunities were limited. They became skilled at investing in the United States and in Japan, and at managing the foreign-exchange risks attendant on such transactions.

After 1980, as a result of the deregulatory actions described, large U.S. financial institutions became international investors.

Pension funds, liberated from excessive corporate control in the United States by the landmark Employees Retirement Investment Security Act of 1973 (ERISA), became very active investors in common stocks, including, during the 1980s, international stocks. European and Japanese pension funds, following the American example, became fully funded for past-due liabilities and invested an increasing proportion of their rapidly growing cash flow in international securities. The total of international investments of private pension funds in the eleven largest industrial countries, as estimated by InterSec Research, a leading pension-fund consulting firm, was $19 billion in 1980. By 1985, these had grown to $83 billion; by 1990 InterSec forecasts them to be over $300 billion, which would represent an incredible ten-year annual compounded growth rate of 32 percent. These pension funds, already a major factor in domestic markets, are becoming an equally powerful force in international markets.

Both the pension funds and the investment companies are managed by professional money managers. These managers have to look for places to invest large amounts of money competitively. They require liquidity, trading support, research assistance, and a steady flow of good investment ideas. Many of these money managers have come to believe that they cannot achieve the best return for their clients by keeping all of their investments in one economy, or in only one type of instrument; they have to diversify. Bond-portfolio managers began to look abroad for higher yields commensurate with risk. Equity managers, looking to invest in the chemical or the automotive industry, for example, learned that they should look for the best of the companies in these industries from all around the world, selecting a portfolio accordingly. Other managers would see opportunities for speculation in discovering "new" foreign areas, that is, moving into (and out of) the Spanish market, for example, ahead of the crowd, and then similarly moving on to another market, Norway, perhaps.

Institutions making the conversion to international investing included many from Europe, the United States, and Japan, which, by the mid-1980s, was bursting with excess domestic savings and a huge trade surplus. The investment of even a small percentage of the total funds under the control of these giant financial institutions around the world represented a very large increase in the total cross-border investment pool. Institutional commitment to international markets is bound to be a continuously growing and powerful influence on global capital markets for many years to come.

As institutions became more of a factor in the globalization of markets, the markets, especially the Eurobond market, expanded. In the early years, Eurobonds were mainly denominated in dollars. Investors were almost always from countries other than the United States, and therefore the foreign-exchange component of their incentive to invest had been a constant factor that differentiated them from U.S. domestic bond-market investors. During the period when the dollar was strong, or perhaps more accurately, when it was not weakening, the Eurobond investors preferred dollars over other currencies and would, on those occasions, take a somewhat lower return than U.S. investors in order to get hold of a dollar-denominated bond of a well-known American company. But the investor, typically an individual whose money was being managed by a bank, did not want to buy the bond in the United States, where it was then subject to a withholding tax on interest payments and where, further, the bond would have to be registered in the investor's name, for tax collectors all over the world per chance to see. Eurobonds provided the same quality investment, but were offered on a basis exempt from withholding taxes and in bearer form (not registered by name but only as payable to the *bearer* of the security), assuring the owner's complete and much-preferred anonymity.

Today Eurobond investors are a more diverse group, which includes institutional investors from many countries whose knowledge of the markets is exact and whose bargaining power with the market is considerable. As a result of their participation, alongside the old Euromarket mainstays, the Swiss banks acting on behalf of their individual clients, the Eurobond market has grown prodigiously. From a highly respectable base of $49 billion of new issues in 1983, for example, the Eurobond market more than tripled by 1986 to a new-issue volume of $189 billion, before falling off to $140 billion in 1987 because of the sharp and sudden weakening of the dollar. At these volume levels, the Eurobond market compares quite surprisingly well to the enormous U.S. corporate-bond market, which produced $68 billion of new issues in 1983 and $220 billion in 1986.[1]

Like those of the individual investors, the appetites of institutional investors for Eurobonds can change quickly because of changed foreign-exchange, political, tax, or regulatory conditions in any of the countries from which they come or wish to invest. The Eurobond market, accordingly, can be more volatile than its U.S. counterpart—that is, prices for issues can change a lot more as a result of a comparatively small change in the things that move the market. As a result, a number of innovative features have been devised to either reduce or enhance this difference in volatility in order to sharpen issuer and investor interest.

Among these innovations was the *bought deal,* in which a single underwriter agreed on the spot to purchase an entire issue from a corporation at a set interest rate. Marketing and syndication (if any) came after the underwriter had made the deal with the issuer, who had locked in his rate and was no longer subject to interest-rate volatility during an extended marketing period. The underwriter had taken all of this into account in offering his price to the issuer. The market risk during the selling period had been transferred to the underwriter in full. There was no longer the need carefully to select syndicate participants or to have extensive road shows, such as those described in Chapter 1. Though some Eurobond deals were done in accordance with the old ways—especially issues for unknown borrowers—the bought deal became the standard.

The popularity of the bought deal among U.S. issuers of Eurobonds led to introduction by the SEC in 1984 of its Rule 415, which permitted the same practice in the United States. Much controversy surrounded this Rule 415 at the time, but it was fairly clear that U.S. corporations would abandon the domestic bond market for the Eurobond market if the SEC did not allow for similar, instantaneous issuance procedures in the United States.

Other innovations originating in the Eurobond market included the *zero-coupon bond*, a highly volatile security that pays no interest until maturity, at which time all interest due over the life of the bond is payable along with the return of the original principal amount. For a relatively small investment now, for example, 20 percent or so of the principal amount, one could make one's children rich on the bond's maturity (when, and if, it paid the 100 percent due on maturity fifteen years or so later), especially as the bonds were available in bearer form, through which many investors were able to avoid taxes on their earnings. Further, those investors who liked to trade bonds found that by paying so little down, the investment was similar to buying bonds on heavy margin. The investment in the bonds was highly leveraged, one that attracted many more speculative investors. If interest rates declined by .5 percent, for example, an ordinary fifteen-year bond with customary amortization would increase about 4 percent in price, whereas a zero-coupon bond would increase by 7 percent.

This innovation too became popular in the United States; however, U.S. tax authorities would not allow investors to defer income taxes on accrued interest being earned on the bonds until maturity, so zeros faded from the scene. They were replaced by an ingenious American arrangement in which U.S. Treasury securities were "stripped" of their coupons, or interest payments, and both the coupons with no principal amount at the end, and principal with no coupons, were sold separately to investors with different preferences for volatility in their fixed-income investments.

The Eurobond market was also the origin of another volatility-enhancing invention, that of bonds packaged with warrants, or options, to buy more bonds at a fixed interest rate in the future. Many other innovations emanated from this fertile and imaginative marketplace that has greatly helped advance integration with domestic markets in other countries.

The growth in the Eurobond sector, in new issues and in secondary-market trading, has fathered similar developments in money-market instruments such as Eurocommercial paper, note-issuance facilities, and floating-rate notes, in which a variable rate of interest based on the London Interbank Offered Rate (LIBOR) is paid. It has also stimulated growth in the Euro-equities markets, which includes transactions involving common stock, convertible debentures, and debt with warrants attached to purchase shares of the issuer.

Most recently, it has fostered a huge increase in *synthetic securities,* in which Eurobonds are combined with interest-rate or currency swaps to create an entirely different payment obligation on the part of the borrower from that originally undertaken.

In 1981, the World Bank wanted to borrow low–interest rate Swiss francs to lend to its developing country clients, but because of overuse of the market and some Swiss skepticism about Third World repayment prospects, investment bankers in Switzerland decided that to place an issue for the World Bank an interest rate of about .2 percent above the Swiss government bond rate would be required. The Swiss, however, were more enthusiastic about lending to IBM, which they thought of as equally creditworthy (or perhaps more so) than the U.S. government. For IBM they were willing to offer a Swiss franc bond issue at the same rate as Swiss government bonds. Meanwhile, IBM, who had been looking to borrow in the U.S. market, expected to pay about .4 percent over the U.S. Treasury rate for its money. This was the rate the market then placed on AAA-rated bonds in the United States (the highest quality bonds available other than U.S. Treasuries), but it was about .05 percent higher than the rate that the World Bank would pay for a dollar bond. The World Bank too was AAA-rated, but it carried an implied obligation of the U.S. and other large governments to support it, thus making its bonds exceptionally creditworthy, regardless of what the Swiss may think.

At this point, a banker proposed that IBM borrow Swiss francs and World Bank borrow dollars, each maximizing its comparative

GLOBAL BANKING TODAY

advantage, and the two of them swap their future debt-service obligations on the issues, and World Bank pay IBM an annual fee of .1 percent of the amount borrowed. IBM thus ended up borrowing dollars for which it paid, net, .25 percent over U.S. Treasuries (the World Bank rate less the .1-percent fee), savings of .15 percent over its own comparable cost of dollar financing, and the World Bank borrowed Swiss francs at .1 percent over the Swiss government bond rate, a savings of .1 percent. Each party has benefited by exploiting different rates for the same security in different markets through the swap of payment obligations, in this case called a *currency swap*. The same thing can be and very often is done between the long-term and short-term debt markets in the same currency, in a process known as an *interest-rate swap*.

With increased liquidity and turnover in both the market for new bond issues and the secondary market that develops after such issues become available for trading, the Euromarket had become large and diverse enough to be able to attract a significant global daily volume of activity. This activity, involving both investors and issuers, takes place in the United States and in Japan in addition to its native activity in Europe. Thus the prices for certain instruments popular among the three markets have become substantially the same. A ten-year U.S. Treasury security will trade at approximately the same price in London or Tokyo as it does in New York. Similarly, a seven-year Eurobond issued by General Motors could be expected to trade within .1 percent or so in each of the three cities. The Euromarket, however, does not do this good a job all across the board; there is virtually no market in Europe for bond maturities beyond fifteen years, and often not beyond ten. Similarly, there is no market for low-grade, high-yielding junk bonds, or indeed even for those rated by the rating agencies at the lower investment grade level, Baa/BBB. Mortgage and other asset-backed securities have been issued in the Eurobond market, but only intermittently. Conversely, there is virtually no market in the United States for floating-rate notes or note-issuance facilities.

Despite a number of areas where markets exist only in one center, international-capital market activity involves a lot of arbitrage between market centers in addition to the traditional capital raising

59

and investing functions. Arbitrage activities are the buying of a security in one market and simultaneously selling the same security in another market at a slightly higher price. Arbitrages, including those involving synthetically created securities involving swaps, exploit market imperfections, but each trade tends to reduce the imperfection, until finally it disappears. When there are no imperfections, markets are completely integrated.

In the real world, however, underlying conditions are constantly changing. Just as it is difficult to manage a cup and saucer while standing on the deck of a ship in a storm, it is difficult for markets to maintain perfect alignment; indeed they may only achieve such perfection while passing from one misalignment to another.

To pursue misalignments, which all arbitrageurs do, one must be able to spot them early and then to act quickly, and often, until the profit potential is fully exploited.

Without the silicon chip, world financial markets, like a lot of others, would not be nearly so active, and globalization and market integration would only be concepts, not realities.

Technology has made several things possible in international financial markets that would not have been otherwise: the ability to settle and deliver large numbers of transactions involving different instruments, currencies, and locations; the ability to keep track of more numerous and more complex trading positions than anyone ever imagined only a few years ago; the ability to hedge positions with financial futures and options; the ability to maintain a portfolio of swap transactions; and the ability to transmit large volumes of market and competitive data internationally at comparatively low cost.

It has also made superior and much faster analytical work possible, which has encouraged market operators all around the world to develop new financial products and ideas. Hence the emphasis on innovation, and more innovation. The newest products are the most profitable, but no successful new product remains uncopied for more than about forty-eight hours after being announced. After IBM and the World Bank did their currency-swap transaction, several others were lined up to come next.

Systems for applying computer, communications, and financial technology to create, execute, and settle transactions have become very effective. As transactions increase, so does market liquidity, and as liquidity increases, markets can be more efficient. When markets are more efficient, orders are filled quickly and smoothly without requiring a major change in the market price. When these conditions exist, market volatility as a whole is lowered. With reduced volatility, investor confidence tends to increase, resulting in yet more transactions.

Not everyone believes markets work like this. Many think that greater activity attracts more profit seekers, or speculators, who jump in with both feet and make the market more volatile and unsafe for decent people. Program traders were seen to be acting in this way by the members of the Brady Commission (and the other commissions), which studied the October 19, 1987, crash and who found it preferable to curtail trading to cool the passions of a panicky market than to maintain a continuous market for the panicked and other investors to use if they wished. The market itself doesn't care whether a participant is a program trader or an arbitrageur (both apparently deemed by the public to be flagrant profit seekers) or an *end-user* (a decent person who is using the market the way it ought to be used—to raise or place investment funds); it only cares that trades that come to the market in an orderly way get completed as close to the price of the preceding trade as possible, and then get out of the way for the next trade.

Perhaps Mr. Brady and the others overlooked the fact that in global markets, if a trade is obstructed from being completed in one marketplace, it can be completed in another. Indeed, Sir Nicholas Goodison, chairman of the London Stock Exchange at the time of the crash, is reported to have commented that if the New York Stock Exchange was thinking of closing or limiting trading to curb speculation or program trading, or for other reasons, the London Stock Exchange would be glad of the business. As our predecessors in the 1910s and 1920s knew full well, global markets can offer many advantages, but equally they can be perverse when one is trying to prevent something from happening or to shelter the homefolks from the unpleasant behavior of foreigners.

The principal positive consequence of financial-market globalization is the pooling together of market capacity around the world to provide a larger common financial marketplace that provides users around the world with a dynamic, cost-effective environment for making selections among an increasingly expanding array of attractive financial alternatives. The principal negative, however, is an irrevocable marriage to coldhearted, free-market behavior, for better or for worse, and for richer or for poorer. The effect on market users has been very beneficial on balance, but there have been some disadvantages.

The World Bank and IBM found their currency swap to be very worthwhile. So have the thousands of others who have followed in their footsteps. "Mostly," they say, "it's a no-lose proposition. If the net cost of borrowing isn't better than what the company could do without swapping, we don't do it. Of course there are risks that the swap might come apart, and have to be replaced more expensively, but the benefits should be sufficient to offset those risks." Swaps not only can be used to save financing costs, but they can also be used by the company to create an arbitrage itself and profit from it directly. This is done by financing at a cost below the market through swaps like IBM's, and reinvesting at the market to pocket the savings.

In the early 1980s, British Petroleum (BP) had been a well-capitalized company without much need to approach financial markets. However, when a bright, energetic petroleum engineer in his early thirties named John Browne was appointed treasurer of BP, things changed. Browne had never had a financial job before; he had spent his career on oil rigs and worrying about exploration. Still, tagged for upward potential, he was sent to the company's headquarters in London to apply his mathematical and engineering skills to learning about finance. In time the workaholic Browne figured out that BP had unused borrowing capacity. If a portion of this excess capacity were used to borrow funds at bargain rates, by accepting some of the suicidal bought deals offered by underwriters hungry for BP's business, or through swaps or other opportunities, the company would be collecting funds that it could reinvest in Treasury securities, bank deposits, or high-grade obligations at a profit. If the company could average .3 to .4 percent reinvestment profits over and above the savings associated with skillful management of its own maturing liabilities, then it could begin to generate some significant banking revenues of its own. In time such revenues would permit Browne to argue that the treasurer's department at BP was a profit center. Browne also saw that there were many markets in which BP might be able to finance advantageously, but unless it was active across the board the treasury staff would never learn enough about the markets to make the most of even their regular duties of managing BP's liabilities. You learn by being active in the markets, Browne reported to his superiors, and the more active you are the more you learn. In the end, Browne incorporated his treasurer's department into something called BP Finance International, a kind of externally recognized in-house bank.

Browne's move was bolder than it may seem because many companies, especially large companies, do not believe that it is their function to speculate in either foreign exchange or interest rates, where risks are well known to be high. Those engaging in such activities do so at their peril. The canyons of international finance are littered with the broken bodies of treasurers who borrowed unhedged Swiss francs at the wrong time, or who came forward with an issue of synthetic securities that didn't work.

"I don't get paid a bonus for saving an extra eighth on some clever deal," one such treasurer said, "but I sure as hell can lose my job by getting one of these wild-assed things wrong." Browne challenged this conventional wisdom and was supported by the BP board. The company now has one of the world's best internal-finance departments, has saved millions in interest expenses and fees, has hedged many of its risk exposures in the futures and options markets, and has sent Browne on to greater things. For Browne, and for BP, globalization has been a plus.

General Electric (GE) doesn't do financings just for the arbitrage. It doesn't need to. At General Electric Capital Corporation (GECC), GE's giant wholly owned finance subsidiary, one kind of financing or another goes on almost every day. Originally intended to provide customer and dealer financing, GECC has long since outgrown GE's own requirements. It now comprises one of the largest finance companies in the world, an insurance company, and an investment bank, Kidder Peabody. GECC is continually in the market, as one of the country's largest commercial-paper issuers, as an issuer of medium-term and long-term securities, as an arranger and underwriter of issues, as an investor in loans and securities of many other companies, and as a manager of a very large pension fund.

GE has also been extremely active on the merger-and-acquisition front since its young and very dynamic chief executive, John F. Welch, took over in 1980. The company has been both buyer and seller of major business on many occasions. Few transactions, however, tested GE's financial acumen as much as its $6-billion acquisition of RCA in late 1986.

The RCA deal was put together very quickly by Welch and RCA's chairman, Thornton Bradshaw. After a thorough but necessarily quick review of RCA's business, GE was ready to negotiate a price. Before doing so, Welch retained and consulted John Weinberg, Goldman Sachs's senior partner and a longtime Welch confidant. The deal was struck and announced. The task of financing the $6-billion purchase price fell to GE's chief financial officer, Dennis Dammerman, a career GE finance man in his early forties.

Dammerman is a product of the GE management training and development system. He has held all sorts of different financial jobs within GE and GECC; he has been to all the obligatory courses at the GE management institute; he spearheaded GE's participation in financing and investing in leveraged buyouts; and he had been involved in all the many merger-and-acquisition transactions that the company had entered into over the past several years. Cool, bright, and tough, there were very few financial tasks he had not performed with distinction at GE.

Dammerman arranged the RCA financing easily and indeed brilliantly. More than a dozen separate financings were involved; markets and investors in the United States, Europe, and Japan were accessed separately without causing congestion or indigestion. Of the $6-billion purchase price, nearly $4 billion was raised on a long-term, fixed-rate basis at an average spread over U.S. Treasuries of

.19 percent, well less than the .3 percent spread above Treasuries that GE would have had to pay for a single domestic offering of $500 million. The entire program was completed in less than twelve weeks.

To accomplish this formidable financing, Dammerman didn't do anything special, that is, he didn't do anything he would not have done in managing the ordinary financial affairs of GE. These ordinary financial affairs, however, had changed a lot in recent years.

Up until the late 1970s, GE had been a company typical of many in the United States: it did not finance often, and when it did it called in its commercial and investment bankers and asked them to recommend what ought to be done, and then did it. The financial staff was much more focused on internal matters than on markets. GE had traditional bankers upon whom it relied heavily and, accordingly, was not receptive to competitive approaches from other banks. Until Welch and Dammerman. Their approach was different, more in keeping with the market capabilities of the times.

Dammerman was guided by three basic principles in running the vast financial empire at GE. First, the GE and GECC staffs would be first-rate, totally market competent and trained by the doing of deals, as in John Browne's BP. Second, though GE would preserve, and continue to value, its traditional banking relationships, it was committed to using multiple financial suppliers and would offer relatively easy access to any reputable banker with an idea. Its bankers would have to compete with each other for GE's business on the basis of the lowest net cost of funds to the company; advice and counsel would play only a limited role. Third, GE would encourage the development of its financial capability in markets all over the world. It wanted to be able to secure the cheapest source of funds from wherever they came. GE's financial resources would be supplied by the best and the brightest, competitively and globally.

The result in the RCA case was proof of the pudding. In the thirteen financings that Dammerman and his team completed, four involved conventional issues done at aggressive rates, six involved international financial markets, two were done with currency swaps and synthetic securities, and one included warrants for the purchase of GE shares. The financings were led by nine different banks and investment banks. Before the exercise began, Dammerman had invited fourteen of GE's closest banking relations in for consultation and advice. Most wanted to serve as GE's financial general contractor for the project, a role Dammerman was unwilling to cede and did not.

Some of the deals were priced too tightly and did not trade well in the aftermarket. "Too bad," said Dammerman, "but those guys who bought the deals from us are big boys, and they know what they are in for when they bid aggressively. I'm here to take advantage of their eagerness and to get the lowest cost funds I can. The markets will recover, and GE isn't going to be prevented from further financing because one of these deals was priced too tightly." He was right. The RCA financing not only depended on having developed a globalized financing capability but also required a major change in the way GE worked with its bankers. Without the later change, to multiple sourcing, the best of all of the globalized market opportunities could not have been realized.

Meanwhile, in a different part of the world, another modern financial drama was unfolding. Years before, Walt Disney had agreed with partners in Japan to construct a Tokyo Disneyland. When it finally came on-stream in the mid-1980s, it was a fabulous success. Walt Disney's share of the profits was considerable, but it was paid in yen. "What are we going to do with all of these yen?" they asked themselves. "If we sell them forward in the foreign-exchange market, we can only do so for about one year—the market doesn't go out any longer—so if the foreign-exchange rates change after that, we are not hedged—we just have to take what we get."

"Why not borrow long-term in yen?" said a bright young financial officer. "We could borrow from Japanese banks at yen prime rate or in the bond market and pay it back out of our share of Tokyo Disneyland's yen profits. That way our long-term yen position, coming from our steady revenues in yen, is hedged, and we can simply convert the proceeds of the yen borrowing to dollars and bring them back to the States."

"But," said a colleague, "yen financing is relatively expensive in Japan. Why not try the market for Euroyen, where it is possible to do a fixed-rate bond issue without all the waiting time we would have to put up with for a domestic yen issue?"

They tried Euroyen and got a quote from a Japanese investment banker. However, the market for Euroyen was only so-so at the time, and the Japanese banker had to do two other deals before he could confirm anything for Disney. Another banker then approached the Disney team with a different idea. A major French government agency had recently borrowed yen in Japan in the domestic bond market. It had had to wait six months to do so, but when its number came up, rates were very attractive, and the issue was done very close to the Japanese government bond rate. "But the French don't really want the yen," explained the banker. "They think the yen/French franc rate will probably increase, making their yen debt more expensive to service. The French want to get rid of the yen. What they want are ECUs."

"ECUs, what are ECUs?" asked the Disney people.

"Well," the banker replied "they are *European currency units,* a financial unit of account used by the member countries of the EEC, the Common Market, that consists of each of the countries' currencies taken into the unit on a trade-weighted basis. Europeans like the ECU because it is the only currency unit without the dollar in it, and therefore not a bad way for Europeans to manage their foreign-exchange positions. For example, if they think the dollar is going to be strong, they can borrow in ECUs, which they can repay less expensively. The French have been the principal backers of the ECU and want to see it become a major capital-market currency. Why not issue Walt Disney ECU bonds in the Euromarket and swap the proceeds with the French for their yen?"

That was the deal that Disney did. The future payment obligations of the ECU financing were swapped for low-cost yen payment obligations. The cash proceeds from the ECU financing were repatriated, converted into dollars, and reinvested at a substantial profit.

As Disney's international business increased, so did the attention paid to them by European and Japanese banks and brokers, who often had useful propositions to make. For Disney, globalization had been a plus too, although later, when it became the target of a takeover effort by corporate raider Saul Steinberg, the Disney people were not so sure that all its international presence didn't simply make it easier for Japanese and other foreign banks to lend large sums to those who might try to take the company over.

Such might have occurred to Sterling Drug, which was the target of an unsolicited, unfriendly tender offer from the Swiss pharmaceutical giant, Hoffmann–La Roche, which ultimately caused the sale of Sterling to Eastman Kodak, which came in as a higher bidder. Both the Hoffmann–La Roche and the Kodak bids were backed by financing from foreign as well as U.S. banks. For those seeking to avoid excessive international financial attention, in order to maintain a lower profile in the acquisition context, globalization is not really a plus at all. However, once they find themselves on the griddle, global markets make it possible to bring in higher bidders or to try other maneuvers that could not be accomplished in the United States.

Firestone found this to be the case when Pirelli, together with Michelin, tried to take the company over, but Bridgestone of Japan came forward with a much higher offer financed by Japanese banks.

An example of another maneuver that illustrates the extent to which globalized financing extends is the story of the successful initial public offering in Japan of the B-R 31 Ice Cream Company, in December 1987. This company, a 50-percent joint venture in Japan (before the offering) of the Baskin-Robbins Ice Cream Company, itself a wholly owned U.S. subsidiary of a large British foods company, Allied Lyons, was able to sell shares to the public in Japan at Japanese stock prices. Thus a large market valuation for the company was created and passed back to its British parent, whose own share price benefited accordingly. The offering was completed two months after the October 1987 market crash, at a P/E ratio of 38. At the time its parent's P/E ratio was less than 10.

Whereas globalization has created many new possibilities for those corporations seeking to raise financing, investors too have become accustomed to global opportunities, of which they now see quite a few. Recently, for example, institutional bond investors in the United

68

States were offered the opportunity to improve their portfolio yields by purchasing German government bonds (*Bundesanleihen*, or Bunds) together with a currency swap or forward foreign-exchange contract that converted the investment into dollars. In December 1987, a Goldman Sachs financial-strategies group reported to its clients: "If you think German bonds will perform as well as (or better than) U.S. bonds over the coming month, you can pick up about 1.65-percent in interest by substituting a covered Bund for a Treasury bond currently in your portfolio . . . without significant exchange risk."

A pickup in yield of 1.65 percent while only substituting the credit of the German government for the U.S. government as obligor is a very large improvement indeed, and many investors followed the firm's advice. The opportunity resulted from a large imperfection in the domestic German bond market that was finally arbitraged away. While it existed, however, the U.S. investor had been presented with an extremely attractive alternative to increase his return for only nominal additional risk through a transaction that he had never before seen, brought to him by an alert U.S. broker that had become well schooled in both the German Bund and the foreign-exchange markets for dollars and deutsche marks.

About the same time, an aggressive portfolio manager for one of New York's leading money-management firms received a proposal from one of its brokers, based on research done by the firm's financial-strategies group, to take a position anticipating a move in the yen relative to the Swiss franc. The yen was still strengthening at the time against the dollar, but the Swiss franc was expected to lag against the dollar. An opportunity existed to bet on the continued movement of the two currencies at minimal risk to the investor.

The broker would arrange for the investor to purchase an option entitling the investor to acquire Swiss francs in the future at today's yen spot price. The investor would pay for this "call" option by selling a call option on the Swiss franc, also to be exercised at today's spot price. The transaction would be self-financing in that the proceeds from selling the Swiss franc option equaled what had to be paid for the purchase of the yen option. If the spread between the yen and the Swiss franc widened, the investor could exercise the options profitably. If he did not, then the options would expire without costing the investor anything. The portfolio manager took the recommended position, surprised at the extensive, if newly acquired, skills of the broker in foreign exchange.

U.S. equity investors have also been looking at international investment opportunities closely for the past few years. During the 1970s, the U.S. markets were unimpressive. The decade closed at a Dow Jones Industrials Average not far from where it had begun. The real action in the 1970s was in Japan, and many European fund managers made substantial profits there. In the middle of 1982, however, the Reagan bull market began in the United States and ran almost continuously until October 19, 1987.

Europeans, especially the quick-acting British and Dutch fund managers, switched back into U.S. securities to catch the rising market, which was accompanied by the additional bonanza, from their point of view, of a rising dollar. U.S. investments became irresistible, at least until the dollar peaked in late 1985 and began its sharp decline. By this time, U.S. investors realized that nondollar investments could be very profitable, especially when they were made in markets that were outperforming the U.S. market. A few U.S. money managers had been significant investors in foreign equities for several years, but by the mid-1980s many more had joined them. By the end of 1987, according to the annual *Pensions & Investment Age* poll, the fifty largest international managers of U.S. pension-fund assets accounted for more than $50 billion of international assets under management. Fifteen of these managers (including four British merchant banks) had international accounts exceeding $1 billion.

Slowly at first, then with greater conviction, the U.S. money managers began to make bold investments in Europe, Asia, and Australia. And they looked to their brokers for the same kind of back-up trading and research support in international securities that they provided in domestic stocks.

Fund trustees and managers realized that there were several reasons for them to emphasize foreign investments. During the last ten years or so, it had been obvious that the fluctuating dollar and U.S. economic performance periodically created superior investment opportunities outside of the United States. It would be prudent to invest a portion of the funds' assets in other markets to hedge against the underperformance of the U.S. market. This also meant that correctly done, a U.S. money manager might be able to outperform U.S. indices (by which his performance was most commonly judged) by spreading some of the investments in his care into markets that would do better than the U.S. market.

International investments have grown as a percentage of total pension-fund assets for the past several years. They now account for about 5 percent of pension assets, a modest amount in comparison to the 20 to 30 percent of total assets that are represented by foreign investments in typical British pension funds. There is still a lot of room for growth.

In the United States, there are approximately two to three hundred institutional investors in international securities, and the number continues to grow. Not all of these investors have committed a substantial portion of the funds that they manage to the international markets, however. Some, of course, have considerably more international investing experience than others, but on the whole, the U.S. institutional investor is still learning.

There are various types of international investment managers in the United States. There are the large investment institutions themselves, such as the College Retirement Equity Fund, Prudential Insurance; several investment companies and mutual funds, such as Capital Guardian Trust, Templeton, and Fidelity International Funds; and banks, such as State Street Bank and Trust, J. P. Morgan Investment Management, and Bankers Trust. There are also specialized money managers, such as Scudder Stevens and Clark, which advise on international investment funds. Finally, there are investment bankers and brokers, such as Morgan Stanley and Shearson Lehman Hutton, which manage funds for their clients.

These various firms compete in the United States for contracts to manage international portions of pension funds and other assets with British merchant banks, Swiss banks, Japanese investment managers, and so on. There are hundreds of non-U.S. investment managers operating in the United States in order to attract part of the lucrative pension-fund business. Competition in the investment-management business in the United States has caused a number of significant changes in the last few years.

The pressure on performance and on fees has caused several players long associated with the traditional domestic money-management business to give it up in order to concentrate on the higher margin international business. Early in 1988, for example, Citicorp announced that it was selling its $21-billion domestic pension-management business, thereby joining Manufacturers Hanover and Bank of America, which had taken similar steps.

Investing in equity securities is more of an art than a science, even when the task is not complicated by choices between currencies and markets. International investing is about as complex a task as can be set for any financial professional. One must weigh many factors in deciding how to allocate funds between different markets, while sticking to the basic approach that persuaded the client in the first place.

Some managers believe, as in domestic investment management, that patience on top of a solid, well-reasoned investment strategy is the best approach. Pick the main areas of concentration, by markets and by industry segment, and stick to them. Truly good investment selections do not have to be changed often. Many savvy portfolio managers invested in Japanese banks and insurance companies in the early 1970s and still hold their original investments, nearly twenty years later, on which their profits have been enormous. Those investors selected the right market, Japan; the right segment, financial services, when they were trading at a very low P/E ratio; and the right time, early in the Japanese stock-market miracle.

Such investors often work with a main theme, such as a long-term optimism, or faith in the Pacific Basin, or in medical technology, or in how people's lives are being affected by social changes. In the international context, however, it is necessary continually to scan the horizon looking for the next Japan. Could, for example, Hong Kong under Chinese rule after 1997 turn out to be the equity market that will reflect China's ultimate power and influence in the world economy; or will it not? The best managers of this type tend to get these questions right enough of the time to be able to reap large enough gains to cover a few mistakes and still look good on the performance charts.

Other managers prefer a more aggressive, trading-oriented approach. These investors look for the next market to experience substantial growth or opportunistic situations wherever they appear. A number of such managers were the first to discover the Korean, or Spanish, or Thai markets, each of which appreciated considerably in the 1985 to 1988 period. They might also invest in international merger-arbitrage situations, for example, by purchasing shares in the Belgian company Société Générale after the Italian entrepreneur Carlo Benedetti had announced his stake in the company and intentions to acquire it. Though often breathtaking, these types of investments can be very rewarding.

Some managers insist on formula guidelines that impose limitations and constraints on the overall balance of the portfolio. Within these guidelines, portfolio managers may trade as they wish, but the principal investment and safety criteria will be preserved by the guidelines. Such guidelines as practiced by the Dutch funds group, Robeco, include limitations on the maximum percentage of the shares outstanding in a single company that the fund may own, a minimum market capitalization for the companies in which they invest, and other similar limitations.

Two governing principles of international investment management that are reflected in the above management practices are the freedom to find the best opportunities, even if these exist outside of one's own country, and the idea that diversification of assets between countries does work. That the juiciest plum may fall in a neighbor's garden is no surprise. The principle of diversification was not so easily accepted; diversified assets may be impossible to repatriate in a hurry if one needs to, and diversification into markets that are less liquid than one's home market may be a mistake when difficult times come. The lessons of the 1930s and 1940s were not altogether forgotten.

The market crash of October 19, 1987, provided a test of the diversification principle. Perhaps, considering the extent to which communications and globalization of markets have developed, it was not much of a surprise to find all the world's principal stock markets participating, sympathetically and simultaneously, in the market collapse in New York. Markets crashed everywhere in tune with New York. But, following the crash, markets recovered at different rates. By the end of October, the FT-Actuaries World Indices* Europe and Pacific stock index (denominated in U.S. dollars) fell by 11 percent as compared to a 12-percent increase for the Standard and Poor's (S&P) 500 index. However, by the end of 1987, the FT-Actuaries index showed a gain for the year of 24 percent, as compared to an increase of only 2 percent for the S&P 500. The FT-Actuaries index is heavily weighted to Japan, which shrugged off the crash and soon returned to precrash highs, but even with Japan weighted less, the diversified international portfolio outperformed a diversified U.S., or German, or Swiss, or French portfolio. Only the Japanese and the British would have been better off in 1987 investing exclusively in their own market.

Most American banks and investment banks would have been better off if they had been able to skip 1987 altogether. Global banking had become difficult and very expensive.

For commercial banks, it was mainly a year of paying the piper by increasing loan-loss reserves substantially to reflect the realities of their Latin American loans. Ironically, the spontaneous industrywide urge to reserve was led by none other than Citicorp, whose chief executive, John Reed, evidently had second thoughts on the Wriston less-developed country doctrine. No doubt such thoughts were prompted by the fact that not only was the principal on the loans not being repaid, which the doctrine prepared him to expect, but the interest wasn't either, even by such promising borrowers as Brazil, in which Citicorp had an enormous stake.

The banks had been digging themselves out from under an avalanche of domestic loan troubles for the past several years. It was not foreign lending that brought Continental Illinois down and Bank of America, First National Bank of Dallas, Seattle First, and too many others to their knees—it was domestic lending. Much of this business had been put on the books, as we remember, in order to make the

---

*A trade mark. The indices are compiled by the *Financial Times*, Goldman Sachs, and Wood Mackenzie & Company, Ltd.

banks' earnings grow, at the Wriston target rate of 15 percent. Third World debt didn't look so bad back then, as the banks were still booking large amounts of income from interest on the loans. Just, however, as the banks had begun to work out of the worst of their domestic problems, by reserving fully against them, the Latin American exposure came home to roost.

Few banks in the United States would feel during 1987 that international diversification of their loan portfolios was much help. For most of the preceding ten or fifteen years, however, they would have; and no doubt the survivors among them will feel so again. Maybe sooner than later—syndicated international bank loans in the Euromarket actually increased in 1987 for the first time in several years.

Banks have instead concentrated on becoming more active in the trading businesses. For many years, they have been very large dealers in foreign-exchange and government securities, and trading profits from these activities have become a very important part of the total income of most money-center banks.

In the fourth quarter of 1987, for example, Bankers Trust reported $338 million in trading profits, an astonishing figure that stunned all of its rivals. The total of all operating profits of the bank for the full year, before transfers to loan-loss reserves, was $637 million, of which foreign-exchange trading income was $593 million. Approximately $300 million of the foreign-exchange profits was reportedly earned by an eccentric thirty-two-year-old star options trader named Andrew Krieger. Krieger left the bank in early 1988, apparently unhappy with the $3-million bonus paid to him for his efforts. In July 1988, however, on orders of the Federal Reserve Bank of New York, Bankers Trust revised its 1987 results to eliminate $80 million of fourth-quarter trading profits, because it had improperly valued its trading positions in foreign currency options. Trading has indeed done much to change the character and excitement of the global banking business.

Most of the major American banks now include in their breakdown of where earnings come from a category called ''investment banking'' or ''merchant banking'' to signify that they are not just worn-out galley slaves of the traditional lending business but modern institutions already engaged in the business they seem most eager to

75

conduct in the United States. Most such banks now report a fairly high percentage of their total income being derived from merchant or investment banking. It may be hard for some to understand how institutions precluded from competing in the underwriting and trading of corporate securities, and which also have a relatively low profile in the merger-and-acquisition and other corporate finance advisory businesses, can do all that much investment banking. The answer is that these banks have transferred their long-standing trading businesses to the investment-banking revenue pool.

However, their trading businesses have been expanding rapidly. Foreign exchange no longer consists of trading in spots and forwards; it now entails extensive activity in foreign-exchange futures and options (which are traded in New York/Chicago, London/Paris, and Tokyo/Hong Kong/Singapore), and in currency swaps. The markets have grown not only in size but also in volatility, the trader's friend. The success that many banks have had in foreign exchange has attracted several of the larger U.S. investment banks into the business, where one or two have made it into the top twenty.

Dealing in U.S. government securities and in noncorporate municipals has also been a similarly expanding, though much tougher, business. Banks on the whole, however, have not pushed ahead all that aggressively in the Eurobond business, or in the trading of non-U.S. securities, though the record shows a few have been successful from time to time. Most banks know that it is very difficult to make any money in Eurobonds without captive placing power or a stable of docile issuers. Periodically banks try, make some progress, then either lose a lot of money in trading or underwriting activities, or otherwise make their superiors in New York nervous, and then back off until the next reorganization of the bank's investment-banking business, after which the process begins anew.

Banks, of course, are restricted in the United States and Japan in terms of participation in trading and underwriting. Such restrictions limit their ability to put together the worldwide distribution network needed to compete with the best.

Several banks have made moves to enter the domestic securities business in the United Kingdom. U.S. banks were among the first to move to acquire U.K. brokers in anticipation of Big Bang. Security

Pacific started things off by acquiring Hoare Govett, a highly respected, profitable firm. Not wanting to miss out on the opportunity to reenter the securities business at last, even if it had to be in London rather than New York, Chase Manhattan and Citicorp followed suit, each acquiring two firms. None of these British brokers was very big, which is the main reason they were for sale (they all needed more capital to compete in post–Big Bang London), so the actual cash outlay wasn't too large despite the fact that very large premiums over the firms' book values were being asked. Acquiring the firms would enable the banks to participate in the expanding U.K. equities market and its separately supervised government- and corporate-debt markets. They could learn the ins and outs of the securities business abroad before being permitted to participate in it at home. One of the banks, Citicorp, was able to engineer a membership on the Tokyo Stock Exchange through one of its U.K. brokers, Vickers da Costa. The London moves were seen as strategic steps in the larger, unfolding game of global banking.

So far the moves have had mixed reviews. Most of the banks have had problems with the U.K. firms they bought, even before the crash, which did not help at all. Earnings have been poor; integration into the larger banking family has been difficult; some of the key people that came with the firms have left; others have proved to have been overrated; many organizational control issues remain unresolved; and, worst of all, the bull market has been suspended. It will take some years for the banks to sort out their investments in the U.K. securities business. Most likely they will end up writing off much of their investments to date and starting over again, sadder but wiser.

The investment banks have done a little retreating too, although they began the globalization era with many advantages over their commercial-banking counterparts. A successful retreater is one who steps back a little when it first becomes apparent that he should. Investment bankers are good retreaters, just as they are good advancers. They pride themselves on their ability to move quickly in response to clear market signals. And to move again when the market signals change.

Most of the major investment banks saw globalization coming

in the early 1980s. At the time, however, more (indeed much more) was going on in their domestic businesses. The international components had to wait until the rest of the firm was ready for its great global advance, which would require opening new fronts in Europe and in Asia. By the time all of the firms were heading this way, competitive factors were tightening and the going became more difficult, especially in some of the trading areas, such as Eurobonds. Most firms were trying to address broader strategic issues at the same time as they were competing for global stock and bond issues. The process became muddy. Then it became obvious that some of the strategic ground to be gained was going to take much more time to become profitable. Hair-trigger overhead alarms went off. The firms reacted. Layoffs of up to 10 percent were effected in London early in 1987, long before the crash. Most firms had pulled in their horns in the trading businesses, especially Eurobonds. The post–Big Bang U.K. business was approached cautiously. Greater attention was placed on international merger transactions and equity new issues. Nevertheless, there was still turmoil among the U.S. investment banks in the international sectors well into 1988. There is always at least some turmoil during retreats.

By the end of 1988, the investment banks were consolidating their global positions. Not too much strategic ground had been given up during the preceding year. Shifts in personnel, top management, and lines of business were still going on, but the turmoil was receding. Most firms had settled into four basic lines of business: trading, brokerage, investment banking, and investment management. All of these could be, and indeed had to be, conducted on a global basis if the firm had any aspirations to industry leadership. Niches would remain viable for some, but the leading firms did not see themselves as being relegated to niches.

On the other hand, to be a major global player in the future, the firm had to survive, a condition that could not be taken for granted in such stormy times, with markets uncertain and predators lurking in the shadows. Survival would depend on the ability to continue to make good money, in most if not every year. Investment bankers are supposed to do that above all; if they don't, the firms fall apart. Key

people leave, for better opportunities or for safety. Capital i
difficult to attract. Clients look for other, more durable sup
Survival and profitability are synonymous for investment banks.
operations have been a drain on profitability during the last few years
of great expansion, however, and may have to be approached more
gradually in the future. Not everything can be done at once, some
things perhaps not at all.

The key, of course, to assuring profitability is assuring good
management. Investment banks have become far larger, and more
difficult to manage, than at any time in their history.

The large firms such as Merrill Lynch and Shearson Lehman
Hutton have 30,000 to 40,000 employees now, most of whom are
commission salesmen. Bankers Trust, by contrast, slimmer than ever
but still a bank, has 12,000 employees; Morgan Guaranty, 15,000.
First Boston, Goldman Sachs, Morgan Stanley, and Salomon Brothers average around 6,500 employees; all are several times larger than
the firms their senior managers joined. All of the investment banks
will get larger still as global activities bear fruit from seeds already
planted. The larger they get, the more important and difficult good
management becomes.

In an environment in which managers are producers, and one's
standing in the firm derives mainly from one's production of revenues, good management is not always practiced or recognized when
it is. The reason, according to Steve Friedman, Goldman Sachs's
vice chairman and co-chief operating officer, is the perception in the
industry that "real men don't manage," they produce.[2] Friedman's
comment certainly reflects a prevailing attitude that has existed in
investment banking, but one also that most observers, including
Friedman, agree is obsolete. Dennis Dammerman's contribution to
GE depends on how good a manager he is. He is a certainly a "real
man" by Wall Street standards, but also a good manager. There
aren't many like him there today, but the future will be different.

Organizations change greatly as they grow. The great passage
awaiting the major investment banks is an organizational one—from
a Seventh Avenue style of hands-on, undelegated management by
personality to something else, something that can adapt the unique,
high-intensity investment-banking culture to one with a higher man-

agement component in it. Some firms are beginning to do this; others will perhaps pattern themselves on those that do so successfully. But unless they do, investment banks may find that competing effectively on a global scale is too difficult. These will then drop out and leave the field to others to exploit.

# 3

# Paul Revere's Return?

It was a big day for Flat Rock, Michigan. It was the ground-breaking ceremony in the autumn of 1985 for a $450-million automobile plant being built by Mazda Motor Company of Japan. Kenichi Yamamoto, Mazda's president, was there, along with a legion of company executives and well-wishers from Nagasaki, where Mazda kept its headquarters. Ford's chairman, Donald Petersen, and many other senior executives from Ford were there too—Ford had a 25 percent interest in Mazda, which it had acquired in the early 1970s. Governor James Blanchard was there and said, "We want to thank Mazda for being part of Michigan's comeback story." The governor was accompanied by numerous representatives of Michigan's economic development office, several members of Congress representing surrounding districts, and many local officials. Senior representatives of half a dozen auto-industry trade unions were present too. So were the media. The event was covered by two television stations and several representatives of newspapers and magazines. The morning's press had included a full-page advertisement by Mazda featuring Mr. Yamamoto announcing Mazda's commitment to the United States and its confidence in American workers.

Crowded in among all these folks, distinctive in their dark suits and pinstripes, was a modest delegation of bankers come to join in celebrating the completion of months of difficult negotiations associated with the packaging of a unique, low-cost financing program for the project. The bankers' platoon was led by myself and another partner from Goldman Sachs and by Tokuyuki Ono, a senior managing director of Sumitomo Bank, the two firms that had put the financing together.

It was quite a ceremony. The Mazda people had gone to a lot of trouble to make it meaningful and, in a kind of bicultural way, moving. They had hired one of the few Shinto priests in America, a very laid-back Japanese American from Los Angeles, to bless the enterprise in a peculiar, homemade, ecumenical service. None of the Japanese present recognized it for what it was, but, nonetheless, the event made the evening news programs. Civic and union officials expressed their appreciation to Mazda for moving to the area, which had been losing jobs regularly for years. Politicians took credit for everything. Mr. Yamamoto warmly thanked everyone present for making this great day possible. Afterward there was a big lunch for the governor and more thanks and congratulations.

The project had begun some time before, as Mazda assessed the market in the United States for its products and the best way to get them there. Mazda was a born-again company, having nearly perished by overcommitment to the Wankel rotary engine, which, being a heavy fuel user, went nowhere after the oil-price rise of 1973. Ford had a small stake in Mazda at the time, which it had previously tried unsuccessfully to increase; when the company started to collapse, Ford backed away. Mazda's principal banker in Japan was the Sumitomo Bank, financial ruler of Japan's Kansai (western) region. Sumitomo rescued Mazda in a classic example of what was then called "Japan, Inc." at work. Sumitomo organized huge amounts of working capital to support the company while it went through an internal reorganization and severe housecleaning. Sumitomo assigned several of its senior officers to executive positions in the company. It leaned on suppliers, customers, other members of the so-called Sumitomo Group of companies, and anyone else it could to support Mazda. They did, and Mazda came through, reverting to traditional combustion engines, good styling, and a reputation for quality. Sumitomo's relationship was cemented for generations if not forever.

Mazda had been successful since its rescue. Its products sold well and were popular in the United States, though they lagged well behind the big three: Toyota, Nissan, and Honda. Mazda decided that there were a number of reasons to consider building a plant in the United States, one that might assemble various models but would center on a sports car that was very popular with U.S. buyers.

One reason for building a plant was the constant concern about trade restrictions and quotas. The concern that the United States might legislate restrictions on Japanese imports had existed at least for a decade; indeed, negotiated import quotas for automobiles had been in place for several years. These tended to favor the bigger companies, and Mazda was not in an advantageous situation to export cars to the United States if the quotas should be tightened.

On the other hand, the prospect of further trade restriction was something of an old saw—the United States had been threatening legislative retaliation against growing Japanese imports for years, and it had always been avoided. The U.S. government of whatever party had been for free trade and had always been unwilling to support tariff or related barriers. Quotas could be sorted out through the rough and tumble of negotiations, and so far the Japanese had done pretty well. Concern about trade restrictions was not to be ignored, but the Japanese knew it was largely a bluff, a public-relations posture set for both countries.

Economic considerations were more important, but figuring these out was extremely difficult. The big Japanese automakers had nightmares over them. Japanese cars had taken over a large share of the U.S. market. During the 1981–85 period, the U.S. economy was experiencing the first phase of Reaganomics—the twin deficits were building, *real interest rates* (that is, nominal interest rates minus the inflation rate) were extremely high, but inflation was dropping fast so therefore nominal interest rates were too. The dollar rose against the yen and all other currencies. The U.S. economy, stimulated by the large fiscal deficit, was growing faster than the economies of Japan or Western Europe. The United States was the place to sell to—the demand was there, and because of the high value of the dollar, prices on imports, including Japanese cars, were low. The current situation favored more of the same: make as many cars as possible and ship them to California. But the current situation could not go on forever.

When the situation changes, the Japanese thought, the yen will start to rise again, and our exports will become more expensive: we will lose market share. To regain it, we will have to either lose money or invest heavily, very heavily, in large U.S. factories to compete with the U.S. companies on equal terms. This will be extremely expensive for us, not just because of the capital costs, but because of the thousands of difficulties we will have to overcome to manufacture efficiently in the United States. Just to build a plant according to our standards, and incorporating our manufacturing procedures, will require the services of hundreds of engineers whom we will have to send to some remote place in the United States. Very few of these engineers and factory-training people speak English. Few have ever met an American. How in the world are they going to train them to make advanced motorcars using Japanese manufacturing methods? How will we deal with the unions, which are supposed to be so notorious in America? Perhaps we can solve these problems, but it will be expensive to do so, and at least for the moment (1984), it looks like our local costs in dollars will be especially expensive for us in yen terms.

"The worst part," one Mazda executive stated, "is that as soon as we start producing there—in a year or two—we will run into a revitalized U.S. auto industry that will undercut our prices to regain their lost market share. They will push us into the sea, where we will drown in the red ink."

"On the other hand," responded a member of Mr. Yamamoto's "faction" at Mazda, "we have no real choice. Honda has already gone ahead with its plans to manufacture in Marysville, Ohio, and the others are surely going to do the same thing. Toyota is talking to General Motors about a joint manufacturing facility in California, and Nissan's experience in assembling trucks in Tennessee has probably been very instructive to them. We must go ahead, relying on our dedication to success, hard work, and incomparable Japanese spirit. Maybe Ford will help us, too.''

Ford did. Finally Mazda decided to push ahead with a plant in Flat Rock, twenty miles downriver from Detroit, where the costs would be low and skilled labor was available. They asked Sumitomo Bank to advise it on financing, which had to be low cost, they said.

Sumitomo knew they could just lend them the money, in dollars or in yen, whatever they wanted. But, they figured, that may be the traditional way we work with our clients, but in this case it may not produce the lowest cost. If Mazda finances in yen, they will have a huge foreign-exchange risk in the future. If they finance in dollars, the rates will be high. The bank turned the matter over to Mr. Ono, head of all U.S.-based operations of the bank. Ono, a bright, personable, high-flying executive of Sumitomo, was on his second tour of duty in the United States. He had also headed the London operations of the bank before taking up his present assignment. He was a very modern banker, in many senses, but also an extremely traditional one insofar as dealing with his Japanese colleagues was concerned.

Ono's view, which he finally persuaded his Tokyo colleagues to accept, was to explore the possibility of some kind of U.S. financial-engineering solution to the problem of lowering the dollar cost of the financing. "If we find something," he said, "we can perform an important service to our client, Mazda, and learn how to do it ourselves next time. We can ask American investment bankers to come to us as Mazda's representative and present their best ideas. We will then select the best one.''

I heard about the project from the Goldman Sachs Tokyo office executive responsible for calling on Mazda in Nagasaki, a full day's trip from Tokyo. Mazda's people were friendly, but kept our man, Ken Kawashima, at arm's length while they were sorting out how the project should be handled. Kawashima called from Nagasaki and suggested that we contact Sumitomo in New York to get a better reading on the situation. We did, and I went to see Ono, to suggest that Goldman Sachs and Sumitomo comanage a privately arranged leveraged lease in the United States for Mazda.

A *leveraged lease* was ideal for this kind of project because it divided the financing into two parts, debt and equity, and financed each separately with U.S. investors. Mazda would not actually own the plant, but it would have a long-term lease on it and an option to purchase the plant at the end of the lease. Sumitomo Bank, which had a U.S. AAA debt rating, would guarantee the lease payments. The financing costs and depreciation of the plant would provide tax benefits for the U.S. equity investor, which Mazda, anticipating some years of operating losses while things got going, could not use. The U.S. equity investor would pay for the ownership of the tax benefits for the future, further reducing the comparatively low cost of the borrowing (at AAA rates). The net cost of the dollar financing to Mazda would be around 5 percent, much lower than the finance people in Nagasaki had expected.

Ono said it sounded just right. He was sure we could work together, as we had before. He would talk to Tokyo and get back to us.

"Nothing Japanese is easy" is a lesson I had already learned. Still, Ono sounded convinced. A few days later, he asked to see the team that had prepared the proposal I had given to him. I was asked not to bother with this meeting. Our team, headed by my partner David George, an extremely resourceful and unflappable lawyer-turned-banker who was in charge of private financing at Goldman Sachs at the time, was delighted with the news and appreciative that my close relationship with Ono had made things so easy. "Let's wait and see," I said. Later David called to tell me how things had gone.

"Quite a meeting," he reported calmly. "They like our idea, which is good, but they seem to think we've insulted them, and therefore are putting our idea up for competitive bidding with sixteen others. Other than that, it was a great meeting."

I asked for some details. "Ono came in and really beat up on us. He started by complaining that we Americans don't know anything about business, certainly not as compared to the 'merchants of Kansai,' whoever they are. He said he would offer us some advice. He began by complaining that when his people were over at our office last week, the coffee was not served correctly." David continued incredulously: "We should know that in Japan, bankers are the 'slaves' and clients are the 'kings.' Though Sumitomo is a slave to its client,

86

Mazda, the king; in New York, as Mazda's representative, Sumitomo is the king and we are the slaves.''

"As slaves," Ono continued, according to David, "you must subject yourselves to us completely. Whatever we want, you must accept.

"In this case," Ono continued, "our Tokyo [office] wants us to act purely as Mazda's representative. We are not to be directly involved as comanager or anything, except as guarantor of Mazda's lease. We are also required to be sure that Mazda receives the most competitive proposal, so we will be talking to other banks and investment banks about their ideas. You will be permitted to submit another proposal."

David, quite naturally, was horrified by this turn of events. "Wait and see," I repeated.

Sixteen others were invited to make proposals. We won the contest, with pretty much what we had offered in the first place. Ono told us that there were many good ideas, and aggressive quotes, but they thought our proposal, if not absolutely the cheapest, was highly credible and therefore we were selected. Everybody saved face. Everybody was happy, except perhaps some of the other sixteen, who might have thought they were being used as stalking horses. For all of the financial participants, the deal was a very good one. Mazda was delighted with the financing; therefore so was its slave, Sumitomo. The value added in the transaction by the bankers was high, and so was the fee. We global bankers set off afterward to find more business like it to do.

Over the last decade or so, the Japanese have been increasing direct investment in the United States. They have bought up over six hundred factories in the United States, many of which were on the edge of being abandoned by their U.S. owners.[1] Such investments have been concentrated in the electronics and automotive sectors. In other cases, whole companies have been bought, as in the case of Bridgestone's $2.6-billion acquisition of Firestone, Sony's $2-billion acquisition of CBS Records, and Dai Nippon Ink & Chemical's $1-billion purchases of Reichhold Chemicals and the graphic arts division of Sun Chemical Corporation. Japanese banks and insurance companies have

87

also been especially active in acquiring U.S. holdings in banks and financial-service firms. Other Japanese investors, including some unlikely ones such as a Tokyo taxicab company, have also acquired several billion dollars of U.S. real estate holdings.

Still, portfolio investments, in bonds and shares in U.S. companies, have nonetheless attracted about 70 percent of all Japanese investment in the United States over the past several years. Japan's total outflow of capital, consisting of its direct and portfolio investments, has more than balanced its exceptional trade surplus. Such a large amount of capital flowing into the United States, nearly $80 billion in 1986, cannot avoid attracting a lot of attention.[2]

Despite the joy in Flat Rock on that sunny day in 1985, the mood has cooled. After years of courting foreign investors, many Americans are reconsidering it. Clayton Yeutter, the former U.S. trade representative, commented at an elite world economic gathering in Davos, Switzerland, in early 1988 that the pace of foreign investment in the United States is creating a "backlash."[3] No less a "capitalist tool" than Malcolm Forbes proposed a presidential panel be formed for the purpose of approving individual foreign investments. Later in the year, journalists Martin and Susan Tolchin published a scare book on foreign investment, *Buying into America: How Foreign Money Is Changing the Face of Our Nation.* The trade bill that didn't pass in 1988 included, at various stages, proposals requiring the disclosure of foreign investment in considerable detail. Even diplomatic Nicholas Brady, author of the Brady report on the causes of the stock-market crash in October 1987 and currently secretary of the treasury, was reported to have suggested that Japanese dumping of U.S. Treasuries had been an important precipitating factor in the crash.[4] As everyone knew, Japanese investors had been buying up U.S. Treasuries for some time and were sitting on a small hoard of them. Buying them up was seen as a fairly good thing while it was going on. They were, after all, financing our deficit. But dumping the bonds was a different story, a downright unfriendly act, some thought.

Yet, observes Harvard's Professor Raymond Vernon, who has studied multinational corporations for two decades, "When Americans begin frowning on foreign direct investment, one cannot help thinking of the missionary turned cannibal. For decades, U.S. representatives have been the world's principal proponents for the virtues of such investment."[5]

Have things changed so much? Have our economic difficulties caused us to become xenophobic? Is Clayton Yeutter right? Is the country mounting up for a financial Paul Revere's ride? Will we soon encounter a man on a horse warning the countryside of the impending appearance of foreign capital in our villages and towns—capital that will make us subservient to others across the seas? capital that will separate us from the sovereign control of our own affairs? capital that will deprive our workers of high-paying jobs, reducing us to Third World status?

None of these points is easy to justify based on the facts, but the issue probably doesn't turn as much on the facts as it does on emotional appeal. A lot has happened in a relatively short time, and much of this takes some emotional getting used to. Just previous to the beginning of the great Reagan bull market in stocks, bonds, and the dollar in 1982, Commerce Department figures showed foreigners owning about $700 billion of U.S. assets, including securities, wholly owned companies, real estate, bank deposits, and other assets. By the end of September 1987 (just before the bull market–ending crash), foreigners had doubled their holdings to $1.4 trillion.

In five years, foreigners had invested as much in the United States as they had in the preceding 211 years. And the United States had decreased its holdings of foreign assets during the period and instead turned itself into the world's largest debtor nation. What a change from the great days of riches and unchallenged economic power following World War II, or for that matter even following World War I. It takes some emotional getting used to.

The basic laws of economics apply in this case, as in, alas, most others. What goes out (in the form of the trade deficit) must come in (in the form of direct and portfolio investment—payment imbalances no longer being settled by shipments of gold between countries). Either the investment is made by the foreign private sector, or it is made by foreign central banks who buy to support the dollar against their own currency and/or to add them to their foreign-exchange reserves.

Eighty percent of the $1.4 trillion ($1,400,000,000,000) of foreign investment is in government or corporate securities, called *port-*

*folio investments.* Of this about $200–300 billion is invested in the approximately $2 trillion of government securities that make up the national debt—10 to 15 percent of the total, roughly half of which is held by foreign central banks as part of their foreign-exchange reserves. So only 5 to 7 percent, or so, of our outstanding Treasury securities is held by *all* foreign private-sector investors, including pension funds, investment companies, insurance companies, corporations, and banks. Half of *this* amount, maybe somewhat more, is owned by the Japanese portfolio investors we have been hearing about. A lot of dollars' worth of our debt, to be sure, but we have so much debt outstanding that it doesn't amount to very much as a percent of the total, about 3–4 percent.

It is true that during certain periods from 1984 through 1986 Japanese investors had a big impact on the market by subscribing to as much as 30 percent of new issues of certain Treasury securities at auction. Such concentrated purchases had an effect on the pricing of the issues concerned and influenced the market in a general way beyond that. This buying activity, however, was concentrated into a relatively small number of issues and only into subscriptions for new issues. Between auctions, however, Japanese investors would sell Treasuries as they became more active as traders. Japanese securities firms and banks were active bidders at Treasury auctions in order to demonstrate their firepower in Japan and to impress the Federal Reserve with their growing qualifications to become primary market dealers in U.S. Treasury securities, something these Japanese global bankers all aspired to be, and which many of them have since become.

Also, at the end of September 1987, foreigners were shown to be the owners of $445 billion of banking assets in the United States, over $300 billion in diversified stocks and bonds, about $300 million in direct investments in factories, warehouses, plants, and so on, and $100 billion in real estate. The growth has been fast—remember foreign direct investment doubled in five years—but in relative terms, total foreign investment is still small, accounting for 5 percent or so of total U.S. assets. In reality, it is a lot smaller than that because the value of the U.S. assets (the denominator in the fraction) is not recorded at market value, whereas the foreign holdings are. Make the whole thing closer to 3 percent. Not all that much.[6]

Ten years ago the largest direct investors in the United States were the Dutch, accounting for 20 percent of all foreign direct investment as represented by the large U.S. holdings of Unilever, Royal Dutch Shell, Philips Lamp, Akzo (a large chemical company), and by substantial real estate and other investments of Dutch pension funds. In second place, totaling 15 percent, were the British, represented mainly by industrial holdings in various companies (foreign-exchange controls still limited pension-fund investments). Next was Canada, then Germany and Japan at 5.5 percent each.

A decade later, British investment had grown to 25 percent of the total, reflecting years of heavy corporate acquisition activity including BP's $7-billion purchase in 1987 of the outstanding minority interest in the Standard Oil Company (Sohio). The Dutch, remaining at 20 percent of total, were in second place. The Japanese, now at about 12 percent, had risen to third place, displacing Canada and Germany. Europeans as a group, however, controlled more than three times the amount of direct investments owned by the Japanese.

An accurate report to his townsmen by today's Paul Revere would have to state that "the British, and the Dutch, Japanese, Australians, and Germans are coming," but it could also add that "the Canadians, the French, and most of the rest are going."

Despite the fact that the total amount of foreign investment is small in relation to that owned by Americans, perhaps as small as it has ever been in our history, some serious perceptional problems remain. Highly visible companies and products have become foreign-owned. Carnation Milk has been bought by Nestlé of Switzerland, Sohio by BP, Firestone by Bridgestone after a takeover effort by Pirelli and Michelin failed, and the consumer electronics business of GE and RCA by Thomson of France.

The perception is that foreign companies, especially the Japanese, have so much money and their money is worth so much more than ours that there is nothing Americans can do to prevent a massive takeover of their best and most important companies by foreigners. When that happens, we'll all be working for them.

To the extent that this is true, it is no more than the global free market in action once again. Dealing with foreigners, we've learned several times in our history, is not all peaches and cream. If we're going to expect the benefits of free capital flows, then we will have to learn to live with the problems.

Most foreign companies, however, fear that if they misbehave in America, or are perceived as misbehaving, then "the people" will settle the score, by boycotting their products, by tying them up in litigation or regulation, or, if need be, by legislation. Foreign companies abuse the American market at their own risk. Executives of Toshiba were horrified to watch on the morning news congressmen smashing their consumer-electronics products with sledgehammers, after the sale of propeller-making machinery to the Russians by an affiliated company was disclosed. Companies with business in South Africa have been forced to divest because of pressure from the people. Foreigners know this; they rarely misbehave.

Different perceptions of things can produce funny results in the United States. In 1987, the giant Japanese computer company Fujitsu announced that it was going to purchase Fairchild Semiconductor, a major U.S. producer of computer chips. The proposed transaction drew a lot of fire. The Japanese were thought to be untrustworthy and devious competitors in the semiconductor industry. A lot of resistance to the transaction developed. The Defense Department was said to object. The government applied pressure on Fujitsu, which backed away. An American victory? Not really. Fairchild was nearly bankrupt as a result of overexpansion and mismanagement. It had been offered to lots of other buyers, without success. Fujitsu finally agreed to take it, with the understanding that it would invest in its restoration. Fairchild's parent company, Schlumberger, the seller in this transaction, was a French company that was eager to see the sale completed as were most of the management and employees even though Schlumberger would take a $200-million loss on the transaction. Schlumberger had few alternatives to the sale to Fujitsu. But the U.S. government killed the deal. Who lost? Fujitsu? It probably managed to save a lot of money; in any case, it went ahead with plans to invest $150 million in a new semiconductor manufacturing plant in Oregon that had previously been set aside in favor of Fairchild.

Following the forced withdrawal of Fujitsu, the House of Representatives' Subcommittee on Competition introduced a new amendment to the pending trade bill providing the president with new authority to halt foreign acquisitions of U.S. firms in cases where the national security of the United States was at stake. This provision remained in the final version of the trade bill that was signed by the president in August 1988, though no doubt no one has any idea of what it means or how it would be applied.

The perception of the reality in the Fairchild case was far different from the reality itself. The government, according to one U.S. executive involved in the affair, created "a whole lot of emotion, nostalgia, and misguided competitive zeal that are really irrelevant to the merging of two foreign companies' semi-conductor operations in the United States." [7]

Perceptions also disguise racial and other factors that can enter into foreign investments. British investors are considered highly acceptable foreigners; they can get away with almost anything. Sir James Goldsmith, however, is notorious for raiding companies and subjecting them to greenmail demands. Hanson Trust has launched several unfriendly takeover bids for U.S. companies. Many transactions have occurred in which the target company was broken up into pieces by a British acquirer and then sold separately. None of these transactions encountered much, if any, public resistance. Imagine, however, if the same had been tried by a Japanese company, or an Arab one. In the late 1970s, when the Arabs had all the money, they were very careful to keep a low profile with their investment activities so as not to provoke American resistance. It used to be said that Saudi Arabia could buy all of General Motors with just a few weeks' cash flow. Maybe, but there was never any question that Saudi Arabia would try to do such a thing. The Japanese today feel much the same way.

There have never been any legal restrictions on the acquisitions of properties in the United States by foreigners other than on the acquisition of certain broadcasting, transportation, and military businesses and of farmland in some states. Otherwise, anyone who wants to can, and always could, buy companies in this country. As a nation, we have always favored free-market solutions to competitive problems, and we should like to see other countries adopt similar policies.

American companies have bought a lot of companies overseas and indeed are continuing to do so. For years U.S. businessmen protested restrictions on their foreign acquisitions by Japan, France, Canada, and other countries and objected when Europeans, incited by Jean-Jacques Servan Schreiber, began to complain in 1967 about "Le Défi Américain" and its aggressive management style. U.S. businessmen do not want to revert to those dark days. They do not want restrictions imposed on them as a consequence of restrictions being imposed on foreigners in the United States. Neither do the owners of U.S. companies want to be restricted from selling them to foreigners.

The real objections to foreign investment seem to be those of hurt national pride. It is not so much the fact that foreigners are investing in the United States; it is the speed and extent to which it has occurred. Few Americans consult the tenets of economic law to explain the phenomenon, the laws of consequences following a quick change from creditor to debtor nation. Mainly, Americans see that foreign financial clout has increased, apparently at the expense of ours. We seem weaker as a result, our national character somehow debauched by too easy foreign money. According to one critic, "The family jewels are being sold to pay for a night on the town."

Most Americans over forty began life comfortably aware of their citizenship in the world's most powerful and important land. Others looked up to us, whether or not they wanted to, for our wealth and power. Some admired our wisdom or generosity or personal freedoms, but they didn't miss the fact that these were packaged with great financial resources. Americans over forty have a hard time watching the sun set on the mightiest days of the American Republic, which is the way many of these people see what's happening today in the shadows of the twin deficits.

Those under forty, perhaps, are better prepared for the future— American power has not during their lifetimes been unchallenged, or indeed undefeated. The future requires give-and-take, cooperation with others, recognizing our limitations. It also requires hard work and competition if one is to hold one's own. This picture of sharing power and wealth in a peaceful global economy may be the more realistic one. There is more than a little irony in the legacy of our oldest, most conservative president becoming the principal, if unintended, cause of our accepting the economic reality of power-sharing, something the young seem already to understand how to live with.

Descending from the philosophical heights of the issue, where heavy traces of the backlash identified by Mr. Yeutter are visible, we can inspect some of the more focused, practical objections to foreign investment.

The first point that often comes up is our vulnerability to sudden withdrawal of foreign investments in U.S. Treasury and other money-market instruments. If foreigners dump Treasury bills (T-bills) in the market, will that bring on another crash, as Mr. Brady perhaps believes it will, or push U.S. interest rates through the roof? The latter concern was often ascribed to Paul Volcker while he was chairman of the Federal Reserve Board. The answer in both cases is "Maybe, but . . ."

As I have indicated already, the foreigners only hold about 10 percent of our national debt, and about half of this is in the hands of coordination-conscious central banks who at least think about the consequences of dumping large amounts of T-bills before they do it. The rest is divided among private-sector investors from many countries who usually have very different motivations for buying and selling U.S. securities. The Japanese are the most potent holders of Treasury securities, and their behavior in the market can be very influential, especially on the margin, which reflects the price effect of their actions over the last five minutes.

If Nippon Life Insurance Company has been a large seller of thirty-day T-bills over the last five minutes, the market in such paper may drop very sharply, despite the fact that Nippon Life may own less than 1 percent of all outstanding thirty-day bills. Seeing Nippon dumping its holdings may prompt other Japanese or U.S. holders, or dealers, to do the same, and the market decline could increase. Sooner or later, Nippon runs out of T-bills to sell, and by then other investors will begin to find the price level to which T-bills have dropped attractive, and the process begins to reverse. Usually it takes more than one or two big sellers to crunch the market, unless something else is going on.

If the dollar is dropping, or interest rates are soaring as a result of bad economic news or a change in monetary or fiscal policy that the market doesn't like, then incentives may exist for lots of investors to sell and markets may react sharply. When they do, the same rules apply: sellers run out of supply and buyers are attracted by the new price levels. But if the market's normal price corrections are amplified by simultaneous dumping by foreigners, short-term intervention by the Federal Reserve and its allies at other central banks may be necessary to help stabilize prices. Such intervention may be too little, or too late, and with enough selling a market crash could occur in one country that could trigger crashes in other countries as happened to stock markets on October 19, 1987. Certainly there are dangers of market collapse that can be heightened by large, potentially unstable foreign holdings.

It is true that foreigners could liquidate their holdings of U.S. securities, and this could have a big, though probably not a long-lasting, effect on various markets. In time the markets correct themselves. But what do the foreigners do with the proceeds of the sales of their dollar assets? They have to put them somewhere. The Japanese could buy yen bonds, for example, which would raise the price of both the yen and the bonds. The Japanese don't want the yen to be all that strong, for fear of making it impossible to export Mazdas and other goods. So the Bank of Japan, in support of the dollar (that is, in support of keeping the yen down), buys the dollars in the foreign-exchange market that Nippon Life sold there a few minutes before. So, in fact, the dollars are still held by Japanese, just different ones.

Of course, Nippon Life may decide that it would rather avoid getting back into low-yielding yen bonds and instead buy German bonds. Sooner rather than later, the German bond market, which is much smaller than the U.S. or Japanese bond market, reacts, setting the same forces in motion. The Bundesbank buys dollars in the market to offset the rise in the deutsche mark and in German bond prices. The net effect is that the Japanese dollar holdings have been transferred to Germans. There are limits on the magnitudes and the timing of central-bank intervention, and sometimes even coordinated efforts by central banks to restrain the forces of economic adjustment cannot succeed in doing so. Prices are supposed to change as a result of these adjustments, and they do.

The important thing is that markets impose discipline on policymakers, whether they like it or not. If policymakers are cavalier about fiscal and international deficits, then the markets will reprice financial assets accordingly. Something will have to give; at the moment that something is the foreign-exchange value of the dollar, which has declined by more than 50 percent against the yen and certain other currencies since its high in 1985, and the flow of foreign investments into the United States, which have doubled since 1982. A 50-percent change in the value of the dollar over three years is extraordinary. Such a big change has made direct investment in the United States much cheaper for foreigners. The $750-million acquisition of the Union Bank of California by the Bank of Tokyo in early 1988, for example, cost the Japanese Y97.5 billion at the time; if the purchase had been made two and a half years earlier, the cost would have been Y165 billion.[8]

After the 1973 oil-price rise, I made an unsuccessful effort to promote the idea that the Saudi Arabians needed us more than we needed them, financially if not geologically. Suddenly, they appeared on the scene with all the money, and those who wanted the use of it had to go to them hat in hand. However, as gargantuan investors, they needed to be sure of access to markets offering safe and sound investments, which they did not want either their commercial or their financial activities to disrupt. The Saudi Arabian Monetary Authority (SAMA) was for some years the world's most important financier. The SAMA staff was supplemented by hired experts, one of whom was David Mulford, who was seconded to Jedda and later to Riyadh by Merrill Lynch. Mulford remained in Saudi Arabia for the incredible period of ten years before becoming assistant U.S. secretary of the treasury for international affairs (now under secretary), an excellent position for someone with his very special background. While at SAMA,

Mulford was in charge of reviewing investment proposals and putting together an investment strategy satisfactory to his ultraconservative and financially unsophisticated employers.

This was a difficult job, because the Saudis didn't want to invest in anything that wasn't top quality, and usually pretty short-term. Considering the enormous cash flow passing through the place, SAMA found itself limited by acceptable investments. At first they put a lot into bank deposits and government securities. They worried, though, that the banks would not be able to handle so much money and limited their deposits to a handful of those who qualified, like Walter Wriston's Citibank. They also worried that their transactions would move the government securities markets against them. Mulford and his colleagues devised a number of ways to live with these problems, but they continued to have them until the oil-price decline in the 1980s.

My idea was that as much as the United States wanted Saudi money to flow back to the United States to offset the effects of the sudden trade deficit, a reflow that was ultimately inevitable, the Saudis needed the United States to invest in. The rest of the world wasn't big enough to absorb all the money being pumped into Saudi Arabia and out again into acceptably safe investments. We could, I thought, suggest to the Saudis that if they wanted to invest in the United States they would have to do it our way, that is, provide a certain amount on a regular basis to be managed by the Federal Reserve or the BIS, or somebody like that, in a spread of maturities, ranging from short to long, in a variety of different securities from corporations and government agencies. In exchange for this gracious and convenient access, the Saudis would also consign a certain portion of their cash flow to investments, perhaps through the IMF or the World Bank, in the capital-deficient Third World. Generally, this was considered a stupid idea because the investment world was already throwing themselves at SAMA, which had a lot of bargaining power, the U.S. markets were already open to SAMA and everyone else, and, perhaps thanks to Mulford's financial skills, SAMA didn't know it had a problem. At this writing, one of Mulford's jobs at the Treasury is devising a U.S. position on the Third World debt problem that won't end up bankrupting some of our largest banks. Perhaps there are times when he considers that his present job might have been a lot easier if he had never had his previous one.

A common complaint raised about foreign direct investment is the fact that "we" are losing control of so many of "our" great companies. The complaint was not aired during the early days after wealth was discovered in SAMA and then in Japan, when the great companies themselves were among the first to offer investment opportunities to foreigners, but after more sober, morning-after reflection, it has been. I always wonder, however, who we mean by *we* in this context. Do we mean *we* the institutional investors of America, the Prudential Insurances and the Dreyfus Funds? American households have been net sellers of common stocks for the past twenty years, so we probably don't mean them.

Do we mean *we* the management of great American companies? Is Roger Smith, the embattled chairman of General Motors, one of "us"? Would we be seriously distressed if a foreigner, that is, some-one foreign-born, were to become chairman of General Motors? The chairman of H.J. Heinz, whose principal stockholder is a U.S. senator, is Tony O'Reilly, an Irish-born dynamo who continues to maintain residences and investments in Ireland. Jack Welch's predecessor as chief executive of GE, Reginald Jones, was British-born. There are many other examples. We don't seem to care too much what the nationality of top management is.

We do seem to like it, however, when the management of large U.S. subsidiaries of foreign companies is American, as it almost always is. The management team at BATUS, the U.S. holding company of BAT Industries, one of Britain's largest companies, which owns Brown and Williamson Tobacco, Marshall Field, Saks Fifth Avenue, and Farmers [Insurance] Group, among other companies, is run by Hank Frigon, a smart, tough, energetic American. There are no large foreign-owned companies in the United States with predominantly non-American management except for Japanese banks and securities firms and they are changing. Foreign companies value American managers highly and feel dependent on them for coping with uniquely American problems, of which, they will tell you, there are many. So, it's not like Paul Revere's time, when armed foreigners

99

were quartered among us. The foreign managers in America are over-whelmingly American. If "we" are losing control of some of our companies to foreigners, do "we" get credit for the Americans that are controlling foreign-owned companies in the United States?

"No," a critic replies. "It's not who shuffles the paper that counts, it's who makes the key decisions."

Here it is true that the top management group and the board of directors of foreign companies are almost always foreign themselves. There are a few British companies, BOC Group and Beecham, with American chief executives, and there are a few Americans who are directors of foreign companies (I am one), but not many. Most of the key decisions that companies make regarding their foreign subsidi-aries concern commercial issues—their objective is to run their busi-nesses successfully, to make money. Key commercial decisions made by foreigners are virtually the same as those that their U.S. compet-itors make. If a meat-packing plant in Iowa has to be closed, they close it. The Iowans may object, and indeed consider anyone with a head office outside of Des Moines to be an unsympathetic foreigner, but these are the kinds of key decisions that key decision makers make.

There can be a problem if noncommercial issues get involved. During World War I, many European companies were required to liquidate investments in the United States in order to raise money for the war effort. The companies did not, however, close their U.S. operations down, as many feared; instead they sold them to Ameri-cans not affected by the same regulations. Some managed to mort-gage the companies and repatriate the proceeds of the loans that were repaid after the war. But despite the intensity and the scale of the problem at the time, it did not result in closing down healthy U.S. companies and throwing people out of work.

Similarly, during World War II, concerns were expressed about German-owned companies such as Bayer and GAF continuing to op-erate for the benefit of the enemy's war effort. A simple solution was quickly found: their U.S. subsidiaries were interned, that is, they were taken over by the U.S. government and operated under the terms of the Alien Property Custody Act.

Concerns about foreign "exploitation" of American companies

through manipulation of transfer prices or violation of labor or tax laws or those protecting patents and other proprietary matters have been assuaged by the legal system in the United States. Complaints can be filed by the government or by citizens in objection to the behavior of others, including foreigners, who are alleged to have violated laws or agreements in this country. A court can issue an order to enjoin actions by parent companies regarding subsidiaries, and of course, it can find against a defendant. Being a foreigner is no protection in itself in civil litigation. The foreigner may never set foot in the United States in order to avoid legal action, but his U.S. assets can be tied up forever if he does not submit to the court. On the whole, foreigners complain about the ferocity of the U.S. legal system far more than any affected Americans might complain about their exploitation by foreign countries. If foreign companies are going to own American properties, they themselves become American under the law.

However, when different national regulations are imposed on parents and subsidiaries, complex issues of corporate sovereignty emerge. Key foreign decision makers under such circumstances may be forced to decide things differently than would indigenous decision makers.

"Right," says our critic, "and that's just when Americans lose control of their companies."

Perhaps some control is lost but not always to one's disadvantage. During the 1973 Arab oil embargo, the producing countries shut off shipments to certain consumer countries, including the United States. Companies from certain other countries, however, were free to purchase oil. The situation was a mess for some time, but oil purchased by subsidiary companies in the free areas was resold to purchasers in restricted areas and thus oil continued to flow to where it was needed. Multinational corporations have coped with these sorts of problems for many years, long before the time when Americans began to worry about losing control to foreigners. In fact, some of the early lessons learned by these companies involved how to operate internationally despite the passage of laws in some areas to restrict their operations because they were large, unfeeling foreign companies.

101

Our critic is not convinced. "That may be fine for big oil and multinational corporations," he responds, "but the fact is that many industries have been wiped out by Japanese competitors in particular, with their low wages and government subsidies, and now they are coming in here to buy up the pieces of what's left in the electronics, steel, and semiconductor industries.

"How can anyone claiming to have American interests at heart believe that this predatory behavior should be allowed to continue?"

Well, our critic has now come to the heart of the matter. Foreign direct investment isn't so bad in itself, most Americans appear to believe, but when it is tied to the end of the unfair foreign competition issue, then it is.

This discussion, however, is not about free trade, which I support, but about free investment. The two go together, of course, which is why the issue gets thorny. Americans hate the idea of having gone soft, of having allowed themselves to become noncompetitive. We are, after all, the champions of the free-enterprise system. Our national heroes include characters from the stories of Horatio Alger; Henry Ford; and his grandson's great nemesis, Lee Iacocca, America's best-selling author of autobiographies. We believe in competition, true grit, and the survival of the fittest. Except when the other side isn't playing fair; that we just can't stand for.

The only way we seem to be able to reconcile our fall from economic preeminence, while preserving our honor and national values, is by assertions of cheating. That this is not often true creates a lot of the problems that now include a swelling sentiment of undisguised protectionism, most recently articulated by Congressman Richard Gephardt in his unsuccessful 1988 campaign for the presidential nomination of his party.

The economic profession, with unusual solidarity, opposes protectionism, believing that we are all better off in a freely competitive system in which to survive one must stay sharp, regardless of whether the other side cheats or not. Economists also believe that neither the Japanese nor the Europeans are big cheaters, on the whole. They have too much to lose by so doing and are too vulnerable to retaliation. The U.S. government has found very few cases of actual dumping of goods in the United States despite hundreds, if not thousands, of investigations.

World trade is far too important to all countries today to turn our backs on it, even if we could. The Smoot-Hawley Tariff Act in 1930 guaranteed the coming of the Great Depression. Trade now represents between 10 and 40 percent of the gross national products of all of the major OECD economies, including our own (20 percent). Moving in the direction of protectionism would be self-destructive. Economists know this. So do most businessmen, excepting those under competitive attack at the time, and so do global bankers.

Economist and management expert Peter Drucker recently noted that "wage levels for blue-collar workers are becoming increasingly irrelevant in world competition. Productivity still matters—indeed it matters increasingly more. Quality, design, service, innovation, marketing, all are becoming more important. But blue-collar wages as a direct cost are rapidly becoming a minor factor." The reason for this, Drucker explains, is that "blue-collar labor no longer accounts for enough of total costs to give low wages much competitive advantage."[9] If wage costs have declined to 15 to 20 percent of total costs, from 50 percent or so, as he suggests, we can see his point. What follows from this point is interesting. If wages are less important, then other things will determine the competitive edge. Those other things, in a nutshell, amount to good management, which encompasses competence in lowering labor costs further, managing the human "software" of business, handling new technology, and dealing proficiently and competitively, like John Browne and Dennis Dammerman, with finance. These are things American companies are good at, or need to be good at in order to survive. If foreigners don't have an advantage in labor rates, and aren't able to get away with cheating all that much, then maybe the playing field is a lot more level than we thought.

Naturally, the poorly managed companies will sink to the bottom, like the Russian fleet in 1905, and though we may grieve for them, their fate is only the result of their failing to stay fit enough to survive. Foreign investment in the United States will help.

Many foreign companies have invested in industries that Americans, for whatever reasons, have let slide. Consumer electronics, a high-tech business that we invented, was allowed to deteriorate. We thought our position was so secure that it was invulnerable. We hardly noticed the presence of the Japanese in the market with their minia-ture TVs and stereos. Before long, however, the superiority of Japa-nese products won the market for them. Where were the great Amer-ican companies during this incursion into their territory?

One important television maker, Motorola, decided to get out of television manufacturing in 1974, even though it had been in the business since the beginning and had a well-established brand name, Quasar. It chose to alter its corporate strategy to escape the killing ground of consumer electronics and move up into the cool highlands of semiconductors. The company decided to put its money-losing television division on the block. Goldman Sachs was asked to help sell it. Matsushita, the giant Japanese electronics company whose Panasonic brand was well established in the United States, was inter-ested. However, the Justice Department thought Matsushita would present an antitrust problem by acquiring Motorola's share of the market. So we went out to about a hundred potential buyers, trying to find someone else. There were no U.S. companies who wanted the business, which employed several thousand employees and controlled a substantial franchise in the television market. Finally, Motorola went back to the Justice Department and said they had been to all of these companies without success, so they ought to be allowed to sell it to Matsushita for $100 million rather than liquidate it. They were.

Matsushita took over and for all practical purposes sent the bull-dozers in to clear out all of Motorola's obsolete manufacturing equip-ment, which was replaced by their own at the cost of another $100 million.[10] After two or three years, the plant was operating efficiently and making money for Matsushita, which has become the largest consumer electronics company in the world, with a huge, healthy business in the United States.

In 1984, Nippon Kokan, a large Japanese steel maker, acquired a 50-percent interest in the failing National Steel Corporation and installed new steel-making equipment to improve production effi-ciency. Several other Japanese companies have done the same with other U.S. steel companies in which they have taken minority interests. New investment was flowing into the sickest parts of the U.S. steel industry through foreign investment. Foreigners have taken much of the ailing tire industry off our hands too. Altogether, these transactions have resulted in the infusion of capital into businesses that are not attracting American investment, and in the introduction of advanced process-engineering technology, which is operated by American workers.

Skeptics say that the foreign companies keep their "really good technology" at home where they manufacture the "really important parts" so these can be shipped to America for assembly there by comparatively unskilled workers. "The really good jobs," they say, "the highly skilled, high-paying blue-collar jobs, are being lost to foreign manufacturers. It may take three times as many Americans to sell over here the cars that Japanese make over there, but the Japanese make more money from doing their job than all the Americans do from doing theirs."

This might be true, at least to some degree. The best technology may be kept at home, but what the Japanese, in particular, excel at is process technology, which they are sending to the United States as they begin to manufacture here. It would make no sense for them to try to compete in the United States with out-of-date manufacturing methods. Matsushita didn't. Nippon Kokan didn't. Mazda isn't. If they make the investment, then they support it as best they can.

Many of the new knowledge-based industries that are coming along (for example, biotechnologies) are being developed in the United States and in Europe. However, the technologies we need to manufacture these new products competitively come from Japan. These, the *manufacturing process technologies,* are the ones being transferred into the United States by foreign investment.

It is also true that skilled engine makers earn more than new-car salesmen working on commission. In the United States there are millions of people in sales and service jobs working for American subsidiaries of foreign companies. Those people would not be in those jobs if they had something better they could do. Perhaps before becoming salesmen they had been farmers or soldiers or laborers. Few, if any, had given up better jobs as machine-tool operators to make a living selling Subarus.

105

American highly skilled blue-collar workers have found plenty of work, their unemployment rate is low, and their pay high. They continue to attract young people to their ranks. As more foreign manufacturing develops, presumably the market for such jobs will expand further.

In any case, Peter Drucker says, it doesn't matter. The competitive edge will be gained on the basis of management skills, which Americans are good at, and are able to improve at low cost. If he is right, what we really need to succeed in the next few decades is tough, healthy competition to sharpen our management skills. No more lazy industries taking their future for granted, no more resting on past laurels, now long faded. We need not be so paranoid about foreigners taking us over as a result of unfair play, unless, of course, we prefer to remain uncompetitive, in which case we should be paranoid that the foreigners might not take us over.

"But, what about the banks?" asks our devil's advocate. "Aren't the banks the next great industry to be overrun by the Japanese? Aren't they guilty of dumping money all over the place?"

Japanese banks have been very active in the United States for many years. During the past decade, many of them have acquired U.S. banks in California, in Illinois, and in other states. The Bank of Tokyo acquired the Union Bank from its distressed British parent, Standard and Chartered; Sanwa Bank bought Lloyds Bank of California from *its* British parent, Lloyds Bank; the Industrial Bank of Japan acquired a majority interest in J. Henry Schroder Bank and Trust Company from *its* British parent, Schroders. These transactions, aggregating over $1 billion, displaced British, not American ownership of U.S. banking assets.

Also, however, the Sanwa Bank acquired the industrial lease-financing business of Continental Illinois when it was forced to liquidate various nonbanking businesses. The Fuji Bank acquired Walter E. Heller, a troubled commercial- and consumer-financing company. The Mitsubishi Bank purchased the Bank of California; and the Mitsui Bank, the Manufacturers Bank of Los Angeles. These transactions, which totaled another $1.3 billion, were largely distress sales by U.S. parents. Most of the leading Japanese commercial banks have

106

bought something in the United States as a base for further development of their business here.

What they gain by these acquisitions is a sizable book of business, a client base, an infrastructure, local employees, and a local presence. It takes quite a while to develop these things for a business being operated by them outside of Japan; it is a learning experience as much as an investment. The return on the investment may be low for many years, but they can afford to be patient in the long-term interest of the bank. Besides, relative to the size of the parent banks, all of these acquisitions by the Japanese have been very small. In that context, they have not been bold moves, but cautious experiments.

The Japanese have not been the only ones to take this route to acquire banks in the United States. The British were the most active, first in acquiring banks here but then the first also to sell them. Significant acquisitions have been made by most of the large British banks, the most aggressive in this area being the National Westminster Bank, which has purchased the National Bank of North America and the First Jersey National Bank of New Jersey.

The most celebrated British bank acquisition was the purchase of a 51-percent interest in Crocker National Bank by Midland Bank in 1981. Midland had troubles with Crocker right from the beginning, but relied on local management to work them out, which it failed to do. Midland increased its ownership of the bank as it was entitled to do under the terms of the original investment and finally took over management control itself. By then Crocker was in a real mess. Large write-offs followed which had to be reflected on Midland's books too. Both banks were then reflecting Crocker's troubles. Midland finally decided to acquire all of the rest of the Crocker shares and to rebuild the bank as best as possible preparatory to selling it. This they were finally able to do by selling the better-quality Crocker assets to Wells Fargo in 1986. Midland, which suffered financially and psychologically from this traumatic experience, was further humiliated by having to accept a new chief executive from the Bank of England for the necessary rebuilding of the bank.

By the late 1980s, it had become very difficult to sell an American bank to a foreigner, even as an experiment. Foreign banks-as-buyers were fully aware of all of the problems being experienced by banks-as-sellers. As more and more banks began to report problem loans, foreign and domestic, and always, it seemed, insufficient reserves against them, only the bravest bank acquirers would step forward.

Instead, foreign banks continued to build up their U.S. assets through the operations of their U.S. branches. This has been the preferred route for most foreign banks in the United States, and total branch assets of foreigners greatly exceeds the total of assets acquired. At least, they thought, we know what we are doing in our own branches.

There are approximately thirteen thousand banks in the United States, a legacy of the American rural past and distrust of large concentrations of financial power, which nonetheless have occurred. Large money-center banks now control about 50 percent of all U.S. banking assets. United States branches and affiliates of foreign banks (including those U.S. banks acquired by foreigners) account for about 12 percent of total U.S. domestic money-center banking assets, or about 6 percent of total banking assets in the United States.[11] This is not a terribly large penetration of our banking business by foreigners, though it is growing. In many European countries, the penetration of local banking by foreign banks is much greater.

To the extent that we are going to worry about the presence of foreign banks on U.S. soil, we should worry about their branches, not their acquisitions. This condition should exist well beyond the end of this decade, as U.S. bank acquisitions are frightening to foreigners, and they seem to have done better than they originally thought they could building up their own business in the United States.

The Japanese have been especially successful building up their business in the United States, and so have the British, the Germans, the Swiss, the Dutch, the French, the Swedes, the Austrians, the Italians, the Australians—practically everyone. They do this by modest injections of capital from the parent, which are then leveraged with deposits or bankers' acceptances, which can be sold in the New York money market. The proceeds are then lent out to clients from their home countries operating in the United States, or more recently, to U.S. companies.

In 1987, total Japanese direct investment in the United States

was about $35 billion, about 20 percent of which represented invest-
ment in the U.S. activities of Japanese banks and financial-service
companies.[12] These banks have hired Americans to serve as marketing and lending officers and have made significant progress in pene-
trating the market for conventional bank loans by offering low-cost
financing and services to people like George from our introduction.
George said he was delighted to have them do so.

Foreign bank branches are subject to regulation by U.S. banking
authorities. The SEC has ruled that there is no difference between the
securities of a U.S. bank and a U.S. branch of a foreign bank. Deposits in U.S. branches of foreign banks are *not insured by the FDIC*,
but they are backed by the full faith and credit of the foreign bank of
which the branch is a part. The branches are subject to reserve requirements. Many of their parents have AA and AAA bond ratings
by Moody's and Standard & Poor's credit-rating services. No American bank holding company has an AAA rating anymore, though ten
years ago nine of them did.[13] Many of the foreign branches enjoy a
lower cost of capital (because of the cost of the capital funds that are
brought into the United States) and often lower funding rates (the cost
of deposits and other AAA funding instruments sold by the banks)
than some U.S. banks, which means they can offer lower cost financing to their customers, while retaining the same profit as their
U.S. counterparts. Some foreign branches are trying to build up their
market share in the U.S. and are willing to reduce their profit margins
below what American banks require in order to do so.

On balance, many are fit and eager to do business in the United
States, and the deteriorated condition of U.S. banks makes it easier
for them to do so.

We are reminded that George Moore and Walter Wriston were
the first to get banks to fight back against the restraints on them in
the securities area and, most important, in interstate banking. Citibank has challenged these two sets of regulations ever since, while
going on to build a growth bank based on product innovation and
overseas expansion. Moore and Wriston might also add, however,
that U.S. banks have been unfairly and absurdly treated by regulators
relative to foreign banks' operations in the United States. The Foreign Banking Act of 1978 cured some of these inequalities, but it

grandfathered those involved in certain businesses at the time. Foreign banks were more liberally treated with respect to interstate banking and participation in the securities business than U.S. banks are. Fifteen foreign banks, including such foreign giants as the Deutsche Bank, the Swiss Bank Corporation, Paribas, and others, are grandfathered to act as investment bankers while still participating in commercial banking. Current bank laws remain very unequal with regard to major international players.

U.S. banks, hemmed in by regulatory factors and suffering exceptional difficulties with their loan portfolios, have been weakened sufficiently to create attractive—almost predatory—opportunities for foreign banks to compete with them in the United States. Not many other countries offer such easy access to competition in the banking field, but the large number of U.S. banking operations abroad signifies that enough do so as to provide plenty of room for U.S. banks to get involved in banking abroad, though perhaps less so in certain high-margin, protected national markets.

In the United States, it is arguable that foreign banks have stepped into those areas in which U.S. banks feared to tread—feared only in the sense that they didn't want to match the rates. The 6 percent of U.S. banking assets now on the books of foreign branches and subsidiaries are where they are because of competition.

Competition in banking was not something their supervisory parents had in mind when they set up the regulated public-utility banking system in the 1930s. Moore and Wriston led the banks out of this desert, where they have provided competition for others for many years. The presence of foreign competition in our home markets, however, may become the catalyst that the banks need to free themselves from what they perceive as unfair U.S. regulatory burdens and complexities. It may also provide the incentive for expanded and invigorated banking services in the United States on the part of those who will survive the reshuffling many banks, because of their recent history, are now undergoing.

Foreign investment that is subject to all appropriate domestic regulations is not a bad thing, but whether it is or is not, it is unavoidable for those countries with large trade deficits. The money flows back from the surplus countries somehow, and direct and portfolio investment is the principal way. In flowing back it actually helps diminish the previous trade deficit by providing local manufacturing that can replace imports and by financing new investments to improve the productivity of the industrial sector, making it more competitive internationally. In time, the trade deficit reduces and the foreign investment recedes with it.

In our case, the dollar has been sharply repriced by the market since 1985; our weaker areas are attracting competition and aggressive corporate restructuring; we are exporting again; and, as Peter Drucker has reminded us, we are lowering the relative cost of labor in our manufactures, which shifts world competition in the future to a different game, one of managerial competence.

These things promise a brighter future, one in which our national self-esteem can be rebuilt. The United States is not going to be turned into the Mexico of tomorrow, no matter what happens in the interim. There is room, however, for reflection on some of the lessons this unusual time has taught us. We cannot allow ourselves to fall into a self-satisfied, noncompetitive state again; the speed at which the global economy will apply its corrective energies to any such situation is too great. No one can escape this ruthless condition, neither manufacturers nor traders nor bankers; we must all come to terms with it or leave the scene.

It is also useful to be reminded that money is not patriotic, that economic laws are economic, not sentimental. If good investment opportunities exist in Brazil, or Korea, or Alabama, money will flow to them. A well-managed company today needs to look throughout the world for the best markets for its products and the most economic areas from which to supply them. The textile industry survived by purchasing goods abroad; the auto industry by acquiring parts from all over the world to be assembled in a particular location, perhaps to be reshipped to other parts of the world. Businessmen refer to this as global marketing and sourcing. It makes sense. It helps economies grow. The United States may have the highest trade deficit ever, but in mid-1988 it reported the lowest unemployment rate in fourteen years.

We Americans need never fear devastation from the coldhearted workings of the global economy. We have every reason to expect success, not failure, from our participation in it. But if we are ever to face such ruin as an economic power, it will not be because we tried to accommodate ourselves to the ever-changing world economic forces, but rather because we tried to oppose them by protectionism and by providing shelter to uncompetitive businesses.

This being true, how can it matter who owns what? Iowans have accepted for years that their economy will grow and prosper more if it is allowed to be integrated with the economy of the rest of the United States. If the meat packer runs into trouble, for whatever reason—fair or unfair—Iowans will have to accept its fate, whether the company is owned by Iowans, Californians, Koreans, or Yugoslavs. Likewise, when the citizens of towns like Flat Rock, faced with heavy unemployment, learn of a new Japanese automobile plant in their area, they can be expected to be glad about it.

There is a trade-off between increasing our wealth through participation in a larger economy and not doing so in order to preserve autonomy. America has accepted this trade-off throughout its two centuries of existence; as we enter the third century, somewhat shorn of our relative economic supremacy, we will have to accept, if not always gracefully, the burdens along with the benefits of the global economy.

In the future, the ownership of corporations will become as much a mishmash as the marketing and sourcing of their products. IBM, already owned by pension funds and institutional investors all over the world, may find some day that as much as half of its stock is owned by foreigners. Sony has been in this unusual situation from time to time over the past ten years.

Many companies would like to see the ownership of their shares distributed among the major capital markets of the world and into the countries in which they do business. They want to increase the number of investors bidding to buy their securities and to develop a broad base of shareholders all around the world. Few oppose this notion today, and fewer still actively attempt to prevent their shares from being acquired by foreigners.

Over the last several years, many British companies have successfully increased their share ownership in the United States and

Japan. Equally, when British companies, subject to a takeover bid from an unwanted foreign suitor—as in the case of the acquisition of Rowntree by Nestlé—try to persuade authorities to block the sale, the authorities have refused to do so, something they did not so readily do a few years ago. We are all getting used to the free-market idea that being foreign does not equate to being excluded.

The free flow of capital across borders is coming to be considered today's financial equivalent of avoiding salt and red meat and eating more fiber. It is accepted, if grudgingly, because more and more people are coming to understand that it is healthier for our collective economies.

Some xenophobia remains with us, as do those with old-fashioned diets, but nevertheless, more international transactions are occurring than ever before. A recent study by DeAnne Julius and Stephen Thompson of the Royal Institute of International Affairs in London pointed out that foreign direct investment had increased since 1983 by 105 percent in the United States, by 40 percent in Japan, by 31 percent in West Germany, by 29 percent in France, and by 21 percent in Britain. They expect foreign direct investment to continue to grow at about 13 percent after inflation through 1995, perhaps indeed until the millennium.[14] Such a pace cannot avoid complaint, anxiety, and aversion, but nonetheless the investment flow should continue.

One wonders whether Paul Revere, if astride in the 1990s, wouldn't be leading a brass band playing a medley of European and Oriental tunes, gladly announcing a sushi and crumpets party at the village Oktoberfest. Certainly, any message announcing foreign investors would have been well received both before and after the midnight ride in '75; if and when the foreigners do get out of hand, we can assume that Mr. Revere's neighbors will know what to do.

# 4

# Hostage to
# the Bandidos

On January 7, 1988, Morgan Guaranty Trust Company on behalf of the United Mexican States requested approximately six hundred banks who were lenders in existing loans to the Mexican government to waive conditions in those loans that prohibited Mexico from issuing new debt that would be collateralized by Mexican gold or other reserves. The banks, in past loans, had insisted on these conditions in order to conserve for themselves all such assets that could be used for the repayment of their loans if necessary.

Since 1982 the banks had had difficulty in collecting the payments of interest and principal on their Mexican loans. The country had suffered a severe foreign-exchange crisis in 1981, which had left it unable to meet the schedule of debt-service payments of prior loans. Partial payments had been made to cover interest obligations, and massive debt-rescheduling programs were arranged in 1983 and 1984, which included about $6 billion of new loans in the form of capitalized interest, but still the problem remained. Despite these efforts and others aimed at reforming the economy, Mexico was still short of foreign exchange with which to retire its foreign loans.

Though an oil-exporting nation of 85 million people with a gross national product (GNP) of $170 billion, the second largest in Latin America, Mexico had fallen upon very difficult times. Efforts by the government to provide for orderly growth and further industrialization were frustrated by falling oil prices, mismanagement, and corruption. A new president, Miguel de la Madrid, a former finance minister, took office in 1982 and attempted to bring the economy under control. The results of his austerity programs were inflation, decline in real wages, unemployment, a reduction of about 15 percent in the per capita income of its citizens from the levels of 1981, the flight of more than $40 billion of Mexican capital into investments abroad,[1] and the first appearance of serious political opposition in fifty years to the governing Partido Revolucionario Institucional.

Over the preceding decade, the Mexican government had relied heavily on loans from foreign banks to close the gap between the funds it wished to spend and those it had available. By the end of 1988, these debts had accumulated to approximately $78 billion, which made it the world's second-largest borrower from banks (Brazil at $82 billion being in first place). Each year approximately 30 percent of Mexico's foreign-exchange revenues from exports, tourism, and investments abroad were required just to make payments due on the debt.[2]

Borrowing money in dollars was a risky business; Mexico had to repay its loans in U.S. currency, the value of which it could not influence or predict. The greater its debts, the more dependent it became on events and circumstances beyond its control. The growing claims of foreign banks on Mexico's scarce reserves would ultimately affect the local economy too, forcing it out of control.

The vast majority of the loans had been made willingly by the banks who were competing aggressively with one another in the mid-1970s to win profitable "mandates" from the Mexican government to arrange large syndicated bank loans. Mexican loans carried attractive interest rates, paid good fees, were available in large amounts, and promised various forms of deposit and other collateral business.

Banks, in the midst of recycling oil dollars from the Middle East, had lots of money to lend and few customers as voracious for it as the Mexicans and other Latin American countries. A substantial number of these loans were made by European and Japanese banks also, but the American banks, profiting from their proximity and long association with the region, were by far the most aggressive. Few among them recalled the experience of Latin American bond issues in the 1930s or worried about the problem of getting their money back in dollars, a problem that George Moore had solved in Germany only by selling "travel marks."

By 1982 the first rumble of the coming storm was heard, and the banks began to show greater concern for collecting what was owed them than for making new loans. The climate changed suddenly. Mexico and the other countries in Latin America were no longer able to sustain the pace of continual borrowing of the past few years, and their economies began to shudder from the deprivation. The flight of private capital to safer regions abroad aggravated the problem severely. Mexico and the others suddenly became net *exporters* of capital instead of substantial net importers, a condition their fragile, inefficient, and capital-dependent economies could not endure.

Mexico, of course, was not alone. Its problems were shared by Brazil, Argentina ($47 billion in bank debt outstanding in December 1988), Venezuela ($35 billion), and most of the rest of the countries of South and Central America.[3] They owed so much to the banks that they were about to prove again the old adage: "If you borrow a million dollars from a bank, the bank owns you, but if you borrow a hundred million, you own the bank." Because of the magnitude of the borrowings, the banks had become hostage to the borrowers.

If the Latin American countries should default, or declare a moratorium on debt repayments, the banks would suffer enormous losses. Several American banks in 1982 had loans outstanding in Latin America well in excess of the bank's total net worth. If the countries failed to pay interest within ninety days on their loans, such loans would be classified by the banks' regulators as nonperforming, thus prohibiting the banks from taking accrued interest on the loans into income. Only interest actually received on nonperforming loans could be counted as income.

116

Just as suddenly as the Latin Americans realized the borrowing climate had changed, the banks too discovered that many would be dangerously exposed were the loans to go bad. Even only having to classify loans as nonperforming would have a significant impact on the current income of all of the major U.S. banks lending to Latin America.

The extent of the problem was considerable. All of the top money-center banks in the United States were heavily extended in Latin America. A collapse or fear of collapse of one or more of these banks could have devastating consequences on the U.S.- and world-financial markets, reminiscent of the failure of the banking system in the United States and in Europe during the 1930s. So the system froze. Latin Americans could not repay what they owed without sizable additional borrowings that banks were not willing to make. Bank regulators wanted the banks to prepare for the worst but did not want to frighten depositors and other sources of finance on which the banks depended. Congress was sympathetic to lending a hand to Mexico and the other countries, but not at the expense of using taxpayers' money to bail out the banks. If the banks failed because of the inability of the Latin Americans to repay the loans, then the government had to bear much of the cost because it insured bank depositors against losses. It hardly seemed fair that U.S. citizens should have to be the ones to pay the price of Latin American profligacy and recklessness on the part of the banks. But, no one else in the picture had any money.

The Reagan administration, already attempting to cope with Castro and the Sandinistas, was aware of the hazard of political revolt in those debtor countries that were suppressing their economies to repay their loans, but it neither had any money in the budget (by then in massive deficit) for aid nor did it wish to retreat from its free-market principles to assist either the borrowers or the lenders.

In September 1985, at the annual meeting of the World Bank and the IMF in Seoul, Secretary of the Treasury James Baker proposed a plan for resolving the debt problem. His proposal, known thereafter as the Baker Plan, was original in that it did not insist on resolution of the situation through reliance primarily on the traditional measures of the IMF that would cleanse the defaulters' economies of

117

inflation and other diseases by the fire of austerity. Instead, he proposed a program for growing out of the difficulty. The countries themselves, international agencies such as the IMF and the World Bank, and the banks were to cooperate in an effort that would provide new capital to the countries so they could finance additional growth from which to repay outstanding loans.

Put simply, into one sentence, it sounds implausible: Latin American debt will be reduced by increasing it. Well, perhaps, but only if the new money going in generates higher foreign-exchange revenues than it takes to repay the new debt, and some of what's left over is used to repay the old debt.

The Baker Plan was meant to attract $10 billion of new money annually for three years, $3 billion to come from the international agencies and $7 billion from the commercial banks. The recipients of the funds would be the fifteen major debtor countries, ten of which were Latin American. The U.S. government would encourage all the other parties to adopt the program, but it, of course, would not be directly involved or put up any money itself except for the occasional bridging loan, or advance, in anticipation of an IMF distribution.

Though polite, the banks took a dim view of the Baker Plan; they had no intention of throwing more good money after bad (though, individually, they would be quite happy if other banks wanted to do so, providing that such participation did not suck them in).

The Baker Plan was greatly praised at the beginning, but without the participation of the commercial banks, it could go nowhere, which is what happened. Indeed, many banks decreased their exposure to the countries involved through the end of 1986, and these would have decreased further but for the Mexican, Brazilian, and Argentinian reschedulings in 1987 and 1988, which involved the capitalization of past-due interest into new loans as well as some minimal new-money concessions. Though the Baker Plan continued as a symbol of multilateral cooperation to save the world from Latin American bankruptcy, its effectiveness was only political. The banks, upon whom so much of the plan was to be based, were not participating.

Meanwhile, others took the view that the banks ought to call a loss a loss and write off, or forgive, part of the loans that were not being repaid on time, so the rest could be. Professor Jeffrey Sachs, a

118

Harvard economist and an adviser to Bolivia and Venezuela on debt-related matters, has proposed a program of partial debt relief for the truly hard-hit economies, such as Bolivia. Sachs claims that the banks have already suffered the penalties of their Latin American portfolios in the decline of their stock prices. His research "has indicated that the market value of the claims on the Latin American debtor countries is already much below par value (i.e., 100 percent), and that the stock market valuation of the major commercial banks reflects that discount." In short, he is saying that if the real market value of Latin debt is fifty cents on the dollar, and a bank has a portfolio of such loans equal to half of its net worth, then the stock market will reflect the reduced value of the Latin debt by reducing the stock price to a level equal to a 25 percent discount from the bank's net worth. This is indeed what the stock market does, as best as it can approximate the numbers. Sachs then goes on to propose that the banks write down the basket cases to the market value of the loans and cancel the amount of the debt written off. Such actions would give the countries breathing room and not really cost the banks anything they haven't already lost.

Senator Bill Bradley also proposed a plan in 1986 through which the banks would forgive 9 percent of the face value of their Latin American loans, and 3 percent of the annual interest (for example, charging 7 percent instead of, say, 10 percent required to be paid) over three years. The forgiveness would not be mandatory, as such would be confiscatory, but voluntary, out of the kindness of the hearts of the banks involved. The senator would encourage all the parties to cooperate in the program, but would not, of course, provide any government funds to assist the parties.

Senator Bradley's idea is that the present situation is intolerable because it is "unfair and kills growth worldwide." Latin American countries, he claims,

> have cut imports to the bone and subsidized exports. U.S. industries and farms find it impossible to sell their products in Latin American markets, and are under fierce attack from Latin America's desperation exports. U.S. exports to Latin America shrank by one-third during the first year of the debt crisis. Altogether, one million U.S. workers have lost their jobs as a direct result of the debt crisis.

119

The Bradley Plan seeks to exchange trade reforms in the debtor countries for debt relief, and proposes a new round of multilateral trade talks at which bankers would be represented as principals. Using the modest amount of debt relief proposed as a pump primer, Bradley suggests that the sought-after trade concessions will enable the system to repair itself. If everyone were to go along with the plan, it would "enable banks to contribute $42 billion of debt relief, compared with $30 billion of new loans under the Baker Plan." The banks, he suggests, can afford the investment in curing their clients of their chronic financial ailments.

The proposal describes the problem pretty well, and it correctly links the debt and trade problems. But his solution, like the Baker and the Sachs plans, places almost the entire financial burden on the banks, who are not direct beneficiaries of trade concessions. Multilateral trade talks are the stuff of governments, not private-sector institutions with well-defined fiduciary obligations.

Bankers, as Senator Bradley knows, are responsible to their shareholders, their depositors, and their regulators to collect every last dime owed to them and not to create uneconomic precedents that will have every other client of the bank insisting on similar treatment. There is no way that the banks are going to step forward voluntarily to provide the additional loans or the debt relief proposed in the various plans without substantial evidence that other parties are contributing proportionately to the effort, and Congress is not going to try to force them to do so.

The debt-relief plans also suffer from the assumption that, once a bank has voluntarily, or otherwise, written off a loan as uncollectible, it will be willing to resume lending to the borrower on the same basis as before because its credit rating has been improved by the overall reduction in debt outstanding and better trade arrangements. If the Bradley Plan were to be adopted, it would be a very long time before the banks made additional loans to those whose debt they had forgiven. Such write-offs may come to be forgotten, but they can never be forgiven. Forgiving banks do not have a great future in business.

On the other hand, the banks were under no illusions. They knew that they would have to take some losses on their Latin loans. Government assistance would be welcome but only if the government did not interfere with the banks' ability to work out of the situation through country-by-country reschedulings, parsimonious use of new loans, and devising new techniques for debt reduction. They were applying skills learned and relearned over many years for collecting bad debts, though this time on a massive scale. They were not primarily concerned with the future of Latin American political and economic stability, a condition only rarely achieved anyway, or with more balanced trading practices, but only with getting their money out as cheaply and as quickly as possible.

John Reed, Walter Wriston's young technocratic successor as chairman of Citicorp, was a sullen hostage at best. He was prepared, in the best Wriston tradition, to cast a blind eye on the Latin American scene as long as the countries were paying interest on their loans and trying to make good on principal payments. He had come to be concerned, however, by increasing rhetoric from Latin American leaders that threatened a moratorium on payments or even outright renouncements of the debt.

President Alan Garcia of Peru, forty-one, a charismatic socialist elected in 1985 with strong support from the Marxist left, has effectively renounced the country's foreign debts, having made no payments except for a token one in 1986 despite an announced willingness to pay up to 10 percent of Peru's export earnings to service the debt. Garcia's aggressive approach appeared to be paying off for a while; the economy grew at more than 8 percent during 1986 and 1987, which added to his popularity and gave him a leg up in his continuing struggle with the Maoist guerrilla group, the Shining Path. Peru continued to be able to obtain short-term trade credit from European and other banks. The economy was performing fairly well despite the cut-off of foreign funds, including those from the World Bank on which Garcia had also defaulted, though its prospects were turning sour by the beginning of 1988. The short-term burst of economic energy caused by the reversal of the country's foreign-exchange cash flow had been spent, and bleak times lay ahead.[4]

Garcia's popularity plummeted as the country's economic performance sharply worsened. He became increasingly isolated from his own revolutionary party, whose other officials began to demand that Garcia take a more realistic attitude toward debt repayment. By the end of 1988, Peru had turned again to the IMF and the World Bank to develop a program for regaining foreign financial assistance, without which some were beginning to suspect Garcia would not be able to remain in office.

Still, said Garcia's prime minister, Armando Villanueva del Campo, after these negotiations had opened, "The dead don't pay." In order to pay back past debts, he added, "We have to survive. For this reason we shall cover our own needs first and then meet our obligations."[5]

Nonetheless, the Peruvian experiment with default has been closely watched by all the other Latin American governments.

"Why not default and get it over with?" many politicians and academic economists have asked. "Then there would be a sharp readjustment, a lot of tears and recriminations, and finally a big cleanup that would enable the countries to go on borrowing and the banks to go on lending. Just like the defaults in the 1930s." The thought was tempting to some perhaps, but, unfortunately, it was based on a wrong reading of history. A study published in 1988 by the Center for Economic Policy Research in London concluded that, contrary to popular views, the resolution of the Latin American bond defaults of the 1930s took up to twenty-five years to conclude. "Default in the 1930s," the study reports, "was often partial and intermittent. Settlement was not achieved in a way that really permitted countries to put the debt crisis behind them." Foreign investment, trade, and international politics were then, as now, all interrelated in the defaults. At first, creditor-country governments applied pressure on the debtor countries to extract repayment, but as World War II approached, the pressure was shifted to obtain concessions from the creditors so as not to push the debtors into the enemy's camp.[6]

Peru's experiment with default was not the only one. Bolivia wasn't paying interest either, and others among the smaller countries were frequently in arrears. Worst of all, Brazil, the biggest debtor at the time, which owed about $80 billion to the banks at the time, had declared a moratorium in February 1987 on interest as well as principal payments until unspecified conditions favorable to Brazil were met; it was upping the ransom for the hostages. Though its foreign-exchange reserves were comparatively high, Brazil argued, in essence, that its economy was being strangled by its foreign-debt burden. Austerity measures proposed by the IMF would only make the problem worse, they said, and in any event were seen as a serious affront to Brazilian sovereignty by its proud people. Brazil's president, José Sarney, had but a tenuous grip on his office and political control, and as in Peru, it was politically essential to take a hard line with Brazil's foreign creditors.

Citibank, with about $4 billion in loans to Brazil, would be much affected by the discontinuation of Brazilian interest. Not only would it lose several hundred million dollars in interest income, but it would have to reduce the value of the Brazilian loans on its books, taking the difference as a loss.

Reed and his counterparts at other banks waited for the second shoe to drop. They had faced these kinds of tantrums before and, after a cooling-off period to calm the Latin tempers, had watched the situation return to what it had been before. But a moratorium on the part of the region's largest debtor had not happened for fifty years. This time the Brazilian maneuver was different. Still, the banks hoped, nervously, that the government would come to its senses and avoid sliding down the slippery slope of default that Garcia had somehow turned into the macho thing to do.

This time, however, Reed had had enough. This hostage was walking out, let the Brazilian bandits do their worst. On May 19, 1987, Citicorp (the holding company for Citibank) announced a $3-billion charge to loan-loss reserves to cover possible future losses on Brazilian and other Latin American and Third World debt. The accounting treatment required that the full amount of the charge be applied to Citicorp's current income, where it would cause a $2.5-billion loss for the second quarter of 1987 and Citicorp's first annual loss since the Great Depression. The net worth of the bank would be reduced accordingly, to $45 per share from $61. Reed indicated that Citicorp would sell other assets to replace the capital written off. The charge was probably the largest single write-down in the three-thousand-year history of double-entry bookkeeping. Citicorp's stock price went up, not down, on the announcement. Jeffrey Sachs had a point; the market had already punished the banks for failing to account realistically for their Latin loans.

But the charge to loan-loss reserves did not mean that Citibank was writing off or forgiving the slow-paying loans. It had added to an existing reserve against such losses in the future *as if* they had occurred. Reserves now totaled an amount equal to about 25 percent of all of the bank's Third World debt. The accounting hit had been taken, but the loans remained due and payable, and Citibank was determined to collect them, or as much of them as it could. Having suffered the loss, Citibank could not be frightened into concessions on rescheduling or forgiveness out of fear of such losses. Brazil had lost much of its bargaining power with the bank, whose actions gave it new flexibility to either hang tough or agree to concessions in future negotiations if it wanted to do so.

Within weeks other American banks emulated Citicorp. Though several banks were very reluctant to take an equivalent loss as large as Citicorp's, virtually all of those with major Latin American exposure did so. Citicorp's Brazilian position was well known, the bank was regarded as very knowledgeable about the country, and others would have to explain why their positions should not similarly be marked-to-the-market. Banks knew that many depositors and investors thought they had been Pollyannaish in continuing to assert that their Latin American portfolios were perfectly sound and correctly valued on their books, and to continue the charade after Citicorp's action would not be credible. Now all of the hostages were in revolt. They had girded themselves for further losses and were adamant that, if Brazil and the others did not revert to the status quo ante in which interest was paid, there would be no more money, period.

Brazilian finance minister Luiz Carlos Bresser Pereira apparently misread the message. Seeing the banks increase their reserves by such large amounts must have implied that the banks were ready for some serious write-offs. Bresser Pereira wanted to take advantage of this apparent movement toward greater concessions, and he wanted to secure a major political victory at home by forcing the banks to bend to his will. In September 1987, he played his hand with a proposal for the banks to exchange their loans for new Brazilian bonds, which would carry a below-market interest rate. The exchange would value the bank loans at their market value, then about fifty-five cents on the dollar. The result would be that the banks would swap the debt on their books at about seventy-five cents on the dollar (after reserves) for new bonds worth actually less than fifty-five cents on the dollar, taking into account the fact that the new bonds would pay below-market interest rates. Understandably, the banks were ice-cold on the proposal, but Bresser Pereira was adamant that this exchange should take place, and he tried to enlist support for the plan from the finance ministers of Mexico and Argentina, whom he claimed were also unhappy with their debt-rescheduling arrangements. He also went to see Secretary of the Treasury Baker, looking for support, but Baker refused to take the plan seriously, calling it a "non-starter."

By the beginning of November 1987, the Brazilians were forced to abandon their plan and enter into negotiations to end the moratorium. Brazil would agree to make a token payment of $500 million for unpaid interest, and the banks would capitalize most of the rest of it. Brazil would come up with another $1 billion, and the banks with an additional $2 billion in capitalized interest during the first half of 1988. Brazil was to resume interest payments on schedule during 1988, and a new repayment plan would be negotiated, to include IMF economic "adjustments," to be in place by June 1988. These arrangements, involving about $60 billion of rescheduling, were announced in Brazil by a greatly chastened Bresser Pereira to much discontent. A former finance minister, Dilson Funaro, helpfully noted that politically the new arrangement was a "disaster." President Sarney bravely stepped forward to announce that "from the moment I took office, I have backed the finance ministers proposed to me by the parties that support me. If their policies work or not, I don't know. I'm not an economist."[7]

The crisis passed, with the solution being the old formula of the interest being paid, partly by capitalizing it into new loans, the principal being rescheduled, and the IMF asked in to propose some new austerity measures. Bresser Pereira was sacked in January 1988 and replaced by a moderate, Mailson Ferreira da Nobrega, who was quick to criticize the moratorium and the heavy-handed tactics of his predecessor. By February 1988, according to William R. Rhodes, chair-

man of Citibank's restructuring committee, "The Brazilian government had concluded that the moratorium was not in the country's
interest. Reserves were expected to increase under the moratorium,
but they did not. Some economists estimate that the moratorium cost
the country at least $1.5 billion" in increased interest payments, lost
investment funds, and capital outflows. Brazil's finally ending the
moratorium in August 1988, after the rescheduling agreement was in
place, was "another milestone in the debt crisis . . . one that dem-
onstrates that moratoriums don't work."[8]

Nevertheless, it had seemed a close call to some of the banks,
who no longer had confidence in the old formula. Wriston and Reed
notwithstanding, they wanted to get out of these loans as soon as they
could, but without taking on more losses.

Even under the old formula, the Latin American debt was drag-
ging the banks down. It affected their credit ratings at the rating
agencies, the interest rate they had to pay for funds, and their stock
prices. It impaired their ability to raise new capital, which most of
the major banks needed to replace the capital transferred to loan-loss
reserves and to invest in other businesses. Continuing to own these
loans also meant that they continued to be exposed to further additions to loan-loss reserves in the future and to the tiresome requirement to keep coming up with new loans in order to get their interest paid.

However, getting out of the loans was much more difficult than
it seemed. The borrowers were not repaying them very fast, and there
were few American banks eager to add to their existing portfolios to
whom they could sell the loans. Some debt-swapping took place in
which a European or Japanese bank might agree to exchange some
Eastern European or Philippine debt for Mexican or Brazilian loans,
but these were transactions with portfolio diversification in mind more
than reduction.

The secondary market in Latin American public-sector debt, upon
which Bresser Pereira based his exchange offer, was much quoted
but not much used. After Citicorp's reserve announcement in May
1987, the prices quoted for most debt issues declined sharply, anticipating no doubt the rush to the exits that American banks might soon
undertake. Brazilian debt, for example, could be sold at about fifty-

five cents on the dollar after the Citicorp loan-loss reserve announcement, Mexican for about fifty cents, Peruvian and Bolivian for about ten cents. If a bank was really determined to get rid of such paper, it could go to the secondary market, but this would be an expensive way out. For a bank that had just written its Mexican loans down to seventy-five cents on the dollar, to sell them at fifty cents would mean taking a further loss of twenty-five cents for each dollar of loans sold, something not many found themselves willing to do.

European banks and some American regional banks, on the other hand, had previously bitten the bullet and written their Latin loans down to thirty or forty cents on the dollar through heavy annual charges against income over the past years. Some of these banks would sell Mexican loans at fifty cents for a loss recovery of ten or twenty cents, which could be recorded as a profit. Some of these banks were in a good position to prosper from the misfortune of their fellows, by going back into Latin loans at fire-sale prices. Such transactions, however, did not reduce the amount of debt outstanding by the borrowers; it just shifted it around.

Starting in 1985, banks had also begun to swap debt for equity investments in Latin American companies and other properties. The idea was to retire dollar debt by exchanging it with the central bank of the country for local currency that would then be invested in a company, factory, or real estate in the country. The result was that the seller of the debt could realize a much higher price than in the secondary market if he was willing to convert the proceeds into a local currency investment that might be more profitable in the long run than the debt, though he would be left with the sticky problem of repatriating his profits into his own currency later on. For some, this alternative seemed preferable to holding on to their dollar loans.

The country involved could retire the debt being swapped at a discount and would receive new capital investment in income-producing and foreign-exchange-creating assets that would add to employment and help to service the remaining foreign debt. Countries were concerned, however, that limits be placed on the number of swaps that took place as the creation of the new local currency to be swapped for the debt could add to inflation. They also wished to avoid criticism for allowing themselves to be forced by the banks to give up the country's best investment opportunities to foreigners. There was therefore only a modest potential for swaps at any given time. In any case, banks too were concerned that swaps only be made into good-quality investments, of which there was always a shortage.

The tempo picked up, however, when manufacturers began to purchase Latin debt in the secondary markets to exchange for funds with which to make new investments in factories and other facilities. Financing such investments with swaps substantially reduced the size of the investments they needed to make.

Debt-equity swap activity continued, though somewhat sporadically, and made an important contribution to debt reduction, especially in countries like Chile and Mexico and, after the moratorium, Brazil. Swaps are estimated to have resulted in the reduction of about 5 percent of outstanding Latin American debt by the end of 1988.

The Mexican initiative launched by Morgan Guaranty in January 1988 was to be a debt-for-debt swap, somewhat similar in concept to the rejected proposal of Bresser Pereira. In this case, however, the proposal was to be designed so as to attract the support of the banks, providing incentives for them to exchange existing debt at a discount for new debt. The new debt would be collateralized by up to $10 billion of U.S. government securities, assuring the investor of its repayment at the end of its twenty-year maturity. Interest would be paid at a variable rate of $1\frac{5}{8}$ percent above LIBOR, a much higher rate than the $\frac{13}{16}$ of 1 percent that had been accepted in recent reschedulings. Banks were invited to tender their Mexican loans at a discount from par value, the amount of the discount being up to the banks. Mexico would then review the tenders and select some, all, or none of them to be exchanged for the new Collateralized Floating Rate Bonds.

The hope was that banks, seeking to lighten their Mexican holdings, would tender fairly aggressively for the new bonds. As the market value of the Mexican loans was then about fifty cents on the dollar, it was assumed that no one would tender below that price. The Mexicans would have no incentive to accept any tenders not including some discount, so the maximum price had to be 90 cents or so. Banks could submit up to five different tenders, to hedge their bets if they wanted to. Mexico would accept all tenders at or below a single price to be decided after the tenders were submitted. It hoped to be able to issue as much as $10 billion of new bonds to retire as much as $20 billion of old loans, a net reduction of $10 billion or almost 13 percent of all outstanding foreign bank loans in just one transaction. This much reduction would require, of course, that holders of $20 billion of old debt would tender it at the absolute minimum price of fifty cents, an unlikely outcome.

The issue was complicated, however, by the structure of the new bonds to be issued. Mexico would gain very little by spending $10 billion of its total of $15 billion of foreign-exchange reserves to purchase U.S. Treasury securities to back the new bonds. Instead, it would purchase up to $10 billion of a twenty-year zero-coupon U.S. Treasury bond (which would be issued directly by the Treasury) at a discount price of about $2 billion. Like a U.S. savings bond, the zero-coupon security would pay no interest until maturity, when the aggregate payment for return of the $2 billion of principal plus all accrued interest over twenty years would total $10 billion. The new Mexican bonds would be secured by the zero-coupon Treasury bond, which is why it was necessary to get the consent of lenders to waive the restrictions in previous loan agreements prohibiting the collateralization of new debt.

The plan was for the $10 billion available at the maturity of new bonds to cover only the repayment of the principal of the bonds; interest would be payable semiannually throughout the life of the loan. The payments of interest, however, were not covered by the U.S. Treasury security; they were strictly Mexican risk. This meant that the new bonds were a new kind of hybrid, part Mexican risk, part U.S. government. It also meant that the new bonds were unlikely to trade in the secondary market at one hundred cents on the dollar, which complicated the problem the banks had in valuing them so they could decide how to tender. The banks also had to give some consideration to the fact that the new bonds would be readily transferable— a liquid trading market in them should develop—and that holders of the bonds would be unlikely to be required to participate in future restructuring programs with their inevitable calls for new loans.

The math was designed to make the break-even point somewhere in the area of seventy cents on the dollar for the old loans to be exchanged. This of course was close to the seventy-five-cent level that many of the U.S. banks had written their debt down to following the Citicorp addition to loan-loss reserves. Those wanting to reduce their Mexican exposure would have to take a modest additional loss to do so. The Mexicans saw little reason to proceed with the scheme if they could not get the banks at least to realize the full amount of the reserves they had set aside for the Mexican loans. The question became one of how many banks wanted to reduce further their Mexican positions by paying a little more.

The answer was some, but not many. The Mexicans accepted all tenders at 69.77 cents on the dollar or lower. Altogether tenders totaling only about $3.6 billion of old loans were accepted, $400 million of these having been tendered by Morgan Guaranty itself. Two and a half billion dollars of new debt was issued, resulting in a somewhat disappointing reduction of only $1 billion. The issue was widely called a flop in the market, but at the same time it received wide-ranging acclaim for the effort and the imaginative approach. Most observers agreed that more initiatives like this one would be welcome, and in time substantial amounts of Latin American debt could be retired through securitization efforts like Morgan's. Mexico said it would try again.

Indeed, Chile, which has been especially aggressive in debt-reduction programs over the past few years through debt-equity and other swaps, came up with a variation on the Mexican deal fairly quickly. In the spring of 1988, Hernan Somerville, Chile's chief debt negotiator, persuaded more than three hundred bank creditors to allow the country to issue up to $500 million of new debt to be 100 percent collateralized with Chilean goods, 30 percent of which would be exports. One expert felt that such an issue might be exchanged for upward of $800 million of outstanding debt. Through swaps and aggressive transactions like this one, Chile has been able to reduce its foreign indebtedness by nearly 40 percent. At the end of 1988 Somerville resigned announcing that his "mission had been accomplished" and that Chile could expect to regain full access to the international capital market over the next two years, following negotiations and restructurings that covered the country's needs through 1991. Though high marks were given to Chile for its performance on the debt-reduction front, most observers felt that little would have been accomplished had not the government been firmly, if undemocratically, in the grip of military strongman Augusto Pinochet.

Efforts to find ways to convert existing bank loans into securities that could be sold in the market have been under way for the past several years. The huge potential for this sort of business, which could rival the $300-billion corporate junk-bond market, has attracted the attention of many Wall Street investment bankers. In fact, the onetime junk-bond king himself, Michael Milken of Drexel Burnham Lambert, had long been enthusiastic about the opportunities to do for Latin American debt what he had done for high-risk, low-quality corporate bonds in the United States.

In 1983 Drexel targeted Third World debt as a business opportunity. It hired Jerry Finneran, a former Citibank Brazilian expert, and a Morgan Guaranty senior vice president for Latin America, Antonio Gebauer, joined the group in 1985. Gebauer was an unfortunate choice, however. He knew his clients just a little too well, which encouraged him to expropriate their funds for his own use, a matter for which he subsequently was sent to jail.

For several years, Drexel has been working on various ideas for selling Latin American debt backed by oil or silver reserves or by a portfolio of bank loans or other assets. So far, however, the firm has not been able to find the map to the buried treasure of Latin American debt securitization. The economics are very difficult—a Latin loan might have an interest rate of 1 percent over LIBOR, or, say, 8.5 percent. Drexel might be able to buy it in the market for fifty cents, such that it would have a current yield of 17 percent. For a 17-percent return, junk-bond buyers are used to companies with a positive, though only slightly so, cash flow after paying interest, and the legally enforceable ability to claim the business and assets of the firm in the event of bankruptcy. Disputes surrounding events of default are governed by courts of law in the United States. None of these things applies to Latin debt. All but the dumbest of junk-bond investors are going to turn away from new issues secured by Latin American debt, unless an irresistible interest rate of, say, 20 percent or more is offered. And no one would be willing to sell Milken their existing loans at forty cents (which would just barely enable him to sell them for a 20 percent yield) if they can sell them in the market for fifty cents.

Still, Milken has a reputation for single-handedly developing the immensely profitable junk-bond market in the United States, and his efforts to do the same for Latin American debt have been taken seriously by many people. His progress in this direction has, of course, been impeded by his own indictment and his firm's considerable ongoing legal difficulties.

One of Milken's admirers is William Farley, of Farley Industries, a junk bond–issuing and –investing client of Milken's, who knows him well. According to Steve Swartz and Peter Truell, who interviewed Farley for an article on Milken's Latin American activities in *The Wall Street Journal* in September 1987,

> Clients say Mr. Milken's interest in Latin America is an outgrowth of his views of the world, which he has been increasingly anxious to share with them. When Mr. Farley flew out to Beverly Hills (where Mr. Milken runs Drexel's junk bond operations) to discuss a "megabucks" financing Mr. Milken is handling for him, Mr. Milken quickly moved the subject to the global economy. "He said, 'Bill, let me run some thoughts by you,' " Mr. Farley recalls. The two men proceeded to talk for an hour "about the world—conceptually," he says.
>
> "We talked about Europe, Asia, Africa, Russia, and Latin America. We went around the world and said 'who controls what?' " says Mr. Farley. "That's how we got into this Latin American thing: It's up for grabs."
>
> According to Mr. Farley, Mr. Milken told him that Europe is a "standoff" because U.S., Japanese and European interests are established there. Asia isn't promising, Mr. Milken said, because "the Japanese have a natural advantage there." "No one will do business in Russia," Mr. Milken said, according to Mr. Farley.
>
> Latin America is Mr. Milken's area of choice because the United States has historic ties there, clients say. They say that Mr. Milken argues that recent moves by U.S. banks to increase their loan loss reserves for Latin debt and to pull back from Latin America creates a void to be filled—either by U.S. entrepreneurs like himself, or by others, such as the Japanese.

Neither Mr. Milken nor his colleagues at Drexel have yet done a deal involving Latin American debt, but presumably Drexel is still trying and continues to be active in the trading of bank loans in the secondary markets. Other firms, notably Shearson Lehman Hutton and Salomon Brothers, have beefed up their market-making activities in Latin loans, thereby adding to the liquidity of the market. The more players there are, the more activity, and the more activity, the more innovative ideas. So far, however, the major lending banks have held the center stage in the area of innovation, with Citibank having been especially effective in debt-equity swaps, and Morgan Guaranty in debt-debt exchanges.

Such activity has made it possible for many U.S. banks to accelerate their programs for reducing Latin exposures. At the time of Citicorp's announcement of its addition to loan-loss reserves in May 1987, John Reed added that he hoped Citibank would be able to reduce its nearly $14 billion of Third World loans by $5 billion within the next few years through swaps and other arrangements. As of June 30, 1988, according to a Salomon Brothers study, Citibank had reduced its portfolio by almost $2 billion, and a composite of thirteen U.S. banks had reduced theirs by nearly $6 billion. The banks were beginning to work themselves out of the hole that their Latin American loans had put them in.

The governments too have been reducing debt. During 1988, the *Economist* estimates, Latin America's four biggest borrowers have "chopped around $18 billion from the $300 billion they still owe."

As countries have made swaps and other debt-reducing activities more available, banks have found more imaginative ways to trade with them. Morgan Guaranty originated a complicated swap of Mexican government debt for new debt backed by the business and assets of Mexicana de Cobre, a state-owned copper company that the government was able to privatize as a result of the transaction. Other deals have included the funding of a buyout of a Brazilian soccer star's contract by a Dutch team through a swap of outstanding Brazilian debt purchased in the market at a discount for cruzados. Many governments too have entered the secondary markets directly, buying back their own debt cheaply.[9]

Many banks, of course, continued to maintain that their Mexican, Brazilian, Argentinian, and other loans were still sound, and that in the long run they will prove to have been good business. That is not to say, however, that these banks wanted to increase their holdings of such loans, either voluntarily or through reschedulings. Most banks already have all the loans they can take, and their maximum carrying capacity for Latin American loans was diminishing because of their poor performance and because of capital-adequacy pressures on the banks. The situation was very different from what it had been in the mid-1970s; Latin America, however great its continuing appetite for capital, was no longer the source of growth for the banks that it had once been. On the contrary, the Latin loans had impaired banks' capital, frightened investors, and raised their cost of financing.

Further, for the first time since the 1930s, regulations restricting the activities of banks in the United States were being revised. Substantial erosion in laws preventing bank participation in interstate banking and in the securities business had occurred; these laws were now thought to be in their last days. To take advantage of these new opportunities, banks would need to invest large amounts in new facilities and capabilities. To do this, they would have to get their balance sheets and earnings in shape.

In addition, the banks were newly made subject to stricter regulations concerning capital adequacy. Concerned by the rapid pace of development in new financial instruments, methods, and market activities, the Federal Reserve and the central banks of eleven other industrial countries adopted new, standardized capital-adequacy regulations put forward by the BIS in 1987. These regulations would be phased in over five years, but they would require many of the larger American banks to raise additional capital just to maintain their present assets and liabilities.

American banks had already passed through some very difficult times in the early 1980s with respect to their domestic-lending activities. The slump in oil prices, real estate values in parts of the country, and commodities had severely affected Continental Illinois, Bank of America, banks in Texas, and many others. Many banks were forced during this period to sell off family jewels, such as their headquarters buildings, leasing and credit-card businesses, and overseas operations in order to meet the capital requirements of the time, which have since been tightened. To be sure, the last ten years have been anything but dull for the banks.

Most of the banks responded to the challenges of the 1980s admirably. They cut expenses, exited marginal businesses, improved their profits by greater participation in trading government securities and foreign exchange, invested in higher margin businesses such as real estate, mergers and acquisitions, and leveraged buyout transactions. Many have acquired interests in regional banks and developed credible investment-banking capabilities in those areas in which they are permitted to compete. They are leaner and meaner than ever, but as such, they are not likely soon to be turning their energies back to building up their Latin American businesses.

Perhaps Michael Milken's world view is right: the retreat of the U.S. banks will leave a vacuum to be filled by entrepreneurs like him or by the Japanese. Financial entrepreneurs have not yet found the key to growth and profitability in Latin American securities, and the Japanese banks and trading companies, like the U.S. banks, already have plenty of loans on their books and are not aggressively looking for more.

Milken's idea of a role for financial entrepreneurs in Latin America has already been tried on several occasions. In the 1960s, a large multinational consortium of banks and industrial companies created a Latin American investment bank called Adela. This bank was to attempt to develop capital market activity in Latin America, to purchase and trade debt and equity securities of Latin American companies, and to make investments in special situations from time to time. Adela lasted about a decade, then went under. A similar fate met another similar enterprise called Deltec Pan American Corp. The area has been hazardous ground at best for investment bankers, though there have been a few successes by indigenous firms in Brazil, Venezuela, and Mexico, though the scale of their operations has been trivial in relation to the requirements of the region as a whole. Cautious U.S. investment bankers, such as Jacob Schiff, have always feared treading where neither angels nor their customers would, and though Mr. Milken's approach may pay off in time, it is likely to be a fairly long time.

It is clear that the banks are rapidly distancing themselves from Latin America. They have not waited for a Baker Plan or Bradley Plan to be imposed on them, nor have they held out much hope for other plans that would have the World Bank or another such institution buy out their Latin loans. They have gone about dealing with their own problems, by themselves, on a case-by-case basis. They have left the formulation of plans and strategies to academics, politicians, and journalists who are not engaged in the situation as principals. It is also fairly clear that the void left by their departure will not be quickly filled, either by the private sector or by the multinational agencies or by government aid. This means that the Latin American governments may have to face a serious capital shortage in the foreseeable future.

How big a shortage is it? The Baker Plan was looking for $30 billion over three years. Senator Bradley says his plan could provide over $40 billion in the same period. A respected group of former Latin American government officials claims that the Baker Plan is too little by half. In reality, except for new loans forced on the banks through continual reschedulings, which have produced only about a third of the Baker Plan requirements, there has been a substantial net outflow of funds from Latin America. The World Bank has noted, for example, that net capital transfers from the seventeen most heavily indebted developing countries from capital flight, and debt repayments in excess of new loans and investments, exceeded $30 billion in 1988 and has been a continuing outflow since 1983, totaling more than $135 billion during the period. A large part of this capital drain has been from Latin American countries, which are hardly prepared to take on the role of a capital-exporting region.

Still, one observer, conservative economist Anna Schwartz of the National Bureau of Economic Research, claims that the countries are not any more on the verge of collapse than they usually are and that free-market solutions to their problems are best. If the Latinos want more money, they simply will have to create the conditions and pay the price necessary to get it. As George Moore has said, "Capital is attracted to places that appreciate it and remains where it is made to feel welcome." We would all be better off, Mrs. Schwartz suggests, if we turned our attention to something else.

Another influential student of the Latin debt problem, William R. Cline of the Institute for International Economics, claims that the condition of Latin debtors is improving and that "after a severe recession in 1983 the countries [have] reestablished economic growth, and in 1986 growth in non-oil Latin countries averaged 6.5 percent." He maintains that as long as the industrial economies can continue to grow at about 2.5 percent annually and interest rates and commodity prices don't turn sharply worse, then the Latin countries ought to be able to work out of their present situation, in which they are periodically unable to service all of the debt that they owe. Cline continues to argue that debt concessions by the banks are unnecessary (except, perhaps, for the basket cases) and counterproductive. But, he adds, the situation depends on the "renewed capital inflows necessary for investment revival and continued growth" and therefore the "proper policy on debt remains a strengthened Baker Plan." He argues that greater incentives to participate in new Latin loans must be offered to the banks, and he suggests a few that did not strike this reader as nearly enough to do the trick.

Senator Bradley and the others advocating debt forgiveness haven't changed their minds—if anything, they continue concerned that the net outflow of capital from the region will poison its ability to maintain economic growth and avoid debilitating reductions in per capita income. Continued and prolonged exposure to economic austerity in the interest of debt repayment is bound to result in political repercussions, which could be adverse to U.S. interests.

Indeed this is the central issue. The banks aren't going to fail based on their present Latin American posture, and none of the major countries appears to be on the verge of immediate insolvency. Latin American economies have always been more than a little chaotic and, like the Italians in the 1960s and 1970s, manage somehow.

It is true, of course, that progress is made only through a painfully slow process and that backsliding is common. After a fairly good year in 1988 in which Brazil seemed to sort itself out and progress was visible in Mexico, Chile, and some other areas, 1989 started out with much bleaker prospects. Hyperinflation had returned to Brazil. Payments of principal were suspended for the first time by Venezuela and Colombia. The enthusiasm of the governments for debt-conversion programs was going limp in the face of increasing inflation and political resistance. The accumulation of these problems depressed the secondary market in Latin loans sharply. At the end of January 1989, Brazilian debt was quoted at thirty to thirty-five cents on the dollar, and Mexican at thirty-five to forty.

Yet the path to salvation was becoming clearer to all: the countries must turn away from confrontation and renunciation such as had been tried with disastrous results by Alan Garcia; instead they must head down the road of internal reform coupled with continued appeal for cooperation, understanding, and forgiveness (if possible) from creditors and their governments. In other words, make the necessary internal reforms in the hope of attracting a return of flown capital and new investments, but buy the time politically from both internal and external constituencies as best as possible by scaring each a little bit.

Most economists seem to agree that the key is the reattraction of investment capital, which can only be accomplished by controlling inflation, reducing the bloated public sectors of the economies and the gargantuan subsidies, bureaucratic inefficiency, and corruption that are wrapped up in them, and conceding some ground on foreign ownership of assets. Tax collection has to improve too. Once capital starts to find its way back, however, it has to be invested wisely, so as to return more than it costs and therefore contribute to the economic growth that must be generated for the debt problem ultimately to be solved. After all, the real tragedy in the situation is how grossly ineffective the investment of the huge amounts of capital provided by the banks has been. If this is not reversed, which so far it has not been by all the efforts at rescheduling, and so forth, since 1982, the situation cannot improve itself and no investor of sane mind is going to throw more money into the black hole if he can avoid doing so.

The critical question is whether existing governments can do these things without being overturned by those who would promise not to.

Latin America has, of course, never been a stable place politically. Revolutions occur all the time, so why dread another one here or there? When economic problems become excessive, political turmoil follows, which, in turn, is often followed by takeover by the armed forces, which usually, but not always, abide by the old formula, as in Chile, for instance.

138

Clearly the risk to the United States is of the overthrow of regimes willing to coexist on friendly terms with the United States by hostile revolutionaries who more closely identify with Cuba and other communist countries. We have enough trouble with tiny Nicaragua. Imagine what it would be like if one or more of the big countries fell into line with it. What if one of those countries was Mexico, with its 85 million people right on our border?

Peru and Bolivia may not be far from collapsing into a communist orbit. Garcia's actions are closely watched by opposition parties all over the continent, and many approve of his stand on the debt issues. Political parties in Brazil, Argentina, and Mexico, our three biggest debtors, compete with one another to express outrage with the way the banks and the IMF are seen to be treating them and scorn for their own officials, who appear always to give in to the dreaded financial imperialists. Though often just for show, these positions increasingly reflect vital political necessity in their own countries.

The Mexican election in July 1988 drove home the point. For the first time since 1929, the ruling party's majority was threatened sufficiently to precipitate actions by the government that many thought covered up the true results. Cuauhtemoc Cardenas, son of a former populist Mexican president of the late 1930s, may not have won the election, but he certainly earned a credible second place and carried the message that an opposition party can and should become an important force in Mexican political affairs. One of Cardenas's strongest arguments during the campaign was that Mexico's cruel economic stagnation was caused by foreign-debt payments.

In a detailed article on Latin popular opposition to the strain of debt payments, Roger Cohen of *The Wall Street Journal* noted that Mexico is not the only place that is experiencing the problem. "Several other left-leaning politicians," he wrote after the Mexican election, "including Peronist leader Carlos Saul Menem in Argentina, Venezuela's Carlos Andres Perez and Brazil's Leonel Brizola are basing their thrusts for power on a message that debt servicing condemns their countries to more misery. All three are favored to win presidential races before the end of 1989."

Perez was elected president of Venezuela in December 1988, and assumed office on February 2, 1989. An old political hand who had served once before as president, Perez inherited an economy that had fallen into sufficient difficulty to threaten future foreign-debt service for the first time. Perez's predecessor, Jaime Lusinchi, had carefully avoided harsh economic measures during his administration. Just before leaving office, however, Lusinchi suspended payments of principal on Venezuela's $33 billion of foreign debt. Perez, surprised to find economic conditions as bad as they were, acted decisively. On February 16, he announced a tough austerity program, despite his having campaigned on promises to alleviate the burden of foreign debt. A week of rioting followed in Caracas and eight other Venezuelan cities, over which the police and national guard apparently lost control as more than three hundred people were reported killed in the surprise eruption of violence. There had been concern that food riots might break out in Argentina, or Brazil, but no one was watching Venezuela, a relatively quiet country seemingly on the outer fringes of the Latin American debt circle.[10]

The Venezuelan incident, which attracted financial first aid from the Bush administration and other friendly institutions, underscored the political threat that the debt situation poses in the region. The threat, however, is not just that these countries will topple into radical regimes that repudiate their foreign debt, as Castro did. There are serious problems that flow from the expectation of these things happening. Even a Michael Milken would be very cautious about lending into such a political environment. The blockage of capital inflow then becomes worse, and indeed the exodus of both capital and talent from the region can be expected. The flight of the best and the brightest of industrialists, of professionals, of skilled labor, of all of those who could be expected to suffer from a Sandinista type of takeover can be anticipated, leaving behind only those who can benefit by revolution. Such a situation would be especially grim in Mexico. We could awaken one day to find that several million Mexicans had walked over the border, and many of those that didn't had become extremely hostile.

The problem with the problem is how much credence should be placed on one or more of the disaster scenarios, and what, if anything, should be done. So far all of those coming forth with plans have focused on the immediate financial aspects, many suggesting that the banks take up the slack. On closer analysis, the problem is mainly a political one, based on acute economic and financial factors to be sure, but political in the sense that the risk to the United States in the problem is a political one—a radicalized, hostile Latin America that would pose a serious threat to our national security. Surely if Nicaragua is a threat, as our previous government has argued, such an enlargement of the political problem in, say, Mexico would be many times more serious.

Well, if the issue is a political one, the solution to it should be also. So far, the U.S. government has not wanted to face it, perhaps subscribing to Anna Schwartz's view that a laissez-faire approach will work out fine in the end. It is hard to see, however, how either she or the United States can pretend the politically darkening sky in Latin America is not there. It is, and it needs to be addressed, if only initially with preventive measures, which as we all know are far cheaper than cures once the disease breaks out.

The United States ought to take the view that it should be in its interest to assist friendly Latin American governments to reduce their outstanding debt sufficiently to regain access to the capital markets. This does not mean that all of the debt has to be retired, but some of it. The retirements cannot come from twisting the arms of the banks, whose willingness to lend in the future is one of the important objectives to be obtained. And the solution must meet another objective; it cannot cost the United States anything, or at least not much. Congress appears to have neither the will nor the funds to embark on a Latin American rescue program, at least for the time being.

Such a program will have to be worked out on a case-by-case basis, each country being treated as the different economic entity that it is. The government's role can mainly be that of a catalyst, though it may be better in some cases for the United States and the countries involved to enter into mutually beneficial trade and economic treaties, so that some of the clearly interrelated issues raised by Senator Bradley can be addressed.

Using Mexico as a case in point, a 30- to 35-percent overall reduction in outstanding debt may be the right level to shoot for to restart bank lending and capital-markets access. Perhaps as much as half of this amount can be achieved by encouraging further restructuring through debt-swapping and other private-sector efforts already under way. The remaining 15- to 20-percent reduction could be the subject of bilateral negotiations with the United States, aimed mainly at encouraging market forces to assist in the debt-reduction effort. For such an effort, each side should be fully advised by its private-sector participants in Mexican business. New incentives to attract foreign manufacturing to Mexico could be discussed. These could, for example, include a five-year window during which 100-percent ownership positions in new, job-creating Mexican investments could be obtained, along with tax holidays and other incentives. The United States could offer a similar window during which Mexican exports to the United States would not be subject to certain tariffs or quotas. U.S. Export-Import Bank financing incentives could also be involved. Specific Mexican industries could be targeted for vigorous development, perhaps in connection with World Bank or other developmental assistance. Mexico could be asked to increase its sale of minerals, especially oil, under long-term contracts to oil companies to assure the availability of a steady source of foreign exchange. It should also be urged to release more of the Mexican economy from the heavy-handed, inefficient control of the state to encourage more of the fresh air of competition and opportunity to help lift Mexican economic performance. These steps, taken together in a coordinated way, could encourage the inflow of new capital, and the return of the old, and assure the profitable investment of such capital, so Mexico could accelerate substantially the reduction of its borrowings over the next several years.

The burden of sudden, large-scale internal economic deregulation, however, is likely to be a heavy one for the Mexicans, who are already distrustful of their technocratic new president's willingness to look out for their interests in the short run. Whether the burdens can be eased by new opportunities and assistance brokered from the United States is yet to be seen, but the need for a new Mexican balancing act seems clear enough.

In his inaugural address in December 1988, President Carlos Salinas de Gotari indicated that he knew his place on the tightrope.

Reforms were necessary and so was a reduction of corruption and abuse of privilege. Growth, he said, was the key to the future, but

> we shall not grow in any lasting way if we continue to transfer abroad each year five percent of our national product. This situation is unacceptable and cannot be sustained. I shall avoid confrontation; but I emphatically declare my conviction that the interests of Mexicans come ahead of the interest of our creditors. Our priority is no longer to pay, but to resume growth.

Salinas was not content with simply stating his convictions—he was ready to act on them. Almost immediately after his swearing in, he raided the home of the powerful leader of Mexico's oilworkers' union and arrested him for various serious crimes including arms dealing. Despite an outcry from the oilworkers, the arrest was made to stick. In the following weeks further arrests for corruption, tax fraud, and stock-market manipulation were made.

Previous to his assuming office, Salinas had received an offer of a $3.5 billion bridging loan from the United States, eager to show its support for the new administration. The loan was never drawn down, however, as reportedly the return of a few billion dollars of flight capital in the period immediately following Salinas's swearing in was sufficient to make the loan unnecessary. It was also reported that the Mexicans found the terms attached to the bridge loan too restrictive.

The new president was off to a quick start, but many huge problems lay ahead. In the process of addressing these problems in a manner that is suitable to Mexican political realities, he will no doubt become, like his predecessors, a prickly and difficult negotiant. We can expect bilateral negotiations over trade, financial assistance, and drugs to be arduous at best.

The Mexicans, considering their long dependence on non-Mexican financial assistance, have perhaps an overdeveloped sense of national pride and identity, which can make them quite trying, and a long history of distrusting the United States' good offices. The Americans, on the other hand, often come across as insufferably arrogant and overbearing. Still, there has to be room, in the light of their mutual self-interest, for the two countries to hammer out a credible program that would ease the way for the restoration of growth in the Mexican economy and the rekindling of confidence in it on the part of foreign investors, including, very importantly, the Mexican owners of approximately $84 billion of capital that is currently invested abroad.[11] The Bush administration began its term of office with a commitment to give first priority to addressing what could be done to reduce Latin American debt, so there is perhaps reason to believe that the two countries may, at last, have developed the will to find the way back from the brink.

A bilateral treaty of "mutual economic benefit," once worked out and announced, could begin to turn the tide in Mexico's favor even before the cash arrived. The fact of the agreement would be discounted by the market, and some of its benefits would be available right away, as expectations change sharply for the better. If the program is good enough, and is seen as increasing employment and living standards, it might even be supported by the opposition parties; if not, it would certainly blunt their criticisms substantially.

A further action that the United States could take as part of a Mexican assistance initiative would be to ask Congress to amend the law that prohibits the sale of Alaskan oil to foreigners. That oil, currently worth about $10 billion a year, is piped from the North Slope to Valdez, where it is next transferred by expensive U.S.-flag tankers to California, where most of it is then sent by pipeline to Galveston, after which it is transmitted by other pipelines to various destinations in the United States. A much more efficient operation would be to sell the oil in Valdez to Japanese buyers, who would far prefer it to politically unstable Middle Eastern oil, and have the oil companies replace the oil in Galveston with purchases from Mexico (and perhaps also from Venezuela, Argentina, and others). By making such an important and secure supply of oil available to the Japanese, the United States would be entitled to ask Japanese oil companies to purchase, say, a billion dollars or so of five-year Mexican notes each year for five years as a contribution to the debt-reduction problem. Indeed, the Japanese have already offered to help in the reduction of Third World debt, though their somewhat vague proposal made at the 1988 annual meeting of the World Bank and IMF in Berlin was put down by U.S. officials. A number of important political issues are

144

involved, but if the Japanese are willing to help, we should certainly take them up on it before they change their minds.

Many solutions to the Latin American debt problem appear to depend on getting the Japanese to pay for them. The trouble is, however, there is not enough money even in Japan to solve all the problems people look to them to pay for. So, at best, Japanese money is going to be rationed. The U.S. government must decide where it wants them to spend it, and then find a constructive and friendly way to get the message to them that is not counterproductive. Apparently it has.

On March 10, 1989, at a meeting of the Bretton Woods Committee, a private group devoted to studying international financial issues, Secretary of the Treasury Nicholas Brady announced a new initiative aimed at Third World debt reduction. The Treasury plan is to be a voluntary program to encourage banks to write off portions of their Third World exposure through various actions, including, the Brady announcement suggested, swaps for new debt that would be backed up in some way by the World Bank and/or the IMF. Japan, the announcement continued, would have a major role to play in the program, the details of which were not then made available.

Brady's announcement was extremely vague. Coming on the day after the Senate had voted not to confirm John Tower as secretary of defense, there was speculation that the administration had rushed the announcement for political purposes. Still, the report was acclaimed as a major change in U.S. policy from helping to encourage financing for developing countries through the issuance of more debt (the Baker Plan) to one of growth through debt reduction.

Commercial banks, Brady said, would be encouraged to forgive debt by exchanging it, at a price that would require a write-off, for new debt to be guaranteed in some way by the World Bank, the IMF, or Japan. Such a program would be very similar in concept to the Mexican debt-for-debt swap undertaken by Morgan Guaranty in 1988.

Many of its benefits, however, depended on the countries involved accepting IMF approved economic reforms and recovering flight capital from abroad, something not all were sure of being able to do.

The immediate reaction to the announcement was generally positive: Senator Bradley, however, said he saw it as a ''turning of the

145

page," but "for the shift in direction to be material, there has to be enough money." Senator Bradley thus touched the heart of the problem: Who was to pay for the new initiative? The Baker Plan had failed because it depended on the banks to put up the money. The Brady Plan was silent as to who would be required to provide the sufficiency of funding for the plan to be able to work. The World Bank, the IMF, and Japan later announced that the amounts they were prepared to commit would be fairly modest. Everyone assumed that the U.S. government would not make a significant contribution to the effort at all, because of its budgetary constraints, and therefore the plan would be totally at the mercy of others.

The intended debt swaps can work, but only when a sufficient inducement is made to those who have to do the swapping. And swapping new debt for old does not provide badly needed new money to the countries; it provides it to the banks.

There has to be a better, more realistic way to accomplish everyone's objectives. Why not say to the countries, "If you are willing to make the necessary changes in your economy and start to suppress waste and corruption effectively, we can get you some new money and a debt-forgiveness program at the same time. Let's agree on what those reforms ought to be, and on what regulatory changes we might make in the United States to help you export more to us and attract direct investment. Then, let's use the credit facilities from the World Bank, the IMF, and Japan to guarantee new investments in key, productive sectors." Assuming one can get this far, the next step is to tell the banks "if you will actually write off (that is, forgive) the debt of specific countries to the extent—but no more—that you have already reserved against it (that is, taken the economic loss on it), then we will arrange for you to be able to participate in new, generously priced loans for the country that are guaranteed by the one or all of the generous threesome." Such a program ought to retire up to 20 to 30 percent of the old debt, and provide the countries involved with some high potency new investment capital that, if not wasted, could result in a shift in the confidence factor and cause market forces to reverse in the countries' favor.

All in all, however, the Brady Plan is a step in the right direction, one that could comprise an important element in the effort to roll back the debt problem. But without the economic changes and reforms in the recipient countries, no new capital will be attracted, in which case the plan may end up being little more than another government exercise in smoke and mirrors.

The U.S. government certainly should welcome a more active role to be played by the World Bank and the InterAmerican Development Bank in the crisis. The World Bank, in particular, has played an especially listless part in the Latin American problem to date. Partly this is because of continual strife between the U.S. government and the Congress, and the Bank over lending policies, capital requirements, and internal organizational matters. This strife, some of which can be attributed to ideological factors insisted on by the Reagan administration, needs to be and probably can be cleared up by the present administration.

A further dispute continues to rage over whether the World Bank or the IMF has the principal responsibility for "resolving the crisis." The result is a hopeless, bureaucratic entanglement that would put most Mexicans to shame. The World Bank can raise lots of capital. It can invest in projects in Mexico, for example, either together with private-sector projects or alone. It has the capital, which Mexico lacks, and the charter to lend it—once investment conditions are adequately stabilized. There is apparently a lot of room here for some positive contribution from the World Bank that has been so far lacking. The Brady Plan will give it a good testing.

Mexico, of course, is only one country in a subcontinent in which all the countries have the same disease. Each country will have different requirements and things to offer in a trade, investment, and finance context. Each should have a substantial incentive to enter into restructuring deals in order to assure a more promising future. The United States can help. Its influence remains great in all of the areas where such restructuring would be desirable, and its and its counterparts' good offices in changing some laws and offering exemption from certain others could create considerable incentives to reduce debt, increase trade and investment, and prepare these prodigal sons and daughters to reenter capital markets around the world.

147

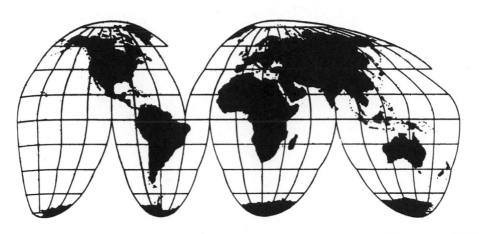

Part II

# Crusades in European Finance

# 5

# Euromoney

"... We appreciate the hard work and excellent teamwork that have gone into our issue, and look forward to working with all of your firms again in the future."

Rising to my feet to respond to these inspiring words, I found my lips almost automatically forming the phrases they had passed so many times before: "Ken, on behalf of all of us in the management group for your issue, I want to take the opportunity to thank you for the kind words and for the confidence you have shown in all of us in selecting our firms to handle this special and unique Eurobond issue for your company. I know we all feel that the market responded extremely well to the company, and we are all pleased that we were able to secure such attractive terms for the issue.

"We did work hard to get it done and we ourselves appreciated the teamwork that ensued. Again, thanks for the business. And now, maybe you Midwesterners would like to soak up some of the atmosphere at Annabel's bar, and although I am not going to be able to join you, my syndicate colleagues will be happy to introduce you to their special corner out there.''

So we concluded another Eurobond issue in the winter of 1984 during the high point of the market. The closing dinner for the issuer's senior financial staff and the twelve managing underwriters was held in the wine cellar of Annabel's, London's most fashionable private club for its most beautiful people. Located in Mayfair's Berkeley Square, midway between the elegant Connaught and Ritz hotels, Annabel's was the favorite late-night watering hole of the Eurobond market's many heroes, arrivistes, aspirants, and camp followers. Its rich decor of fresh flowers, comfortable chintz-covered sofas with too many throw pillows, and hundreds of paintings and prints of landscapes, dogs, women, and cartoons all hung close together on gold-covered walls was meant to be everyone's idea of the fantasy English country house. The place was packed almost every night with an eclectic group of well-dressed Londoners and visitors of all descriptions. Newly prominent among them were the highly paid syndicators, traders, and salesmen of the Eurobond market.

Walking out of the wine cellar into the main bar, I nodded an obligatory greeting to the prince of Eurobonds himself, Hans-Joerg Rudloff of Credit Suisse First Boston, sitting in his usual niche, surrounded as always by protégés and admirers. Rudloff returned the gesture with a slight scowl to signify that he disapproved of closing dinners being held at Annabel's by firms other than his own. I suspected that before long he would be offering his audience a critique of our issue, which had only barely squeaked by in the market, despite the ebullience that the market was then experiencing.

The issue had been won in competition with many other firms, the winner being the one that was thought to have offered the most competitive pricing, the ''most bang for the buck.'' We had proposed a $100-million straight-debt issue with traditional features, called a *plain vanilla issue,* that would be sold together with warrants entitling bondholders to purchase additional bonds with the same terms over the next five years. If interest rates should decline over that time, then the holder of the warrant could exercise his right to buy the additional bonds, which would carry an above-market interest rate, or he could sell the warrant in the market to someone else who would pay a premium in order to be able to buy the bonds himself. Considering the unpredictability of interest rates over the next five years, and the high degree of volatility that had become the norm in financial markets, the warrants were considered to be potentially very explosive little packages, worth quite a bit to the original investor in the bonds.

This investor could not be an American—the bonds could not be sold in the United States because they were not registered with the SEC. The investor would be someone whose home currency was not the dollar, who therefore saw in the bonds and the warrants not only a speculative opportunity with respect to interest rates, but also an opportunity with respect to the dollar. If the dollar strengthened in relation to the Swiss franc, or whatever currency he kept his books in, he would have a further gain, one he would not have if he were a dollar-based investor. In 1984, interest rates were continuing to decline, increasing the value of earlier issued, higher interest rate bonds, and the dollar was still strengthening, providing a double bull market environment for Eurobonds and warrants that might be attached to them.

When we analyzed the competitive situation before submitting our bid, we knew that the company, a well-known, highly regarded American manufacturing company headquartered in the Midwest, would be able to attract financing offers in the Eurobond market that would carry a significantly lower interest cost to the company than anything it could get in the United States, because of the appetite in Europe for dollar-denominated securities of the best quality. European investors would pay more for the same security than American investors would because of the potential profit on the dollar and because high-grade American corporate securities were the best paper available in the Euromarket.

U.S. Treasury securities, which might have been the preferred investment of the conservative Europeans, were not available because at the time they were subject to a withholding tax in the United States on interest paid to foreign investors. Corporations could avoid this

153

tax by issuing Eurobonds through an overseas subsidiary that would be guaranteed by the parent, but the Treasury could not so exempt itself from its own tax.

The fact that the market was very strong and attractive to the company did not ease our problem, which was to find a way to win the business in a competitive bidding without losing our shirts. True, if our bid was too aggressive, we might look to continuing interest-rate declines to "bail us out" of our holdings of unsold bonds at a later time, but our competitors would take this factor into account too. To win, either our bid would have to be so far ahead of the market as to be suicidal, or we would have to come up with either an innovation or a different marketing approach so we could shave off a few more basis points (.01 of 1 percent is a *basis point*) of interest cost to the company. Alternatively, we could use our close relationship with the company to try to talk it into accepting the best bid that would still result in "a successful offering" rather than simply the lowest cost, irresponsible kamikaze offer.

My preference was for this latter approach, but we all knew that it would be difficult to be credible when so many other reputable firms, including perhaps the formidable Credit Suisse First Boston, were the ones that would be likely to be offering the impossibly low rates. To win, our effort would have to combine aggressiveness, innovation, and persuasiveness. In this case, we were successful. Our warrant pricing was more aggressive than some of the others. We had lined up a Japanese co-manager who had agreed, in exchange for the co-managership, to precommit to selling at least 20 percent of the issue to Japanese insurance companies, trust banks, and individuals.

We had also had our salesmen make some discreet calls to line up some investors in Europe and the Middle East who were very aggressive buyers of warrants, especially if we would buy back from them some similar warrants we had sold them several months earlier, which would yield a good profit to them. Warrants were often detached, or "stripped," from the bonds they were packaged with and sold separately to speculative investors. The plain vanilla bonds, without the warrants, would be sold to conservative investors at a discount. In this case, we had priced the package at 100 percent, about 94 percent representing the value of the plain vanillas, and 6 percent the

154

value of the warrants. We would receive an underwriting discount, called a *gross spread,* from the issuer of 1.875 percent, much of which would have to be rebated to institutional investors for us to be able to sell the package to them. But if we sold the two parts separately, and moved quickly before the rest of the market caught on to the deal, we hoped to be able to market the package for something more than the usual price at which we would expect to sell an issue of bonds only.

Our bid was close to, but higher than, the absolute best, which did not include warrants. We were able to differentiate our deal on the basis that the company, by issuing the warrants, had the best of both worlds—if interest rates, hereafter, should rise, then the warrants would expire without being exercised and the 6-percent warrant value would remain in the company's pocket, where it would be used to lower the effective cost of the bonds issued. If interest rates drop in the future, the warrants would be exercised at the same rate as today's bonds, but without any further underwriting or issuance costs. Besides, we argued, the distribution of warrants and plain vanilla bonds at the same time would bring the company to the attention of many more and different types of investors, including those in Japan and the Middle East, than would an issue of plain vanillas only.

In the end, we won the assignment, but were required by the company to include the lowest priced bidder as a *co-lead manager,* which we did. The lead managers receive a larger portion of the management fee and more visibility than the co-managers do. *The* lead manager, or *book runner,* however, by controlling the allocation of the bonds among the other managers (including allocation of bonds to his own firm) stands to make (or lose) by far the most money on Eurobond issues. Having a co-lead manager eats into this potential profit quite a bit, but at the same time it provides someone to share the difficulty with if things go wrong. In this case, the two of us would be co-lead managers, between us underwriting about 30 percent of the issue. In addition, there would be ten or so *co-managers* who would underwrite about 50 percent of the issue, and a small group of twenty of so general *underwriters* who would take up the last 20 percent. In addition to the management fee, applicable only to the first two groups, all of the underwriters would receive an *un-*

*derwriting fee,* against which all of the unreimbursed costs of the issue (including the lavish dinner at Annabel's) would first be applied.

The allocation of bonds for sale by the underwriters is the sole responsibility of the lead manager. Thus the lead manager gets to determine which other underwriters get to earn the large sales commissions that are allowed on Eurobonds. The lead manager is also solely responsible for any efforts to stabilize the price of the issue through the purchase of bonds in the immediate aftermarket.

In the Euromarket, money is made by the firm capturing the lead manager's slot, but only when the issue can be sold profitably. In competitive bidding situations, the margins are tight. Success goes to those who can spot changing appetites among investors and provide them instantly with what they want when they want it. This requires good knowledge of markets and investor preferences all over the world. It also requires the ability to design innovative securities and to use syndication and lead-management skills to their fullest.

Once we had won the deal, we proceeded immediately to firm up our Japanese connection and to select and invite the co-managers, who would underwrite 50 percent of the issue between them. This meant that they would agree to buy that much from ourselves and our co-lead manager, who had bought the entire issue from the company at the time that we had been awarded the mandate. Strictly speaking, any co-manager who accepted would be helping the co-lead managers to share their risk rather than helping the company, whose risk exposure ceased when we shook hands (over the telephone) on the deal.

One co-management slot was filled by the Japanese firm, but nine others had to be found within the next hour or so. Our salesmen and those of our co-lead manager were already flogging the issue throughout Europe. Soon the fact that the issue had been created would be known in the market and a price for it would be quoted. We wanted to be able to announce the successful completion of syndication (with first-rate underwriters) as soon as possible to set the right tone for the issue. Also, we wanted to fix the laying off of our underwriting risk.

We hurried. A major Swiss bank attracted by the opportunity to place the high-grade bonds with managed accounts accepted. A major German bank that represented the company in Germany had complained strenuously about the low interest rate on the issue but was still thinking about it. A major French bank, eager, like many at the time, to improve its standing in the rankings of Euromarket underwriters, accepted on the wire. Another French bank was called. A Dutch bank declined, and so did one major U.S. investment bank, another one accepting. The German called in to accept, but asked us not to allocate any bonds to it. A major U.S. commercial bank that specialized in warranted transactions declined on the grounds that our warrant pricing was wrong. Another U.S. bank accepted with pleasure. The second French bank accepted. A Japanese commercial bank's merchant-banking subsidiary accepted appreciatively; it had been invited because we were soliciting it for another transaction in Japan and wanted them to know we cared. A Finnish bank was invited for the same reason and accepted.

In the middle of this two-hour process, however, the issue appeared on the television screens of one of the bondbrokers at "less 1.75." *Bondbrokers* are wholesalers who broker trades in bonds between bond dealers and underwriters. If one of our co-managers, for example, had wanted to be seen in the deal but had nowhere to place bonds and wanted to get out of the bonds he thought would be allocated to him, he might call a bondbroker and offer bonds to the market, anonymously of course, for sale.

The bondbroker's first call would be to the lead manager of the issue. The lead manager would be told that the broker had a certain number of bonds for sale that he would sell to the lead manager at the official offering price (for example, at 100 percent), less a discount of, say, 1.75 percent. If the lead manager did not want to buy the bonds back, then the broker would be forced to offer them elsewhere at a lower price, which would soon be shown on his screen for all the market to see. The selling co-manager, who did not participate in more than a little of the management fee, would be selling his bonds at, or slightly below, his cost in order to eliminate the risk of holding them until buyers could be found.

This informal, unofficial market between participants is called the *gray market*. It is well known to all in the underwriting and trading of bonds because it is so influential. In 1984, the gray market was only about three years old, but its impact was considerable.

Through the gray market, underwriters of issues could dispose of their expected allocations with comparative impunity by "dumping" them on the market. No one would know who did it, and the bondbrokers would never tell. The loose bonds would have to be picked up by the lead manager, and redistributed, or they would be marked down for sale elsewhere, depressing the market for the bonds that the underwriters had not yet sold. As a result, the market for new issues in those days became very unstable.

David Watkins, Goldman Sachs's Eurobond syndicate manager at the time, felt we should buy back the bonds the bondbroker was offering to us. The deal was still being syndicated, he said, and if the gray market went way off, we might spook some of the firms we had not yet heard from. We definitely wanted to get the whole deal syndicated as soon as possible, as any unsyndicated portion was strictly for our own and our co-lead manager's account. We bought $2 million of bonds, which indicated that the sale had come originally from a member of the co-management group.

Our head salesman reported that the market was timid. The buyers, he said, "think the deal's too tight, that they will be able to buy it cheaper tomorrow."

"Is Crown interested?" asked David, inquiring about one of our largest Eurobond customers, The Crown Agents, a British government agency that managed investments and other affairs for Commonwealth governments. The sultanate of Brunei, an independent oil-rich Commonwealth state on the northwest coast of the island of Borneo, was one of Crown's principal accounts at the time; it was a big buyer of high-grade plain vanillas.[1] "Yeah," said the salesman, "but not in the warrants, and they'll probably want us to take some Swedish," meaning they would want to swap some Swedish Export Credit bonds they owned for any of the new issue that they bought.

"Look," David said, "we just bought back 2 million at less 1 and ¾ [that is, at 100 less 1.75, or 98.25 percent]. Strip the warrants for 6 and offer the bonds to Crown at 92¼. And tell him we can offer him 2 million more at a ½ [that is, at 92.5]."

David went on to advise me of his plan. "We're working on a

158

big sale of warrants in Hong Kong, and Rudi's got a big nibble from [the] Rothschild [bank in] Zürich. I think we'll be okay on the warrants, but the bonds worry me. We've got to lighten up our own position, and I don't want to buy the whole issue back from the bondbrokers tomorrow morning. If we can get the bulk of it done overnight in the Far East and then we let the gray market go, then the bastards [that is, our other co-managers] will dump bonds like crazy, but we won't buy them back, and we won't allocate any either. They will have to cover their short sales in the market, which will bid the price up again."

Overnight sales were fairly good, and more warrants had been sold than David expected, many at $6½. By the opening of business the next morning, we had sold most of our own underwriting position of $10 million and had resold about $3.5 million bought back in the gray market, which now, however, was quoted at "less 2." More gray market offers came in, $10 to $12 million more, which David rejected, and the market fell to "less 2.25 to 2.50." The gray market quote was telling us that our other underwriters were losing up to half a point just to get out of their positions.

"All these sellers into the gray market can't be just our underwriters," said David. "We've got other folks dumping this deal, trying to screw it up. There's another AA American name coming in the next couple of days—could be the lead manager is trying to hedge the long position he expects to have in those bonds by shorting our deal." Being both long and short Eurobonds for similar issues is a much better hedge against interest rates rising suddenly than being long Euros and short a different security, like U.S. Treasuries, in a different market, that is, the United States.

"Either that, or some sour grapes bozo is trying to make us look bad in our client's eyes." By short-selling bonds into the gray market, a speculator could hope to crater our deal, then cover his position cheaply after the market collapses.

At noon, David sent word to the syndicate that allocations would be approximately 25 percent of their underwriting obligations, not the 100 percent that most had expected. Between our own, our co-lead manager's, and our Japanese co-manager's sales, about 50 percent of the issue was done. The co-managers and the underwriters between them were to underwrite about 70 percent of the deal, only 18 percent of which would actually be allocated. Together with our 50 percent that was sold, 68 percent was therefore spoken for. However, 32 percent of the deal was not sold; if the gray market did not respond to our tactic, we would own $32 million of bonds with a market price at least .625 percent lower than what we paid for them, a loss of at least $200,000.

Most of the underwriters were relieved, because, like the Germans, they really didn't want the bonds. Others railed at David for hours (thus he had some idea as to who the culprits were who had to cover short positions), but he held fast. All the loose bonds in the gray market were bought up at rising prices, as those who had sold more than they had been allocated searched for bonds they could buy to deliver to their customers.

When the price reached less 1.75, David began to sell some of the bonds he had not allocated. The price settled at about less 1.50. Soon David had sold out the remaining position. All in all, the firm made about $200,000 on the deal, and everyone else much less, but it was a fairly close call—we could easily have lost it all, especially if the market had turned down while we were executing these various maneuvers.

The client liked all the action and the results. The immediate drop in the gray market confirmed that the price at which we had bought the bonds from him was not "too low," and the recovery indicated that the market had recognized the high quality of the paper on offer and stepped in to buy it when it appeared cheap.

Our best estimate was that about half of the issue ended up in various Alpine vaults, where the bonds will never again see the light of day as they will be held until they mature. Perhaps six or seven Swiss banks acquired these $50 million of the bonds and placed them with perhaps hundreds of individual accounts and some commingled bond-fund accounts. Perhaps $20 million ended up in Japan as agreed, most of which will be traded again. The rest were bought by British and continental banks and institutional investors for accounts they manage, and by a few Middle Eastern names.

•   •   •

160

Hundreds of these deals were done in 1984 and 1985, the peak years for dollar-dominated, fixed-rate Eurobonds, of which more than $19 billion were issued. In late 1985, the dollar began what became a protracted decline, and the investor enthusiasm that had so characterized the preceding three years subsided. U.S. issuers found the U.S. market equally or more accommodating, as by 1985 the SEC had emulated the Eurobond market by adopting Rule 415, which permitted the bought deal to be done in the United States. U.S. corporations were then less likely to be able to finance more cheaply in the Euromarket, so they financed at home.

Their place was taken in the Euromarket by a huge increase in the issuance of floating rate notes (called FRNs), mainly by banks, which were sold to other banks hungry for marketable dollar assets to replace illiquid bank loans and to provide investments for the abundance of funds available to them. FRNs were generally five to ten years in maturity with interest payable every six months at a given margin, or "spread," over LIBOR. A good-quality bank, for example, might issue five-year notes at a spread of 25 basis points over LIBOR, with underwriting commissions totaling about .35 percent, or .07 per year. The sum of the spread and the fees would add up to 32 basis points over LIBOR in this case. Most banks, at the time, could secure deposits from their long-standing customers, or purchase them in the market at about 12.5 basis points below LIBID (the "bid" side of the London Interbank Rate for deposits, normally about 12.5 basis points less than LIBOR). The bank's cost of funds, therefore, was about 25 basis points less than LIBOR. Investment in the FRNs would provide a total return to the bank of 57 basis points (32 on the FRNs plus 25 on the funding). The FRNs would be marketable, which bank loans generally were not. FRNs were also extremely flexible in design; a different interest-rate exposure could be created by having the interest-rate reset occur every three months instead of every six or by having it payable every three months, but based on resets every thirty days. Some Third World governments issued FRNs too, and these were issued in bearer form, which meant that purchasers would not be identifiable for any future new-money rescheduling exercises.

Maturities could also be extended, in exchange for a bit more

161

spread over LIBOR. Twenty-year FRNs appeared in the market, some designed to be subordinated to deposits and other obligations so as to count as "capital" for the issuer, something many banks were eager to raise.

Finally, in 1986 the first "perpetual" FRN was introduced in an issue for National Westminster Bank, the leading British clearing bank. Principal would never be repaid; when you wanted your money back you could sell the security in the market. The British government had issued similar undated securities, called *Consols,* for years, though not recently. Interest on the NatWest perpetuals was payable at 37.5 basis points over LIBOR, a comparatively healthy premium over the 12.5 basis points or so that the bank would have had to pay for a conventional five- or seven-year FRN. The issue was a big success; it was sold far and wide, apparently with Japanese banks taking up a large amount of it. These banks had plenty of comparatively cheap financing available to them, but few loans to make with it. They needed high-grade assets that could be available in large amounts that would carry a higher than usual spread. Except for the high-grade aspect, their purchase of these securities resembled the recycling of oil dollars to Latin American borrowers by U.S. banks in the late 1970s.

Trouble came when rumors appeared in Japan that the Ministry of Finance was considering a regulation that would not allow Japanese banks to invest in securities that represented capital of other banks. The Japanese were thought to be dumping or about to dump their perpetuals in the market, but there were few buyers. The market fell into a rout, then went dead. There were virtually no buyers or market-makers in perpetual FRNs; then the disease spread into the conventional FRN market too, substantially shutting down further new issues. In 1986 nearly $38 billion of FRNs were sold; in 1987 only $7.8 billion.[2]

The combination of problems in the FRN market and a reduced volume of fixed-rate issues in 1987 resulted in the first significant drop in Eurobond new-issue activity since the market's beginning. Total volume in 1987 fell to $162.3 billion from $191 billion the year before. The decline would have been much greater except for extensive new issues of bonds denominated in currencies other than the

162

dollar. Many of these issues, which accounted for 60 percent of the market in 1987, as compared to only 18 percent in 1982, were swapped into dollar obligations which still offered savings over domestic financings, as Dennis Dammerman and others had already learned. By the beginning of 1988, after the dollar had stabilized again, new issues began to reappear, though sporadically.

Prior to 1963, the method used to raise long-term capital from international sources was to float a bond issue in some other country, denominated in the currency of that country and issued in accordance with the laws and procedures of the bond market there, usually at a premium interest rate reflecting the exotic nature of the borrower and/ or the possibilities of difficulties in collecting payments due. Such issues have been called *foreign bonds* since Jacob Schiff's day. Today, when foreigners issue dollar-denominated bonds that are registered with the SEC they are called *Yankee bonds*; those registered with the Japanese Ministry of Finance and denominated in yen are called *Samurai bonds*; those involving sterling issues in the United Kingdom, *Bulldog bonds*; and so on. They are all foreign bonds.

The total annual volume of foreign bonds averaged $2.6 billion during the period from 1964 through 1974; then, after the removal of U.S. capital-market controls in 1974, it jumped sharply to average about $16 billion annually for the rest of the 1970s. In the 1980s, reflecting much greater use of the Swiss and Japanese markets, foreign bond issues have been averaging about $30 billion each year.[3]

The Eurobond market, by contrast, is not regulated by any government or official body. It is a completely ad hoc creation that has perpetuated itself and grown because it avoids regulatory impediments and rewards initiative and innovation at the same time as it polices itself informally through a clublike atmosphere among the principal and most powerful participants.

In the Euromarket, for example, issues are not forced to wait their turn in a queue set up to protect the market from excessive usage or to explain their newest ideas to a regulator whose approval must be secured before the issue can be made. Nor do issuers from one country have to convert their financial records to the format or style of another country or to disclose facts about their businesses that may seem irrelevant to investors if not to regulators. Nor do investors in Eurobond issues have to worry about their identities being revealed to tax authorities in various countries in which an investor may have assets.

Each country, to be sure, has regulations concerning the distribution of securities within its borders, and Eurobonds must comply with these, and they do. They are treated as foreign securities in each of the countries in which they are sold. There are laws restricting the sale of such securities in all countries, but almost always sales to sophisticated and knowledgeable investors, such as banks and institutional investors, are permitted. No country claims jurisdiction over the Eurobond new-issue process, however, which therefore continues to exist as an unregulated, stateless capital market.

Between 1964 and 1974, the total annual volume of Eurobonds averaged about $2.3 billion; from 1974 to 1979, it averaged about $11 billion, less than that for total foreign bonds, but from 1980 to 1986, the average annual new-issue volume soared to $80 billion, with almost $190 billion done in 1986 alone—almost five times the volume of new issues of foreign bonds in that year.[4]

After 1979, markets everywhere had to adjust to the landmark change on the part of the U.S. Federal Reserve in the way in which monetary policy was managed in the United States. Interest rates were allowed to float to whatever level the market required in order to maintain a certain level of money supply, the reverse of the previous policy. Interest rates thus became free to adjust themselves to the extremely high levels of inflation that existed in the United States and elsewhere.

Perhaps no single event in the postwar period contributed more to the volatility of interest and foreign-exchange rates than did this action by the Fed, but whether so or not, the volatility of both increased to several times what they had been in the 1970s. The markets, then, had been put in charge of controlling U.S. inflation and reacting to the uncertain economic policies of the new Reagan administration, but markets can be, and usually are, fickle, changeable, and, for brief periods, irrational.

Such increased volatility placed a larger premium than before on the ability to act quickly, before markets changed, and to take advantage of different investor preferences around the world through innovative new securities. The Eurobond market was the only market in the world that could provide these features. National bond markets required waiting periods of three weeks to three months to enter the market; they required extensive, expensive, and time-consuming disclosures to conform to national requirements; and these markets were still oriented to charging foreigners a premium for their use of them. Euromarket issuers could avoid all of these impediments to efficient financing, and, as indicated earlier, many could borrow in the Euromarkets at substantially lower rates than they could at home. At such times, the Euromarket naturally became the market of preference for those borrowers whose size, recognition, and strong balance sheets qualified them to use it.

At the end of 1982, Goldman Sachs's partners were astonished to learn that the London office of the firm had sold more corporate bonds at new issue than had the New York office. During the period from 1981 through 1984, top-rated U.S. companies such as General Motors, AT&T, General Electric, Ford, Coca-Cola, Citicorp, IBM, Sears Roebuck, Xerox, Kimberly Clark, and many others issued more bonds in Europe than in the United States. The impact of globalization on these companies, and their bankers, was considerable.

In the early days of the Eurobond market, certain practices were established that were unique at the time, but were effective in making this exotic market function efficiently. Among these was the practice of announcing a deal, then marketing it, then pricing it, as was done in the United States. More common European practice had been to announce a deal that had been already priced and underwritten, then let people subscribe to it, with any unsubscribed portion being left with the underwriters, who were substantial institutional investors. The American system emphasized the marketing of the issues, through syndications and through the efforts of well-trained bond salesmen. Such intimate contact with the market that salesmen provided gave underwriters, especially lead managers, the sense of the market they needed to fine-tune pricing and to detect trends.

Initially, the Eurobond market was a system for organizing the portfolio management departments of leading European banks for the purpose of distributing bonds into individual investment accounts at the banks. There were only two kinds of investors in the beginning:

individual clients served by banks that were in the syndicate, and individual investors who were served by banks that were not in the syndicate. The individual investors were an extraordinarily diverse group consisting of tax-averse, secretive middle-class Europeans, whom the market calls the *Belgian dentists*, wealthy Middle Eastern and Latin American families, and rogue potentates like Ferdinand and Imelda Marcos who have squirreled money away in numbered accounts in Switzerland.

There were only a few institutional investors at the time, but with the increasing involvement of U.S. firms with their sales forces, new ones were uncovered, including corporate pension funds, especially European funds of U.S. companies, offshore insurance companies, merchant bankers and others acting as investment advisers, special investors like the Crown Agents, and others. After 1979, with the deregulation in the United Kingdom and in Japan of foreign-exchange controls affecting overseas investment, many new institutional investors poured into the market, accounting for much of its sudden growth in the 1980s.

Still, at the beginning the greatest accommodations had to be made to the Swiss banks, whose portfolio managers comprised by far the greatest placing power in the market. The Swiss banks were at first restricted in acting as a lead manager of issues, lacking at the time affiliates abroad that could do so freely. The Swiss in any case preferred situations in which foreigners came to them, on a case-by-case basis, to ask them to serve as co-lead manager and to take a large number of the bonds being offered. This they would do as long as the economics were right.

"You must understand," they would say, "our clients are only interested in the best-quality names, but even so, our portfolio managers, who act quite independently, must be satisfied with the issuer and the rate.

"Further," they might add, "it is very expensive for us to approach all of these portfolio managers and to give them information about bond issues. When they do buy, they tend to do so in small amounts, so we must bear the expense of placing the bonds in perhaps hundreds of accounts."

For this, they would need to be well compensated for their participation in the issue. "Well compensated" meant gross spreads, or underwriting commissions, of about 2.5 percent for long-term bonds, 2 percent for medium-term bonds (that is, ten years), and about 1.875 percent for shorter maturities.

As co-lead managers, they would receive a generous part of the management fee for the issue and underwriting and selling commissions based on the amounts underwritten and allocated to them. With their considerable in-house placing power, they never had to go outside the bank to sell their allotment. They could access the demand for an issue from among their portfolio managers before deciding to underwrite it.

When they did participate, their rewards could be considerable. For example, on a $100-million seven-year issue, a very typical offering in the 1980s, they might earn $50,000 from the management fee, $38,000 for underwriting $10 million of bonds, and $250,000 for placing close to $15 million in their clients' accounts; a total of $338,000 with no risk to the bank. This is in comparison with the $200,000 Goldman Sachs managed to earn on the transaction described earlier despite possibly a fifty-fifty chance that the firm would have lost $200,000. For the Swiss, the whole process was a piece of cake. Other Continental banks followed the same practice, though on a lesser scale. There was no way around them—between them, the Swiss and the other Continentals controlled the market.

Later, as institutional investors came to be much more active in the market, a *two-tiered pricing system* arose: one price for the Swiss banks' customers, who were sold bonds at the official price after the gross commissions had been paid; and another price for the institutional investor, who would only buy bonds at market rates, that is, at investment yields consistent with comparable investments available to him in the secondary markets. What this meant, in practice, was that the underwriter would have to pass on some or all of the gross spread in the form of a discount to the institutional investor.

An issue priced at 100 will be purchased from the issuer by the underwriters at 100 less the gross spread (say, 1.875 percent), or a net price of 98.125. The company will have sold the deal to the underwriters on a basis in which the total cost of funds to the issuer is calculated and compared to other offers. The winning bidder will select an interest rate, or coupon, for the issue that, at the purchase price, will yield the agreed cost of funds.

If the underwriter is a Swiss bank, he will sell bonds to his accounts at the official price of 100, taking the difference above the 98.125 purchase price for his profit. If the underwriter is selling to an institutional investor, however, the latter will not buy at 100, but instead will insist on a discount such that his investment yield will be fairly close to the issuer's cost of funds. In the tale at the beginning of this chapter, Crown was offered bonds at "less 1.75" after the warrants had been stripped. The total gross commissions on the deal were only 1.875 percent, so the Crown sale by Goldman Sachs earned the firm only 0.125 percent. Most of the seemingly large gross spread has to be rebated to institutional investors in order to get them to purchase bonds. So they buy at one price, and the retail investor being advised by the banks buys at another price.

The sale of Eurobonds to institutional investors is little different from the sale of domestic bonds to U.S. institutional investors—the margins are thin, the risks are high, and the profits are made on the volume and the trading opportunities created by the high level of volatility in the market and on the ingenuity contained in individual transactions.

In the mid-1960s, a few U.S. brokerage firms had taken the trouble to staff sales offices in Europe to sell U.S. stocks to European investors. Most of these operations were small. One of the larger ones, however, was White, Weld & Co., an old-line "white-shoe" Wall Street firm. The firm was then headed by a second- or third-generation member of the White family, August (Augie) White, who liked to travel abroad, especially in Japan, where he was well known, and was a committed internationalist.

About the time of the first Eurobond, White Weld was in a good position to benefit from the event by virtue of its European sales offices. The firm was among the top dozen U.S. underwriters at the time, but certainly one of the very few who were both leading underwriters and distributors in Europe. Using their many excellent connections, especially in Switzerland, the firm became very knowledgeable about Eurobonds and made a market in them in New York. They were one of the few firms to do so.

One of my early assignments for senior partner Stanley Miller was to prepare a daily list of the market quotes for the top U.S. Eurobond issuers. The only way I knew to get such quotes was to call up the increasingly irritable Eurobond trader at White Weld and read him my list, which was lengthening every day, hoping he wouldn't catch on that we were using the quotes to demonstrate market knowledge so we could compete with his firm. Maybe he did know and gave me phony quotes, but if so, none of us at Goldman Sachs ever knew.

White Weld jumped off to an early lead in share of the market for U.S. Eurobond issues, which was fairly quickly equaled by Morgan Stanley through its subsidiary Morgan et Cie. in Paris, then jointly owned by all of the corporate blood descendants of J. P. Morgan & Co.: Morgan Stanley, Morgan Guaranty Trust, Morgan Grenfell, and several prominent European banks.

White Weld's European operations were run by a Swiss named Robert Genillard. Over the years, under his direction, the firm invested in a number of European activities, and these were ultimately consolidated into a holding company, White Weld Limited. This company was partly owned by its management and at least one outside investor, the billionaire shipowner D. K. Ludwig. The White Weld Limited group operated a U.K. merchant bank, a joint venture with Sumitomo Bank in Japan, a French money manager, and a small Swiss investment adviser that it had bought from Crédit Suisse, one of the three big Swiss banks, for an 18-percent interest in White Weld Limited. White Weld Limited specialized in servicing blue-chip issuers and investors in Europe.

White Weld's U.S. investment-banking business had, however, headed off in a different direction from serving the largest companies, into one specializing in smaller company deals and venture capital. Its European colleagues, wanting to preserve autonomy over the high-class business they had created in Europe and to preserve their potency in the fledgling Eurobond market, succeeded in persuading White, Weld & Co. in 1970 to reduce its holdings in White Weld

Limited to a minority by spinning it off into a new trust, called WW Trust. The business of White Weld Limited and WW Trust would continue to be run by Genillard and his principal lieutenants, John Stancliffe, John Cattier, and Stanislas Yassukovitch. In 1974, recognizing the importance of Swiss placing power, Crédit Suisse was encouraged to increase its interest in the holding company to 47 percent, and White Weld Limited was renamed Credit Suisse White Weld.

John Craven, a very bright, laid-back former director and head of international corporate finance at S. G. Warburg, had joined White Weld Limited at the age of thirty-one, impressed by its youth, energy, international focus, and the financial package he was offered. He had become involved with the Sumitomo–White Weld venture early on and was sent in 1973 to Tokyo to take up residence there as the man in charge of the marketing and execution of deals that would take place with Japanese companies. He had found an office and an apartment and had begun to get them decorated when a telex arrived asking him to return immediately to London, where he was appointed the first chief executive of Credit Suisse White Weld. Genillard had decided to retire, feeling that "the frontier time is over." The others had either left also or were not considered sufficiently "administrative," as Craven evidently was.[5]

Craven was a natural for the job. He forged a kind of brain trust of Euromarket wunderkinder who were eager to do new things and to test their ideas on the market. The link to Crédit Suisse gave them extremely valuable advantages: market insight, credibility, and certainly tame, if not submissive, placing power. The firm rose quickly to prominence as the only bank in all of Europe that was focused on market transactions of all types, not just Eurobonds.

Unfortunately, Augie White's successors at the New York firm had failed at turning it into a powerhouse of smaller company transactions, and in 1978 it was sold, in distress, to Merrill Lynch. The principals of Credit Suisse White Weld, however, wished to continue in business but without the entanglements of being part of the Merrill Lynch colossus. So instead, the White Weld & Company interest in Credit Suisse White Weld was sold to First Boston, also in trouble at the time, who paid for the investment with the proceeds of the sale of a 30-percent interest in First Boston to the new company, now

170

called Credit Suisse First Boston (CSFB).[6] In other words, First Boston now owned approximately 30 percent of CSFB (which was still 47-percent owned by Crédit Suisse), and CSFB owned a 30-percent interest in First Boston, an incestuous arrangement that would cause many problems for both firms in the future. These problems ultimately had to be resolved by the merger of First Boston and Credit Suisse First Boston into a new holding company in late 1988.

Craven did not like the arrangement with First Boston, thinking that the firm's freewheeling, innovative ways would be crushed by a bunch of know-it-all, New York large-firm bankers getting into his act, and decided to leave. He returned to Warburg, where he was made a vice chairman, though the reentry never worked. Merrill Lynch then hired him in 1980 as chief executive of its European operations, and all the worst of his fears about First Boston came true instead at Merrill Lynch. Again, he quit, for the third time in about four years, attracting considerable media attention. "Most people," he once said, "commit their follies in private. I seem to have a penchant for committing mine on page one."

After Craven's departure, the helm of CSFB was taken over by Michael Von Clemm, an eccentric American expatriate who had lived in London for more than twenty years. A senior CSWW executive who had a good technical reputation and was renowned for coming up with new ideas, Von Clemm had an unusual background for a banker: a doctorate in anthropology, time spent living with a primitive tribe in Tanzania, time spent as a reporter on the *Boston Globe,* time with Citibank, time as an instructor at the Harvard Business School. According to one close observer, he was one of the most vivid, self-promotional, and extravagant characters in the Euromarket. He was an owner of the only three-star restaurant in London, Le Gavroche, where he entertained customers lavishly. He was generally considered outspoken, arrogant, difficult to deal with, brilliant, and a great "hunter" of mandates, as he referred to business-getting, something he was very good at.[7] During Von Clemm's time, CSFB built on its base and became the undisputed leader in the Eurobond market. It occupied first place in the annual *league tables* (a term borrowed from the ranking of professional soccer teams, which listed banks by volume of underwritings managed) from 1981 to 1986.

Von Clemm, however, was not considered a very good manager, a distinction shared by many top investment bankers. Morale, and presumably results, were suffering, it was said, from infighting and internal turmoil. In 1982, the shareholders decided to cure the problem by transferring Jack Hennessey, a vice chairman at First Boston, to London, where he would become chief executive of the firm, with Von Clemm remaining as chairman and Hans-Joerg Rudloff becoming deputy chairman. Craven's fears for his old firm were beginning to come true. Von Clemm left CSFB in 1986 to join Merrill Lynch as a "roving ambassador" and has never been heard from since.

Perhaps Von Clemm's greatest contribution to the firm was his hiring, in 1980, of Rudloff from the backwaters of the International department of Kidder Peabody to head CSFB's syndicate operations.

Rudloff is to Eurobonds what Michael Milken has been to junk bonds. He started in the business in the 1960s as a bond salesman covering Switzerland. Being German, he did not fit all that well into the cozy little world of largely unschooled Swiss bond traders. However, he learned to talk to them in "Schweizerdeutsch," the Swiss-German dialect of Zürich, shared their inside jokes, and gradually became one of the boys. He became a fantastic bond salesman, knowing every cranny of the Swiss market and, in time, other markets too. He rose to head the Eurobond syndicate desk at Kidder Peabody, then a low-to-middle-ranking participant in the market. He was bold, however, and enjoyed a number of successes, including the first bought deal in the Eurobond market in October 1975 for the government of New Zealand.

Soon after Rudloff was put in charge of Eurobond syndications at CSFB, he began to show that he was pretty much the whole franchise. He has always been very controversial. Described as arrogant, brilliant, petty, outrageous, treacherous, and ruthless, he has become a vivid figure in the Eurobond market ever since he reached the high table at CSFB. As deputy chairman of CSFB fully responsible for its Euromarket operations, he more than shared power with Hennessey. When First Boston and CSFB were merged in December 1988, in a transaction that sent Hennessey back to New York as chief executive

172

of a worldwide firm, Rudloff was named as head of the European part, a vital component. The chairman of the new firm is Rainer Gut, Crédit Suisse's chief executive and Rudloff's mentor, who invited Rudloff onto the board of the Swiss parent in 1986, where he became an executive director and general manager, a very senior appointment for a non-Swiss. A similar appointment has not been offered to Hennessey.[8]

Rudloff has had many Euromarket triumphs, but perhaps the greatest of these was the issuance for Texaco in March 1984 of $1 billion of convertible debentures in the first coordinated "global" offering.

Texaco had just acquired the Getty Oil Company for $10.1 billion in cash, the largest acquisition ever done at the time, in a transaction that later ended for Texaco in a nightmare of controversy, litigation, and disappointment. At the time, however, Texaco was seen to have made a great deal, having acquired all of Getty's abundant U.S. reserves at a net cost of approximately five dollars per barrel.

The transaction was financed initially with expensive bank financing, which Texaco's finance staff hoped to replace with lower cost long-term fixed-rate financing. Like Dennis Dammerman during the RCA transaction, Texaco called in a number of bankers to ask for their best ideas. Because of an illness, Texaco's vice president and treasurer, Richard Brinkman, was unable to participate fully in the deliberations, which were led by Assistant Treasurer John Hewitt, whose reputation was that of a very tough negotiator who wielded a very sharp pencil.

Perhaps among others, Goldman Sachs had the idea that a billion dollars could be raised through the sale in the United States of fixed-rate bonds that could be converted into shares of Texaco common stock at a later date. Such convertible debentures were fairly common, but not in such a large amount. Also, we proposed that the price of the Texaco stock at which conversion could occur be set at a level substantially above the present stock price. Institutional investors in the United States were looking for convertible debenture issues that would pay an interest rate prior to conversion that was higher than the common stock dividend, but where conversion into straight equity could occur if the stock price rose in the market over the next few years.

Most likely First Boston had thought of something similar. They discussed their ideas with Rudloff, who had led two prior offerings of plain vanilla Eurobonds for Texaco and who knew Brinkman and Hewitt fairly well. Rudloff saw an opportunity to snatch the transaction away from the American market and to bring it to Europe, where CSFB would run it. He hopped on a plane to New York and went to see Texaco.

When he arrived, he proposed that Texaco issue $800 million to $1 billion of similar convertible debentures in the global capital markets outside of the United States, where he was convinced the issue would be a great success. He argued that the terms were about the same as those proposed for the U.S. market, but a global deal would not use up financing capacity in the United States, which would certainly be needed to complete the refinancing program, and new investors in Texaco securities from all over the world would be attracted by the deal.

To ensure that the issue would be fully underwritten, Rudloff proposed a global distribution comprising four simultaneous offerings in Switzerland, Germany, Japan, and the rest of the world outside of the United States. The most prestigious firms from each region would be asked to lead their respective issues and to effect discipline on its underwriters. There would be an extensive road show during which senior Texaco officials could explain the Getty acquisition, its implications for the future, and their plans for financing it. Banks from all over the world who already did business with Texaco, or who would wish to in the bright days ahead, would want to be in on the deal. The whole thing would be arranged so that banks would feel they had to participate.

Texaco bought it. Despite being assured, at least by Goldman Sachs, that they could raise the same or more money at better terms in the United States, the thought of doing the first global offering was too good to resist. They appointed CSFB as lead manager of the global offering, with Goldman Sachs and Morgan Stanley as co-leaders. The Swiss syndicate would be run by Crédit Suisse, the German by

174

Deutsche Bank, the Japanese by Nomura Securities, and the rest of the world by Morgan and ourselves.

I was in charge of the Goldman Sachs London office at the time. A phone call from one of my partners in New York brought me into the picture. When, finally, I was asked what I thought about the deal, I had to reply, "Jesus!

"It's a great idea, and I'm glad you guys have got us into the thing, but you know doing $800 million of Texaco convertibles is kind of overwhelming. The largest convertible debenture offering ever done in this market was for $200 million. How much do we have to underwrite?"

"About $100 million before syndication," I was told. "Hey, you're the only one who isn't really excited about this."

"Yeah," I responded, "that's because we're the ones over here that have to sell it. If the deal gets hung up, it could be a real disaster."

"Well, the deal's decided, and we've accepted, and we're all happy over here, so you better make sure we do sell it."

I went to see our head bond trader, who assured me that the arithmetic checked out and that it had value to the investor at the terms we were talking about. He thought we could sell a lot. That was what I wanted to hear, but I knew that he'd spent a total of about two months in London at the time and was still learning about the sometimes strange and unpredictable ways of the Euromarket.

The deal, I was convinced, was too big to be sold in the ordinary way. The tranche that we and Morgan Stanley were responsible for was about $200 million, certainly doable under normal circumstances, but here the Swiss, the Germans, and the Japanese were not to be included in our tranche. We had to sell our bit in the United Kingdom, in France, Holland, Belgium, and Luxembourg, which might prove to be very difficult. Whether all the other tranches could take up that much, without dumping them into the gray market and destroying the market for unsold bonds, was highly questionable. The deal could only succeed if each of the tranches were willing, in effect, to hold bonds with the underwriters for as long as it took to sell them in an orderly fashion, perhaps weeks, or months.

175

Some European banks were prepared to do that, especially the Swiss and the Germans, who would take their unsold bonds on their own books if necessary to avoid having them appear on the market. These banks were in constant battles with the "Americans," that is, all of those who did not wish to be exposed to such high levels of market risk with respect to unsold securities and who accordingly did sell them or shorted other bonds in order to liquidate or hedge their positions, such behavior led to outraged complaint about the lack of discipline and backbone on the part of such profit-greedy firms that was destroying the market. By no means were all such firms American, though of course many were.

The secret to the Texaco deal was the discipline that had to be imposed on the participants. Rudloff knew this very well and knew that if everyone believed that his performance was being carefully watched, and the deal went reasonably well at the beginning, he could expect the discipline to hold—otherwise, most likely not.

He used Texaco's president, Al DeCrane, to make some of the calls to the chief executives of the major tranche leaders. His own chairman, Von Clemm, and chief executive, Hennessey, were not much involved, but the chairman of Crédit Suisse, Rainer Gut, was deeply so. Rudloff talked to Gut, a former syndicate partner of Lazard Frères in New York, on the telephone "several times a day, about every little detail." Gut would personally ensure the good behavior of the Swiss, and Alfred Herrhausen, Deutsche Bank's chairman elect and a member of the board of directors of Deutsche Texaco, would guarantee the Germans. All kinds of pressure was put on Yoshihisa Tabuchi, Nomura's president, to ensure that the Japanese bonds, if unsold, would stay put. Similar efforts were made to get commitments from the heads of those banks from the rest of the world before they were to be invited as co-managers of the tranche. Altogether some thirty-three banks were involved as co-managers, so the task was a formidable one.

Meanwhile, news of the transaction leaked to the press and of course the gray market. Rudloff was on the phone constantly, sweet-talking the journalists and the bondbrokers into understanding the grand conceptual design of this, the Euromarket's greatest financing, its finest hour. Rudloff's magical way with the press and the market-makers was similar to Henry Kissinger's during his time as national security advisor and secretary of state. He was the media's darling and its chief expert on his subject, the unquestioned authority. It worked. All the atmospherics were in place, the prestige of the deal established, the commitments ironclad, and the values accepted.

The deal was to be structured in the old-fashioned way, with preliminary terms being announced, followed by a road show, with final pricing to be fixed after a marketing period had elapsed. Before it did, the size of the transaction was increased to a round $1 billion. Demand was fantastic. Everybody bought. It was the deal you had to be in. A triumph like none before it.

Even when the deal was completed, the enthusiasm for it continued to linger. Rudloff, like a successful general after a great victory, sent his cavalry out to finish off the retreating foe. He persuaded Texaco to issue an additional $500 million of the same securities, through the same syndicates, within a month of the original issue. This one, too, was successful, an echo of the earlier one.[9]

Rudloff's talent is for smelling opportunities in the market and for stretching the system to accomplish the undoable, by cajoling, badgering, and threatening all the players into commitments to support the action. It works enough of the time for people to go along with him despite their better judgment.

Thanks in part to the Texaco transactions, Goldman Sachs finished sixth in the fixed-rate Eurobond league tables, and eighth overall after taking FRNs into account, for 1984. We were second that year in issues for U.S. companies. Four other U.S. firms were in the top ten for overall managerships that year, indicating the importance of the Euromarket to U.S. issuing clients. Seventy percent of all fixed-rate Eurobonds were denominated in dollars, and about a third of these were issued by U.S. borrowers during the year. Nineteen eighty-five went on to be the high point in the cycle for Americans when over 40 percent of all fixed-rate dollar issues were by U.S. companies. In late 1985, however, the dollar broke from its exceptionally high levels and continued in decline for the next three years. By 1987, fixed-rate dollar issues comprised only 39 percent of all Eurobonds, the lowest level ever.

•   •   •

Trying to manage a Eurobond business (if you are not Swiss) must be like owning a Ferrari. When it's working right, it goes like hell and surrounds you with youthful, envious admirers. Your superiors, however, think it's unbecoming, too flashy, very dangerous, and, perhaps, un-American. The thing is expensive to run, breaks down all the time, and can only be fixed by some foreigner whom you can hardly understand, and every so often it goes off the road and bangs everything up. There are, no doubt, very few old and experienced Ferrari owners, that is, ones who have held on despite all the danger and grief.

In the period before 1979, there was not enough volume for anyone really important to care about the Eurobond market. Only John Craven and his youthful band of nonconformists took it seriously. Afterward, however, the first problem was deciding to set up trading operations in a location outside of New York. Senior bond traders are poor delegators on good days. They worry about controlling their junior traders and their positions and about all the problems of delivery and settlement, but mostly they worry about trading that goes on outside their immediate reach. Globalization of markets was bad news for most of these people. Their day would begin at the crack of dawn when they first talked to London and not end until Tokyo had said good-night. Any serious trade overseas would mean a call, day or night, to the boss trader in New York. They bought beepers, and then car phones, and thought their lives were ruined forever.

But that was not the worst of it. In New York, the new-issue market bore a logical relationship to the secondary market, where bonds that had already been issued were traded. New issues had to be sold at a fixed price; underwriters could not offer discounts to customers. A new issue would be priced to be sold at, or very near, the bond yield prevailing in the secondary market for issues of like quality and maturity. Not in the Euromarket.

New-issue pricing for Eurobonds did not necessarily relate to the secondary-market yields of comparable bonds, partly because there was much less aftermarket trading in bonds and therefore the secondary markets were not always reliable indicators of what the yield ought to be, but also because underwriters, in competing for the new-

178

issue mandates, would drastically underprice new issues. This meant that a bond trader, driving into work at 7:00 A.M., might get a call on his car phone from a colleague in the London syndicate department (originally, in many U.S. firms, a part of the corporate finance division, not the fixed-income division of which the bond trader was a part), in which the trader was asked for his "consent" to bid for a new piece of Eurobond business.

" 'Morning, Eric," the syndicate man would begin. "We've been asked to bid on a $200-million five-year issue for General Electric. Bids are due in about half an hour. It's going to be very tight because GE is letting five or six firms bid. As you know, John Weinberg [our senior partner] is trying to get us closer to the company, and winning this bidding would be a big step in that direction. . . ."

"GEs here are 8.90 for five years. They might go a little better in Europe," says the trader, referring to the market interest-rate yield to investors for the bonds, knowing what was coming.

"Yeah, well, we think there will be bids around 8.50 to 8.60, especially as the Swiss like the name and are bound to come in aggressively. To win this thing, we'll have to bid about 8.45. That okay with you?"

"What's the secondary market for the last GE five years?"

"Well, the last deal was about ten months ago, and we never see them anymore, but John [the London trader, who works for Eric] thinks they could be sold at 8.60 or 8.70."

"And you geniuses want to buy some at 8.45, you gotta be outta your [expletive deleted] minds." (A five-year bond bought at 100 to yield 8.45 would have to be sold at 99.4 to yield 8.60.)

"Well, we think we can syndicate maybe 80 or 90 percent of it to Japanese banks and lesser Swiss names, and maybe a couple of French banks. We'd try to hedge what was left."

"You can't hedge against buying it at the wrong [expletive deleted] price."

"Yeah, well, Eric, can I say this bid is okay with you?"

"If you investment bankers want to blow your money like that, who am I to stand in the way, but if anybody asks, tell 'em I think you're all crazy."

"Look, Eric, if we don't bid around 8.45, we don't have a chance to win it, and this is one we want to win. We can't go around looking like we don't know anything about this market to big industrial companies like GE. There's more in this than just this deal. If we can get out even, or maybe with a small loss, we've still scored points with GE, might rise in the Euro league tables, gain some favor with the other players over here, and control all the profitable trading in the secondary market for the next six months while the bonds are still around.''

"Whadaya ask me for, I'm the trader. Your deal is way the [expletive deleted] off the market. It's a loser. They don't pay me to tell you numbnuts how to lose money, they don't need to, you know all there is to know already.''

"What did Eric say?" I asked the grim-faced syndicate man after he had hung up.

"He didn't say he absolutely refused to let us go with it.''

The Eurobond market evolved with internal conflicts like this always in the forefront. Logically, the investment bankers wanted to find a way to build up market leadership. Logically, the traders wanted to protect their profits, fragile though they were. Also, the traders tended to see only the tough institutional end of the market. Deals could often be done (that they did not think could be) at lower rates with retail investors or those banks wanting a relationship with the issuer.

In the end, there was little that could be done if the two groups, the originators and the traders, were to remain in separate profit centers. Traditionally, the poor results were blamed on the department that was deemed to be responsible, but which usually was not—if new-issue losses mounted, trading would ultimately be accountable; if market share was lost, investment banking would be. The system had to be integrated. It took a while for this lesson to be learned, the inertia overcome, and the necessary organizational changes made.

In time, the two groups were combined into a fixed-income capital-markets desk that would supervise new issues, secondary-market trading, interest-rate and currency swaps, and hedging. Altogether, profits were indeed possible, but it had to be recognized that different components would, from time to time, represent costs to the system, not always revenues.

180

I believe most of the U.S., and other non-Swiss, firms went through this evolution in one way or another. Those that did not were already integrated, but only into an autonomous Eurofinance unit. They were not, however, immune to the problem. The Eurofinance unit might be losing money, which would ultimately attract head-office attention, and perhaps criticism, reform, and reshuffling. However, unless the Eurobond operations were integrated as well with the domestic bond business, it would be impossible to look at the results of the global bond financing activities of the firm. Again, Eurobonds may reflect a cost to the whole fixed-income system from time to time, but they support the system overall by virtue of their tapping into other markets that periodically are important to them.

Those firms that ignored the Eurobond markets in 1982 to 1986 began to disappear from even the domestic rankings of market-share leaders. You can't tell GE you want to talk to them only about U.S. deals.

Another difficulty we had in managing the business was the fact that vast sea changes that completely upset our strategic positioning would occur every so often. A Ferrari is great on the highway, but not so hot on winding streets in little French villages or in weekend traffic on the Long Island Expressway.

From about 1965 until 1974, for example, U.S. companies had to do Eurobond financings (to comply with U.S. government capital controls), so business was comparatively easy. Then the controls were lifted, and U.S. companies stopped using the Euromarkets, and foreigners started using the U.S. markets for Yankee bond issues. The shift in our marketing focus was total—stop calling on Americans and go out and get some business from foreign clients, who previously hadn't been called on because they were prevented from issuing securities in the United States. Then, after 1979, the Euromarket began to explode and U.S. companies again became the dominant players, but after 1983 on a largely competitive bidding basis that made our syndicate people dread their calls to Eric and the others in New York. During the period from 1981 to 1984, the market experienced an enormous amount of product innovation, when ideas like bonds with warrants and zero-coupon bonds flourished. The zero-coupon issues were very attractive to Japanese investors (who did not have to pay taxes on the imputed interest income), which resulted in an extremely vigorous and profitable period of marketing Eurobonds to Japan. Then came FRNs, where the issuers were not corporations, but banks and governments. Finally, though, in late 1985, the market turned away from the dollar, and financing in other currencies, often accompanied by swaps, became the most important sector.

No firm can have a natural comparative advantage in all market conditions. At the moment, the Swiss are being nudged out of the top market-share spots by the Japanese firms, which, because of extensive rigidities in the Tokyo corporate-bond market, have been able to persuade Japanese companies to issue securities in the Euromarket that are in turn sold to investors in Japan.

All of the bankers active in the Euromarkets, however, try to be as much as they can to all people, but also try to maximize their comparative advantages whenever they can. The U.S. investment bankers rely on their network of clients and their ability to follow markets globally to provide the latest opportunities to hundreds of companies around the world. They rely on innovation and knowledge of market niches to structure unique securities that are attractive to both issuers and investors. At this point, they may be the only banks that are trying to be truly global in their reach, expanding aggressively into nondollar and non-U.S. clients in Europe and in Asia. They are also learning, however, that it is easy to develop a case of eyes that are bigger than one's stomach, and indigestion causes some slowdown in the pace of expansion, even when following a period of exceptional growth.

American commercial banks also rely on their vast network of clients, their ability to take swaps on their books, and their commercial-lending and financial-services capacities to increase their participation in Eurobonds. European, Canadian, and Japanese banks tend to be vital players in their own currencies, and sometimes in the dollar, or when their own clients are involved, but otherwise, they tend to fade into the background. All have learned how difficult it is to make Eurobonds pay off over the long run.

Japanese securities firms are exceptionally aggressive when either a Japanese issuer or a Japanese investor is involved, but much less effective otherwise. They know that they will not be able to continue to hold their recently won market shares without increasing their activity with non-Japanese clients, who require that the firms bid aggressively with their own capital and develop active secondary-market trading and research capabilities, which most Japanese firms have not yet done.

U.S. firms have had quite a bit of success in developing business with non-American clients. After 1974 many had initial success with Yankee bond issues for French, Scandinavian, Austrian, Japanese, and Australian government entities and the EEC and its constituents. Relationships so developed led to a willingness on the part of the clients to talk with U.S. firms about other markets, despite existing Euromarket relationships. The firms had marketing people who were thorough professionals and who knew the markets and their territories well. Still, non-U.S. business could be difficult.

In 1983, we recruited to Goldman Sachs one Peter J. R. Spira, who had previously been finance director of Sotheby's, the auction house. Peter had been at Sotheby's for eight years, during which time the company had gone public, and had then just been sold. Before joining Sotheby's, Peter had been one of two vice chairmen of S. G. Warburg, the other one being David Scholey, its chairman and chief executive today. Peter would like, he said, to get back into the business.

Despite his "advanced age" (fifty-two at the time) and his long absence from the business, we took him on as a vice chairman of Goldman Sachs International Corp., our London-based international subsidiary. Peter was a success from the start—he developed business, provided advice and navigational assistance, and helped to train "our young," in which he included just about everyone, in the ways and manners of the City with great effect.

On Peter's first day, he said he was going to try to get some business from Imperial Chemical Industries (ICI), one of the United Kingdom's largest industrial companies, whom he had known well while at Warburg. ICI was already a client of the firm, but only for U.S. business. We had not been able to break into the business they did in the United Kingdom or in the Eurobond market, which had been the traditional province of Warburg.

A few weeks later, Peter, accompanied by David Watkins, came

183

to tell me that they had been having very promising conversations with ICI about an innovative dollar-denominated bond that could be converted into a sterling-denominated bond under certain circumstances. ICI had about half its business in dollars and half in sterling, so it was indifferent about which currency exposure it would have as long as the market would enable it to borrow at a lower interest rate than otherwise because of the value of the conversion option that was inherent in the new bond. Peter and David were just about to go back to ICI with their ''final'' price thinking, which offered some but not all that much of a bargain on the interest rate.

''Can't you sweeten it?'' I asked. ''Why not offer some detachable warrants to buy ICI stock [effectively] to lower the coupon further?'' The previous week there had been two or three debt-with-equity-warrants deals that had gone very well.

They went back to the meeting and offered their final price thinking, which did not seem to impress ICI overly. So they suggested the warrants, which the company thought was a great idea. After further discussion, ICI decided to go ahead with the deal. We were delighted—our first lead management position for a major U.K. industrial in the Euromarket.

''Can we sent the telexes tonight?'' asked David, eager to get going.

''Well,'' said ICI, ''there is the matter of Warburg.''

''What about them?''

''We must tell Warburg, who is our lead bank in the United Kingdom, about the deal and ask if they wish to co-lead it.''

Warburg was told, and Warburg didn't like it. They were extremely cross that ICI had taken an idea from us that they had not first vetted and also cross that ICI would allow someone like us to take the lead position, especially without hearing from them first to see whether they had a better idea. Undoubtedly, they didn't like a U.S. firm trying to snatch traditional business away from them on their home turf, though if we had been challenged on this directly, we probably would have let them run the show. They also didn't like Peter's involvement in the whole matter. This was, as Peter put it, ''likely to be thought of as behavior unbecoming to, if not traitorous by, a former employee.''

184

Warburg said it would have a better idea the next day. We said time was of the essence, the market could not wait around for Warburg's better idea. It was a very complex security and therefore very perishable: very slight changes in the market could make it unworkable.

"Bear with us a bit," ICI said.

Late the next day Warburg submitted its plan for the financing. ICI did not find it superior to ours and decided to go ahead with our original idea. However, Warburg had been offered a co-lead managership, along with ourselves and Morgan Grenfell (whom we had brought into the deal in compliance with U.K. regulations requiring a U.K. bank to be in the management group), and wanted to think about it. We would know on the next day, a Friday.

The market had not changed much, but David was worried about its "tone." Friday came, Warburg accepted the position we had offered them, but too late in the day to proceed over a weekend. We waited until Monday, a bank holiday in England, to price and launch the deal, which received a fair amount of press acclaim.

The first trades were close to the offering price, meaning they were very profitable. Many investors wanted either the warrants or the bonds, but not both. Bonds went more easily. By the close of business in London, the issue looked in good shape, though we still had about half of it to do. Monday afternoon, however, saw Wall Street drop twenty points, and investors were nervous about the equity markets.

On Tuesday morning, as the British investors came back to work, the market in the warrants began to weaken, first a little, then a lot. The warrants pulled the unsold bonds down too. David bought warrants in the gray market. Then he bought more.

Pricing a common-stock warrant is a work of art, at the best of times. In the Euromarket stock warrants were fairly rare, and warrants exercisable into U.K. stocks rarer still. The pricing had been based on what the market had paid for the two most recent warrant issues, now a week to ten days old. One was for a German company, the other a Canadian. Neither was a close parallel to ICI, and we had priced the ICI deal more conservatively than the others had been.

185

ICI's stock was comparatively cheap, our research analyst thought, and the outlook for the company's business quite good.

Toward the end of the preceding week, and during Monday, the market turned away from the other warrant issues. The effect became that of a very wet blanket on our issue. The warrants were continuing to be offered in the gray market.

At this point, we decided to hedge against a decline in the ICI stock price, which would depress the warrants further. We bought *puts*. If the stock price dropped, the warrant price would too, but, theoretically, by no more than the puts (or options to sell the ICI stock at today's price) would increase in value. This way we could weather the storm that was obviously coming, in which we would have a lot of exposure to the warrants while working out of our position.

We couldn't have been more wrong.

The market for the other warrants stayed about even, but the market for the ICI warrants continued to sink. The dual currency bonds were all right, but the warrants were being dumped, and no one wanted to buy them. Then the ICI stock price started to rise. The marketing attention the issue had received caused investors to buy the stock, not the warrants, which were soon trading at a discount.

The warrants, in which we were long, were dropping in price, and the puts, which established our short position to hedge the long position, were also dropping in price because the ICI stock was going up. This deal was defying the laws of gravity—and we were losing money on both ends of our "hedge."

"How can this be?" we asked ourselves. "If the stock is rising, the warrants should be too." But they weren't. Too many had been dumped by those who had bought the package just for the bonds and stripped off the warrants. The market saw the surplus of warrants on offer and decided to boycott them. The main holders of ICI stock, large U.K. institutions, were not buyers of ICI at the time; a lot of the support for the rising stock price was coming from the United States, where ICI had become popular. U.S. investors could not buy the warrants, which had not been registered with the SEC. The market in ICI stock had become dis-integrated,contrary to current trends. We were left shaking our heads and holding the bag.

Finally, the warrants stopped dropping, but they remained life-less. No one would buy them, even though they knew they were now reflecting almost irresistible values, because they were afraid that the thin market, turning its back on them so emphatically, would never recover.

The stock price continued to rise. We bailed out of the puts, then the warrants. On balance, the firm lost over a million dollars on the transaction, at the time the largest new-issue underwriting loss in the firm's one-hundred-plus-year history. ICI was embarrassed at how poorly the deal had gone and therefore not all that happy with it despite the good deal it was for them. Warburg was gloating. The financial press was lavishly criticizing the deal. And our colleagues in New York were starting to organize for the second coming of the Spanish Inquisition.

"Not bad for your first month on the job," I said to Peter. He was of course paid to get the mandates, not to price the warrants.

In retrospect, we made two major mistakes. First was in not insisting that ICI go ahead when they had accepted our offer. We should have told them that the offer could only be good until the close of business on the day it was made. We would accept Warburg as a co-lead, but we couldn't wait for them. We were afraid, how-ever, that ICI would think that we were bullying them and might drop us to do the deal with Warburg after all. Today, as ICI knows very well, you do not hold up deals to fit traditional bankers in, assuming you care to do so at all.

We learned that when dealing with a non-American client we do not have the rapport that we are used to with our domestic clients (with whom we are able to communicate much more naturally and directly), that they can be easily offended and perhaps a bit irrational when they are, and that when things are tough, we are always the outsider. Despite these valuable lessons (or, I hope, because of them), Goldman Sachs has gone on to improve relations with non-American firms and greatly increase the business done with them.

The second mistake we made on the ICI deal was in panicking with our position. The position, which was logical, was going com-pletely the wrong way for reasons we did not understand at the time. We decided to cut our losses, which, in this case, had the effect of realizing them. If we had stood our ground we would have seen the price of the warrants recover, to parallel the movement in the stock, and we would actually have made a lot of money.

A valuable lesson here was that when dealing in exotic positions in new issues, those in the firm with the greatest experience and skill in managing such positions should be doing so, not the new-issue desk by itself. David had consulted widely on the warrants with equity traders in New York before and during the event, but they were not familiar with U.K. securities; they were sympathetic, but unwilling to get involved in someone else's mess. In those days, if we were going to do exotic deals like the ICI issue, we simply had to figure them out as best we could, take a deep breath, and jump.

Today Goldman Sachs has a global-equity capital-markets desk, similar to its fixed-income capital-markets unit, that manages the new-equity issue process from beginning to end, from the initial proposal, to the marketing, the allocations, the gray market purchases, the hedging and position management. It operates on an integrated global basis, that is, in the United States, in the Euro-equities market, and in the national equities markets of other countries, keeping track of investor appetites all over the world, and devising securities and distribution methods to make the most of them.

Among the group's activities have been the placement in the United States of shares of large privatization stock issues of corporations such as Singapore Airlines, Veba and Volkswagen from Germany, Telefónica de España, and Société Générale in France. It has also coordinated sales in Europe of many large U.S. stock issues and lead-managed Euro-equity issues for Japanese, Italian, Norwegian, Danish, and other companies. The firm was also selected to serve as the U.S. lead manager in large global offerings for British Steel, British Gas Corporation, British Airways, Philips Lamp, Barclays Bank, Norsk Hydro, and, among others, unforgettably, British Petroleum.

In October 1987, the British government announced the sale of 2.1 *billion* shares of the British Petroleum Company (BP), or 31.5 percent of the company, for approximately £7.1 billion ($12 billion) to underwriters for distribution globally in several tranches. Approximately $1 billion of the offering was underwritten in the United States, making it one of the largest equity issues ever sold in the country.

188

The issue was underwritten according to British practice, in which the underwriters were at risk during the entirety of a two-week subscription period. During that period, the 508-point stock-market crash of October 19, 1987, occurred. At the end of the subscription period, there were virtually no subscribers; almost the entire issue was left with the underwriters. The U.S. syndicate lost more than $200 million between them. A large issue had finally gone wrong, and when it did, the damage was considerable.

The bankers had grown considerably since the early days of the Euromarkets, and the profits for the year were good, so the hit to the firms involved, though painful, could be absorbed. More important, perhaps, was that the largest stock financing ever done in Europe or the United States came forward, despite highly distressed market conditions, to be placed in global tranches around the world, and was.

Despite the crash, and the BP issue on top of it, the global markets held, took the strain, and continued doing business as before. The comparatively mild response to the event was due in part to *Euromoney,* or the liquidity in the global financial system that operates without regard to national borders and the maintenance of that liquidity by the central banks of several countries.

Euromoney exists in many forms, as bank deposits, as certificates of deposit, and as bonds, floating-rate notes, and equity securities. It is denominated in many currencies and moves around the world generally in accordance with the laws of economics. It flows to greater returns, improved security, and better liquidity. When the dollar weakened, it moved into yen, deutsche marks, and Swiss francs. When U.S. companies pulled back from the Eurobond market, it moved into FRNs. Then, as the stock-market boom spread to Europe, it went into equities in various different forms, including warrants. After the crash, it moved again into U.S. Treasury securities, then gradually back again into bonds and stocks. It is always moving, as markets change. The only place it never goes, it seems, is back into the country whence it came, at least not permanently.

It has always been fashionable to predict the next great change in the Euromarkets, including, as has often been done, its demise. It was supposed to die in 1974, and again in 1979 after capital-market controls were removed. The dollar sector was supposed to disappear after the massive run on the currency in 1986–87. FRNs have been pronounced dead, along with straight debt issues with maturities of over fifteen years. Such predictions, being based, foolishly, on the indefinite continuation of prevailing conditions, have never been accurate.

Today, some forecast that the Euromarkets will drift off into insignificance because of the growing strength and capabilities of national capital markets, all improved as a result of deregulation, greater liquidity, and integration with markets abroad. As European national markets join each other in an effort to reduce remaining barriers to capital flows, it is said, these markets will strengthen further, rendering the Euromarkets, and Euromoney, in the free state of its birth, unnecessary.

One cannot deny that national markets seem to be developing greater capabilities and that the European market integration is on the march. But equally, one cannot deny that Euromoney, as we have known it, will change again, modify its markets, and continue to exist as a freewheeling, unregulated, free-market alternative to national markets everywhere.

# 6

# The Big Bang

There was a film made in the 1960s starring Richard Burton, who played a dissolute demolitions sergeant who was sent, with a very correct, supervising young English lieutenant, to blow up a giant power dam somewhere in the middle of Europe during World War II. After parachuting in with their high explosives, they made their way to the dam, disguised themselves as workmen, and somehow got inside the dam and down at its bottom. There they affixed a delayed charge and got themselves out and onto a nearby hill to watch the fun. Nervously, Burton's companion searched for a sheltered place from which to observe the blast. "Relax," said Burton, attempting a nap under a nearby tree. The companion could not; tensely he counted down while Burton snoozed. At zero hour, nothing happened, only a little *pffpft*. The officer was at first stunned, then furious, convinced that Burton had botched the job. Burton, kicked awake, shrugged and said something like what can you expect from a drunk who didn't really have his heart in this war anyway. They had to pack up and make their way to the pickup point. The officer planned to shoot Burton, or something, for being such a sorry excuse of an Englishman. Just before he did, Burton glanced at his watch and asked if the officer had heard anything, the plane perhaps. Searching the skies, the officer heard a faint roar, which became louder and louder until finally it turned into the earth-shuddering sound of a giant dam being carried away. Burton grinned and, I guess, flew off into the sunset with his little twit of a lieutenant now suitably humbled.

The little *pffpft* at the base of the mighty dam that was holding back incredible forces did the job. It weakened the structure just enough so it could no longer contain the forces that it was supposed to, and the whole thing, after a while, was carried away in a big bang.

On October 27, 1986, an earth-shuddering event took place in the City of London, the square-mile financial center of Britain that dates back to the time of the Norman Conquest. The event was one that would change forever the lives and the businesses of almost everyone who worked there, and indeed had already done so in its anticipation. City life would never be quite the same. The long-dreaded Big Bang had gone off. The old rules of the stock exchange had been changed; new rules were now in force.

As a result, nineteen of the twenty largest stock-brokerage firms in the United Kingdom had been sold, and so had all of the principal *jobbers,* or dealers, whose function is most closely paralleled in New York by stock-exchange "specialists." Stock-exchange transactions in London would now occur through a brand-new, just-installed electronic market-making system that closely resembled the United States' National Association of Securities Dealers Quotation System (NASDAQ).

The giant U.K. clearing banks and aggressive, experienced foreign banks and securities firms of all types had been allowed to muscle into the U.K. stock-brokerage business for the first time.

Also, the market in government bonds and other fixed-income securities had been stood on its head; it was now a copy of the U.S. Federal Reserve's primary market dealer system. The new system was scheduled to start off with twenty-nine market-makers instead of the two or three jobbers that had handled all the government's new-issue business before.

Competition on the exchange was expected to become ferocious overnight. Volume would soar, perhaps, but as commissions were now to be fully negotiated, profits surely would be hard to come by.

Life was going to be very tough for the British firms, who now had to face, all at once, the total deregulation of their market, fierce new competitors, completely new trading systems for both stocks and bonds, new mergers and corporate-integration problems, critical back-office requirements, and radical changes in the economics of their business and the skills required to succeed in it. Very few British firms had experience in any of these matters. Old dogs would have to learn new tricks, quickly, or get out of the way. Many elected to do the latter.

No wonder the English wags called all of this change and reform, all the *Sturm und Drang,* the "Big Bang." Like much of British wit, the term was ambiguous, however, with several alternative interpretations.

Certainly the Big Bang was explosive. It blew up more than a hundred years of comfortable institutional practices and procedures with one mighty charge. The whole thing was gone, swept into the Thames, destroyed by one great cleansing blast. The old City would be replaced by little green shoots of a new free-market City, poking up through the rubble to blossom into a better, healthier, more competitive place in the future.

Certainly it was noisy, a shot heard 'round the world by regulators, elected officials, users of global-securities markets in other countries, and, of course, potential competitors everywhere. Its coming was noticed and emulated in Australia, and in Canada, and in many other countries within Europe. Few financial events have had greater impact.

Certainly it was portentous—the birth of a new expanding universe and of many yet-to-be-understood changes in financial markets of the future. We all felt a little like we were being hurled, in total darkness, through a great void at the speed of light.

No other country had ever subjected its financial markets to so

193

many changes at once. Indeed few had ever imposed any changes at all. The SEC had nudged the New York Stock Exchange into controversial and wide-ranging reforms in 1975, which were implemented on May 1, or "Mayday." These reforms were essentially aimed at securing negotiated commission rates for stock-market transactions in replacement of the fixed commission rates, which the Justice Department's antitrust division thought might be in restraint of trade. No changes were made, at the time, in the exclusion of foreign members of the Exchange, or in the Glass-Steagall Act, which prevented banks from competing in the brokerage and underwriting businesses. Though considered drastic at the time, and rightly so in terms of the ultimate changes in the U.S. securities business, Mayday was a picnic compared to Big Bang.

What was going on? Did Margaret Thatcher really want to nuke the City, on which the Conservative Party depended for support, both moral and financial? Was the existing financial system so bad that the whole thing just had to be blown away?

Or was it, at least partly, an accident?

It all began in 1976, when the Labour government passed legislation extending the laws dealing with restrictive practices in business to the service industries. The stock exchange, along with travel agencies and surveyors, was asked to present its "rule book" to the Office of Fair Trading (OFT) for inspection. In 1977, the OFT identified 150 rules that appeared to be in violation of the law. In 1979, just prior to the election of Margaret Thatcher, the OFT brought the matter to court in a restrictive-practices suit. The stock exchange regarded the whole thing as a nuisance, though a potentially dangerous one. "The system wasn't broken," its representatives argued, "so why fix it?" In any case, they continued, "The correct place to argue about changes in the long-serving but delicate stock-exchange rule book is certainly not in a law court, where no one understands how the system works."

The suit never got much beyond the requests-for-information stage before the Labour government was swept out and Mrs. Thatcher was swept in, with much applause from the City. The Thatcher regime lost no time in bringing about economic reform. Taxes were cut, foreign-exchange controls removed, and free-market principles followed whenever possible. Strikes were taken and broken, industry was shaken up and made more competitive, and a hard line taken with the EEC. When the Argentine government sent troops to occupy the Falklands, Thatcher led the nation into battle, with leased passenger liners, container ships, and commandos, and got them back, despite several sunk ships and near misses on others.

The Honourable Lady soon became the Iron Lady. Totally self-assured, tough, determined, and relentless, she became the nation's nanny, someone everyone respected more than liked but nonetheless obeyed. Notwithstanding a more than 12-percent unemployment rate, resulting from her free-market economic policies, and a respectable if not inspired opposition, she won reelection in the spring of 1983 in a landslide.

In the early days of her second term, the previous government's antitrust suit against the stock exchange was reexamined by Cecil Parkinson, Thatcher's new minister for trade and industry, whose department had jurisdiction over the suit. The issue was a potential thorn in the government's side. It had had a lot of support from the City, and it had ambitious plans to begin large privatization sales of shares in government-owned industries in the coming year, so it did not want the market upset or weakened by the restraint-of-trade issue. On the other hand, the government had forcefully articulated harsh free-market economic thinking ever since coming into office, and the whole issue of the stock exchange was that it simply wasn't free-market enough. The Bank of England was also lobbying for much-needed reforms of the government-securities market. And, certainly, Parkinson did not want to appear to the press to be "wet," or soft, on the issue of bringing the City into line. The government's policies, after all, had been hard on industry and blue-collar workers. It could not appear to be lenient on the stockbrokers. The issue needed resolving, and it fell to one of Thatcher's favorites, Cecil Parkinson, a handsome, smooth-talking, skillful politician, to resolve it. One can imagine Parkinson's situation easily by recalling an episode or two of "Yes, Minister," a popular British television comedy of the time that has since had wide exposure in the United States.

Parkinson did resolve the issue, within six weeks of his appointment, following discussions with the stock exchange's shrewd and prescient chairman, Sir Nicholas Goodison. The deal was that Parkinson would get the government to withdraw its suit, which was due to come to trial in six months, if the exchange would offer proposals for reforms that would be acceptable to Parliament. These would include a three-year phase-out of fixed commission rates (which the stock exchange had raised in 1982) and greater participation by nonmembers in the affairs of member firms and of the stock exchange's governing council. Goodison was thus considered to have won a solid victory for the exchange, having conceded so little. Parkinson was thought to have "caved" in order to wrap things up quickly.

The only significant concession that Goodison made was seen to be the move to negotiated rates. In reality, all things being equal, this was not thought to be such a big deal. At the time, brokers offered substantial volume discounts (called "bargains") to institutional customers that they did not offer to individual, small-time customers. The difference between the bargains and the undiscounted rates was already approaching a factor of ten. Given three years to implement them, full negotiated rates should not be impossible to handle.

Goodison had conceded nothing regarding the two other key alleged restrictive practices, the separation of the functions of brokers and traders, and barriers to entry to membership, which were supposed to sustain the exchange's monopolylike position in London.[1]

Parkinson, having finished this bold piece of work, did not remain long on the scene. He was embarrassed by a classic British scandal: Parkinson's former secretary had just delivered his illegitimate child and complained in public that Parkinson had promised to marry her, but his wife had got him by the ear and refused to let him. Parkinson, embarrassed and apologetic, resigned from the cabinet, not to reappear until five years later, when his penance was considered done.

Some people, like Richard Burton blowing up the dam with a delayed reaction, knew what was coming. Although he had saved the separation of brokers and traders from the agreement with Parkinson, Goodison had reservations about the future: "At best," he said, "there can only be great uncertainty about the pressure which the abolition of fixed commissions might put on the dealing system within the central market."

196

Added crusty Gordon Pepper, a senior partner of one of the brokers who would figure actively in subsequent negotiations at the exchange, "Once you start these forces, you can't stop them halfway."

Though not especially happy with the deal, the members of the exchange recognized the inevitability of their situation and began to analyze its effects. The principal issue was whether the exchange ought to give up voluntarily the separation of brokers and traders in the interest of protecting the profitability of members.

The stock exchange, along with the rest of the City, that is, the banking and insurance communities, was organized into a rabbit warren of small, exclusive compartments that were interconnected by well-worn paths between them. Stockbrokers, for example, were not allowed to function as market-makers or specialists. They could only deal with the public, and with jobbers, from whom they acquired or sold positions wholesale. Jobbers could only deal with brokers, not with the public. Both had to be organized as partnerships and could not, except to a limited extent, permit nonmembers to become partners. Banks could not become members, neither commercial banks nor merchant banks, despite the fact that merchant banks were the principal underwriters of securities in the United Kingdom and among the principal managers of pension funds and other investment assets. Insurance companies too could not join, nor could nonmembers own more than a 29.9 percent interest in any member firm, a level recently increased from 4.9 percent.

Stockbrokers were the first to get nervous. "With our commissions way off, our revenues will shrink. With greater competition for business, our expenses will have to rise. Our profits will disappear unless we can get into the profitable businesses of the jobbers and compete with them."

This meant that brokers would want to be market-makers, to perform the services of traders, just as brokers in the United States had done. Then the brokers would be able to offer block trading to their clients and other improved services, such as program (or computer-assisted) trading. But to do this, the brokers would have to have much more capital. Never having needed capital in the business before, the brokers had mainly paid it out to their partners. Even Cazenove, the largest and most prestigious of the brokers, had retained very little capital. Altogether, the brokers had little more than £10 million or so of capital.

To get more capital, the brokerages thought about taking on nonmember partners, to the extent now permitted. There were a number of nonmembers eager to acquire an interest in a top-ranked brokerage firm. These were largely British and American commercial banks, which, by comparison, had enormous capital reserves, or so it seemed to the brokers at the time. They would be prepared to pay a substantial premium over the brokers' book value to obtain an interest. Those brokers who sold would in fact be receiving more than their businesses were likely to be worth once negotiated rates began.

The jobbers figured that if the brokers were going to be coming over the wall into their businesses, they had better develop some ability of their own to compete. Though the jobbers did have capital in their firms, they had no distribution or other capabilities. They too began to see more merit in having new partners.

Some of the members, however, were appalled at all the talk about doing away with the rules that restricted members to acting in a *single capacity,* that is, as either a broker or a jobber, but not both. Permitting members to act in *dual capacities* would be ruinous, they argued. The self-regulatory system that the exchange had maintained for years worked pretty well. To change the rules to permit dual-capacity operations would mean that a new system to protect investors against inappropriate sales by brokers of their trading inventories to unsuspecting customers would have to be devised.

"Why change a perfectly good system?" they would ask. "If we have to negotiate rates, we'll negotiate them, but in a single-capacity system there's probably a limit to where rates will get cut to. If people want to trade uneconomically, they won't be in business long anyway."

"Besides," one or two were known to have added, "it's the bigger firms that are getting bought out at a premium in order to compensate their owners for the loss of their valuable franchises. How is such compensation to be made to the smaller firms?" The question was deemed to be a good one by the members, and in time a special compensation fund was arranged to be shared in by those whose firms did not find a buyer. On the final vote by members on the rule changes, it was assumed that the smaller firms would go along quietly with the majority on the move to dual capacity, which they did.

A year before Parkinson's negotiation with Goodison, the first significant acquisition of a minority interest in a broker by a non-member was announced. Hoare Govett, an old and prestigious firm with a substantial client base, had sold a 29.9-percent interest to Security Pacific Bank from the United States, with the balance of the firm to be sold when the rules should so permit. The price paid for the 29.9 percent was not disclosed, but was thought to represent a large premium over the firm's book value. Some of the money paid would go into new capital, to help Hoare Govett build up its overseas businesses, it was said. The remainder was paid out to the firm's partners. In London, the move by Security Pacific was seen in significant strategic terms, but in the United States it appeared simply an opportunistic step involving very little money. Hoare Govett's total book value in 1982 was only £7.5 million.

Many other deals followed the Hoare Govett move, especially after the summer of 1983. The boldest of these, perhaps, was taken by S. G. Warburg, which announced a four-way merger in August 1984 (to be effective after October 27, 1986) through which the firm would combine with the leading jobber, Akroyd & Smithers, one of the largest brokers, Rowe & Pitman, and "the government broker" (in gilt-edged securities), Mullens & Co. Warburg's plan was to assemble the strongest across-the-board team that was possible, in an effort to become, according to David Scholey, Warburg's chairman, "a British equivalent of a Goldman Sachs." The merger was valued at £350 million, the largest transaction of its type ever done. Warburg would spend more than a year working to integrate all the separate companies and businesses into one, smooth-functioning organization. Scholey could not afford a mistake; he had bet the whole firm on the decision to move quickly and grab up the best.

A similar move was made by Barclays Bank, which was also eager to put together an all-in-one powerhouse. In March 1984, it announced plans to acquire the other major jobber, Wedd Durlacher Mordaunt, and a major broker, De Zoete & Bevan. These firms would be combined with Barclays Merchant Bank to create a new investment bank, Barclays De Zoete Wedd. Barclays then recruited Martin

Jacomb, a vice chairman of Kleinwort Benson responsible for its securities business, to head the new entity.

Other transactions reflected differing strategies. Citicorp and Chase Manhattan each acquired two brokers, in an attempt to put together global securities firms with a substantial critical mass. Kleinwort Benson acquired Grieveson Grant, an important broker, and attempted to integrate it into a broad-based securities business with substantial activities in the United States. Hong Kong Shanghai Bank and Union Bank of Switzerland each picked up prestigious brokers, James Capel and Phillips & Drew, respectively, which they left alone to stick to their knitting. National Westminster, Midland Bank, and Morgan Grenfell contented themselves with lesser players. Hill Samuel, after some vacillation, acquired Wood Mackenzie, a prominent Scottish broker.

The buildup to Big Bang resembled the start of a large long-distance yacht race, with different vessels tacking to and fro across the starting line to position themselves for the gun. All the boats were going to the same place, but how they got there would vary quite a lot. For some, a large boat and a straight line would offer the best chances for winning, by surviving the unpredictable and carrying the most sail. For others, remaining light and agile offered the best promise of speed and maneuverability. Some individual yachtsmen would give up their own chances in order to sign on with a stronger crew. Others would be out looking to build up a top crew by recruiting the best talent away from others. Some would ignore all of this and concentrate on training their own people. There were many different approaches, both for firms and for individuals.

One man's approach to Big Bang was especially different. John Craven, late of CSWW, Warburg, and Merrill Lynch, had formed his own firm, aptly named Phoenix Securities. Craven had been doing small deals here and there, and some consulting, when the move toward consolidation on the stock exchange began. Craven was called in to advise one of the firms about to be sold. He was an expert in small acquisitions, clever, sensitive, capable, closed-mouthed, and, perhaps best of all, not part of one of the major merchant banks who were also out looking for acquisitions of their own. It was not an easy time to know whom one could talk to. But they could and did talk to Craven.

Craven and his partners became instant experts on everything to do with selling stock exchange firms. In all they represented twenty firms in Big Bang acquisitions, virtually cornering the market.

Another prominent figure in the City played Big Bang still differently. Jacob Rothschild, a former director of his family's merchant bank until he left to set up his own business as an aggressive financial entrepreneur, was selling out. Rothschild had begun in the early 1980s to build up a wide range of holdings in financial-services businesses, mostly in the United Kingdom. He had assembled investments in a broker, a merchant bank, and several unit trusts involved in investment management. Early in 1984, however, he began to sell off his interests, completing the job a year later. He did not like the outlook for the businesses; Big Bang had changed things too much. "It is legitimate to distinguish between what one thinks is right for the marketplace and what may be right for one's [own] business," he explained.[2]

Nineteen eighty-four was a record year for profits in the City, but Rothschild didn't like what he thought deregulation would do to future profits. Changes in the stock market were bad enough, but he couldn't have liked what he learned about the gilt market, even if it was likely to be good for the market in general over the long run.

The Bank of England had long wanted changes in the way U.K. government, or "gilt-edged," securities were distributed. They now had an opportunity to design a system that would meet their requirements for both market efficiency and regulatory integrity. In October 1984 the bank promulgated new rules for the gilt market.

The new system would be based on the one used in the United States by the Federal Reserve. Securities dealers willing to subject themselves to extra regulation, including regulation of their levels of capitalization, could apply to become "market-makers" in gilts, in which capacity they would be entitled to bid for new issues, trade directly with the Bank of England, and have access to financing from the bank. A market-maker would have to be separately incorporated and capitalized so as not to allow any use of gilt-market capital to support other activities. Approximately six months later, the bank announced that twenty-nine firms, including eleven U.S. and fourteen U.K. groups, had been approved to become market-makers on October 27, 1986, when the system would go into effect. The U.S. firms included six commercial banks eager to demonstrate their capital-market skills, four investment banks, and one financial conglomerate, American Express. Morgan Stanley had decided not to participate, preferring to save its powder for the global-equities markets. Credit Suisse First Boston had joined, counted as a Swiss-U.S. joint venture by the British. No Japanese firms were accepted initially because of the lack of reciprocal rights for British firms in Japan.

Everyone knew the market would be tough at first. Even Eddie George, executive director of the Bank of England responsible for the new system, expected a substantial shake-out period. "It will be like a marathon footrace," he said, "lots of people starting, lots dropping out along the way, and only the strongest finishing." Still a big improvement over the present system, he must have thought.

The new capital requirements for the gilt market, applied across the twenty-nine new market-makers, indicated that the total amount of capital behind the market would rise to $800–$900 million from less than $200 million before the rules were to go into effect. Too much money chasing too little business was the general consensus. Still, all these firms turned out to become market-makers, with their eyes on the marathon's finish more so than its start. One American observer estimated a fivefold increase in daily trading volume in gilts over the coming five years. Only an American, the British thought, would make such an unpredictable prediction.

The equity market was also entirely redesigned and rebuilt during the run up to Big Bang. The old system would be abolished. A new one that provided for appropriate disclosure and investor protection was agreed on after a hurried inspection of trading systems all over the world. A system similar to NASDAQ, called Stock Exchange Automated Quotation System, or *SEAQ*, which involved market-makers showing their prices on screens located in trading rooms and customer offices, was adopted. There was scarcely enough time to order all the screens before the curtain went up, but herculean efforts were undertaken and were successful.

Not all stocks would be treated the same; they were categorized according to trading volume. Not all exchange members would have to act in dual capacities. There was a lot of room to pick one's own strategy or niche, and to prepare for it. One complication, however, was the fact that until October 27, the old rules had to be followed. Single capacity was all there was. It was hard to practice dual capacity beforehand.

In the meantime, however, some British stocks were being traded off the exchange by nonmembers, mostly, but not exclusively, U.S. firms trading in American depositary receipts (ADRs) of British stocks in New York.

ADRs are transferable certificates, denominated in dollars, that represent ownership in particular British shares. U.S. investors have bought and sold ADRs for many years. Their growing interest in international investments, however, had caused the volume in British ADRs to increase sharply since 1982. This volume was handled mainly by U.S. brokers, who were draining off some of the business of the London exchange. An even greater amount of business was lost to the stock exchange, however, when British institutional investors began to deal in ADRs with U.S. brokers, because in doing so they could avoid a stock-transfer tax in the United Kingdom. U.S. brokers were also spoiling these institutions with offers of block trading, research, and other services. Even were they able to meet this competition, which many could, the British firms had their hands tied until the official event of the rules change, and until the tax matter was cleared up. It was considered somewhat "unsporting" to take advantage of the British firms during this period, but there were plenty of unsports doing it.

This issue brought to the public's attention the whole subject of how American firms were handling the transition. There were frequent rumors that one or another firm was on the verge of a big acquisition announcement, which they were not. Only two U.S. investment bankers, Shearson Lehman and Prudential Bache, bought U.K. firms. All the rest abstained for the simple reason that they already had a brokerage business in the United Kingdom covering accounts, and they were already market-makers in many U.K. stocks (through ADRs). There was little for any of us to gain, and none of us was eager to pay a premium to acquire a franchise that, after Big Bang, would most likely be worth a lot less than we had paid for it.

The presumption in the press, and indeed within the City, was that the U.S. investment banks were preparing a major effort to seize significant market share after the new rules took effect. I was called often by the press and by friends in the City to comment on rumors and to explain what we were up to. Our party line on this was that we were in London for the long haul, saw many opportunities in the coming changes, but expected to move cautiously and in a manner calculated not to offend our British hosts. No one believed a word of this.

A popular television show in London in the early 1980s was a financial news show called "The Money Program." One day I received a call from one of the show's producers asking if a camera crew could come into our offices to film our London trading room when the weekly New York money-supply figures came out. These figures were supposedly having a big effect on the bond markets within minutes of their publication. Or so it was thought at the time. They wanted to capture the very moment when the news flashed across the wire and pandemonium would break out, or something.

"That's fine with me," I said, "but the money-supply figures don't come out until sometime Friday afternoon, around nine or ten P.M. London time. No one's likely to be too interested at that hour," I added, "but in any event there are not too many U.K. clients left at their desks for our people to give the information to."

"Well then, Mr. Smith, why don't we just do something else? Could we come in and film you getting ready for Big Bang?"

Somehow this odd conversation ended up with a camera crew outside my office one day. The producer had moved to another assignment, doing a series of documentaries on change in the City, and wanted to shoot some footage at Goldman Sachs.

We worked out a script. I was filmed making some general remarks about our desire to combine genteel profit-making with good citizenship and then walking into our trading room to introduce our head securities sales and trading people. They would identify their colleagues bent over their desks, phones plugged into their ears, talking to clients in the United Kingdom and offering quotes. In the morning, before New York opened, our principal trading business was in shares of Japanese and British stocks, in ADRs. While the cameras rolled, clients would call in for quotes and to conduct their ordinary business with our salesmen and traders.

Then, unexpectedly and on live television, a major U.K. client called with a long list of large block trades he wanted us to quote on. The trades were all in British stocks. The traders were busy fielding the quotes; the salesman was feeding the quotes back to the customer. The head salesman, however, was horrified; he did not want his most sensitive trades with his best customer broadcast on television. He squinched down in his chair and covered his mouth with his hand. Meanwhile he waved at me to get these TV guys out of here.

The producer saw through it all. The microphone was pushed next to the salesman's face. He smiled into the camera and flashed murderous eyes at the producer.

"You just bought ten million shares of ICI," he murmured on national television "five and a half million BPs, and sold four million GECs. I'm getting quotes for you on the others." These were very large trades at the time.

After ten minutes or so of this, complete with the trader, who sat twenty feet away bellowing the quotes across to the salesman, who then tried to cover them up, we turned to another subject. A few minutes later, the cameraman noticed that the salesmen were now offering quotes in Japanese stocks to their British and other European clients on the phone. They thought that was interesting too.

When they came back from lunch, the cameramen watched the salesmen and the traders handle the orderflow in U.S. stocks. Many block trades were executed, most of them modest in size by our standards. When the day was over, the producer told me he had been completely overwhelmed by all the activity.

"It was great," he said, "just great."

"When is it going to be shown?" I asked anxiously, beginning to hope that it never would.

"Third Friday night in July," he answered, "ten P.M."

"At that time," said Peter Spira, who was in the room at the time, "you would have more people watching it if it were on Hungarian television."

Peter was right. We never heard a word about the show. The producer later told me that he had been made to cut out most of the footage of that sour-looking guy who kept mumbling into his phone.

The run up to the big day continued to bring further deregulation of the exchange's rules. Having voluntarily given up the single-capacity restriction, the exchange then decided to open membership to corporations and, from March 1986, to allow a member firm to be 100-percent owned by a single outside business, including foreign-owned firms. In March, Merrill Lynch and Nomura Securities from Japan were admitted as corporate members, and many other foreign firms also joined.

October 27 came and went uneventfully. The whole equity market had been asked to come in on the preceding Saturday to "practice" getting the orderflow right. But, other than this, most of the consequences of Big Bang had already been identified and responded to. On December 8, the British government launched the £5.4-billion privatization of British Gas, at the time the largest such issue ever done. This issue was a big success; the market took it completely in stride.

In an article published just before October 27, Sir Nicholas Goodison proudly noted that the reforms embodied in Big Bang had begun long ago, had been painfully and skillfully worked out over the three years following the Parkinson-Goodison agreement, and were continuing. Further work was being done on the technical side of the exchange and on making paperless transactions possible. The exchange was aiming at becoming the world's leading center for international stock-market transactions.

Having deregulated almost everything connected with the exchange, the government had submitted a financial services bill to Parliament to provide for its reregulation and the necessary enforcement powers. The bill was enacted in 1987 and came into effect in stages over the following year. The new regulatory system was complex, in a number of respects unclear, and controversial (as such bills always are). Perhaps the main difference that the new bill presented as compared to the old system was the lessening of reliance upon totally self-regulated bodies, like the exchange, to enforce the nation's securities laws. Also, the bill threw a net over many of the London-based international activities of financial-services firms, most notably by imposing capital-adequacy regulations on member firms participating in the Euromarkets.

Naturally, so much regulation was resisted by those most affected by it. Some compromises were made, some matters left to be worked out as time unfolded. Concerns were expressed that too much regulation might encourage market participants, especially those mainly involved with Euromarket activities, to move their bases of operations to a more hospitable country. Since then a new head of the Securities Investment Board, David Walker, a former executive director of the Bank of England with considerable understanding of securities markets, has been appointed. Walker has already undertaken efforts to simplify, if not loosen, regulatory requirements. The give-and-take over securities regulatory issues in the United Kingdom continues.

There were a few problems in the wake of Big Bang. Morgan Grenfell's newly hired head of equity trading, Geoffrey Collier, was arrested for insider trading only a few days after Big Bang. The exchange's computers broke down because of the heavy volume. When the back-office and quotation-service computers were designed, the capacity requirements of the system were underestimated. There soon developed a long delay in settling trades, but this was sorted out after a few months. Stockbrokers were working harder, and later, than ever before. Volume in both gilts and in equities started to build, but profits, as expected, were hurt by all of the competition, which included efforts by some to capture a commanding market share through block and program trading.

An odd thing was happening on the floor of the exchange: no one was using it. Traditional trading methods involving face-to-face deal-making were rapidly becoming a thing of the past. Despite the $4-million renovation to the floor that the exchange had put in, and the expense incurred by members to fit out and rent their "pitches," business was fast gravitating to the firms' own dealing rooms. Within a month of October 27, both the established British firms and the foreign newcomers had pulled most or all of their staffs off the floor. Talk was circulating about how the exchange was going to cope with all the empty space. "Rent it out to the financial futures market," some said. "Turn it into a skating rink," added others.

Still, it was a time when all eyes were on the equity markets, which had been inflamed by merger-and-acquisition activity. Big Bang's early days saw soaring volumes and rising stock prices in all equities markets, not just the London market. Records were being broken in New York, Frankfurt, Tokyo, Sydney, Zürich, and Hong Kong.

In the United Kingdom, however, the merger market was off. The government had commenced criminal investigations into the effort by Guinness PLC to take over Distillers Company in the year's most hotly contested deal. The merchant banker advising Guinness, Roger Seelig of Morgan Grenfell, was held to have acted improperly and was dismissed to face charges. The Bank of England was not satisfied with Seelig's head, and before long it secured the resignations of Christopher Reeves, Morgan's chief executive, and Graham Walsh, its head of corporate finance. Takeovers were put on hold temporarily while these matters rolled through the City.

Morgan Grenfell, London's merger powerhouse in 1986, reeled from the blows of the Guinness and Collier scandals. Even its most envious competitors looked on in horror. The firm, which, after 150 years in private hands, had gone public in June 1986 at a share price of 500p, had dropped in value to around 370p. Its board decided on a bold resurrection. The bank's chairman, Lord Catto, would retire, and Sir Peter Carey, its deputy chairman, would succeed him. To replace Reeves as chief executive of the firm, the board selected John Craven. Morgan bought out Phoenix Securities and installed Craven on May 5, 1987. By the end of the year, Craven had reorganized and remotivated the firm and had it functioning more or less smoothly again.

All the merchant banks were suffering pains of adjustment as they accommodated Big Bang, and several attracted unwanted investors. Saul Steinberg, fresh from a successful greenmail raid on Walt Disney, took a substantial stake in Warburg; Gerald Tsai of American Can did the same at Kleinwort Benson; and an Australian investor acquired 14 percent of Hill Samuel.

A deregulated, free-market financial system was one thing to the

Bank of England, but letting their prime candidates for success in the new system get bushwacked by foreign greenmailers was another. At the beginning of February 1987, the Bank of England stepped in and announced that any acquisition of more than 15 percent of a bank's shares would require its approval. The raiders soon sold out.

One bank that assumed it could get the approval if it asked for it was the Union Bank of Switzerland (UBS), whose London affiliate had bought Phillips and Drew, a distinguished firm of U.K. stockbrokers. UBS was interested in Hill Samuel, a merchant bank that itself had acquired the Scottish stockbrokers Wood Mackenzie in 1985.

Hill Samuel during the 1970s had been a tired institution, holding on to a large banking and money-management franchise that was supplemented by unrelated financial-services businesses in insurance and ship brokering. Its greatest successes had been achieved in Australia and in South Africa. In 1980 its board, seeking long-needed rejuvenation, decided to appoint the man responsible for these successes, Christopher Castleman, then thirty-nine, as chief executive of the firm. Castleman's experience in setting up and running the two overseas businesses was a legitimate qualification. Others shared responsibility with him for their success, but he was the principal architect and his the driving energy behind them.

Castleman was a good operational manager. During his five years as chief executive, the quantity and the quality of the firm's profits rose steadily. His original approach to Big Bang was to avoid paying a lot of money for an acquisition. It would be better, he thought, to build what was needed from within and afterward to see what could be picked up from among the rubble.

Apparently, he changed his mind as he saw what Warburg, Kleinwort, and Barclays were doing. A capacity for distribution of securities was going to be more important than he had thought. He began to look around, and he ran into John Chiene.

A bit older than Castleman, Chiene was still young enough to be remembered as an original "enfant terrible" of the U.K. brokerage business. Trained for a while at Warburg, he had joined Wood Mackenzie in Edinburgh and after a decent but short interval became its senior partner. He moved the main office of the firm to London, hired and trained a first-class research team, invested heavily in computer software, and watched the firm rise in the esteem of its customers and its competitors.

I first met Chiene in 1980, when I moved to London to head the Goldman Sachs office there. He invited me to lunch and proceeded to explain to me in his fast-talking, jargon-rich way about the coming problems of the restraint-of-trade suit and the requirements that brokers must meet to accommodate the movement toward globalization that was overtaking us all. I understood very little of what he was saying at the time, but we would meet again often. There was something there, I knew, if I could only figure out what it was.

Chiene was well tuned in at a very early stage to what was going on within the U.K. brokerage industry. He understood the many changes that lay ahead, the need to expand appropriately in New York and Tokyo, and the coming need for capital. Long before many, he was losing sleep over the future of his firm. He thought it ought to be merged with another firm, but a dynamic one, not just some relic of the past that might soon be frozen into the ice as the weather changed. He and his partners wanted a future in the new market, not just a well-paid exit.

There followed a series of negotiations with a firm of financial entrepreneurs and money brokers, Exco, which ultimately were broken off, apparently over issues of autonomy and control. Chiene was not going to abandon his colleagues to someone else's management.

When he met Castleman, Chiene thought he had found a kindred spirit. They were both young, international, smart, stubborn, strong-minded, fast-talking, athletic, and from Cambridge. Chiene was just what Castleman was looking for to head up a brokerage cum corporate-finance wing of Hill Samuel. The engagement was brief, the marriage consummated in the summer of 1985.

It wasn't long before Chiene realized that working for someone else was not the same thing as being the boss, but he was a good soldier and slogged on, preparing for the twenty-seventh of October, which apparently was met by the firm without losing its balance, its shirt, or its head. Early in 1987, however, the raiders discovered Hill Samuel and staked out a position.

Being in someone else's sights can be nerve-racking, but the

Bank of England's position on takeovers eased things a bit. The real question at Hill Samuel was how was the firm going to fit itself into a future made up of larger, stronger, and more competitive players? The firm was somewhat in the middle between the giants like Warburg, Barclays, or Merrill Lynch (whose name, like Nomura's, was often used when conjuring up pictures of competitive monsters) and the niche players like Hambros or Baring Brothers. Chiene believed that the firm, to compete with the larger players, would need more capital, and probably a more streamlined business. Castleman was reluctant to change things all that much; getting the capital that would be needed would most likely mean merging with someone else, which he did not want to do.

The issue came to a poignant head when an approach was received by Hill Samuel's chairman, Robert Clark, from the head of Phillips and Drew, London brokers owned by UBS. UBS, he was told, might have an interest in Hill Samuel's corporate-finance and related businesses. Clark replied that the firm would not be broken up, but if UBS was interested in the whole thing, discussions might be possible. UBS was interested, apparently eager to get on with negotiations. Clark raised the matter with the board of Hill Samuel. Castleman strenuously objected to negotiations with UBS or anyone. After a lengthy and very tense meeting, he resigned over the issue. The sudden resignation forced Hill Samuel to disclose the fact that negotiations with UBS were under way, something they would have preferred not to do at such a preliminary stage.

Chiene supported the negotiations even though he realized that if UBS did acquire Hill Samuel, it would probably sell off either Phillips and Drew or Wood Mackenzie, having no need for two comparatively large brokerage firms. Most likely it would be Wood Mackenzie that would be sold.

After Castleman's abrupt departure, Hill Samuel continued to negotiate with UBS. All had finally been agreed, except the price, which for some reason had been left to the end to be resolved, instead of being brought to a head at the beginning. A small group from Hill Samuel's board was to meet in Zürich in July 1987 with Robert Studer, UBS's young new chief executive. Studer, who some say looks like a Swiss Robert Redford, had headed UBS's investment-banking

activities before rising to the top and had clearly been deeply involved in all that had gone on before involving Hill Samuel.

The meeting in Zürich was a fiasco. Apparently Studer had not
been able to win over the UBS board on the matter of acquiring all
of Hill Samuel in order to get the parts they wanted. He told the Hill
Samuel delegation that he was only interested in buying the pieces
that had been originally discussed. Take it or leave it. The London
group was horrified. They had gone out on a limb on this deal, which
had cost it a chief executive, premature press exposure, and expectations both within the market and within the firm for a deal being
done, only to find that there was nothing there. They returned to
London and made the grim announcement, which then kept them
busy explaining how all of this could have happened to a firm of
investment bankers experienced in deal-making.

They also had to spend some time assuring the public that UBS
had not uncovered something terrible in their investigations about Hill
Samuel that had caused them to turn away.

The Hill Samuel board moved quickly to get a hand on the tiller.
Perhaps Chiene was asked to take over, but everyone must have known
that at that time the independent future of Hill Samuel and of Wood
Mackenzie was much in doubt. Chiene, no matter what other duties
he might be given, could hardly be expected to set aside his concerns
for the future of his firm within the firm, while the future of Hill
Samuel lay unaddressed. Indeed, approaches from others interested
in acquiring Wood Mackenzie had been made. In any event, the board
asked one of its outside directors, David Davies, to take over as chief
executive, which Davies did.

Davies, a former Hill Samuel employee, had, until recently, been
chief executive of Hongkong Land, a real estate company controlled
by Jardine Matheson, which in turn was controlled by the Keswick
family, which resided in Britain. A disagreement with the Keswicks
over policies concerning Hongkong Land led to Davies' resignation
and return to London. Hill Samuel was fortunate that Davies was
available at the time. He took over, righted the helm, steadied everybody down, and conducted a strategic review of the firm and its future. As a result of the study, Davies guided Hill Samuel into a safe
harbor at last, selling the firm outright to Trustee Savings Bank for a

price that must have finally brought joy to the hearts of all of its shareholders.

The sale of Hill Samuel did nothing for Chiene. Trustee Savings was not interested in the brokerage part of Hill Samuel's business. Wood Mackenzie would be sold again, virtually at auction. Chiene's preferences were considered, but Trustee Savings wanted a clean sale at a high price. In the end, Wood Mackenzie was bought in December 1987 by County NatWest, the merchant-banking subsidiary of National Westminster Bank. This was the fourth time in the past three years that John Chiene had negotiated the sale of his firm to others. The Big Bang years had not been idle ones for Chiene.

At least he had been able to deliver the firm as a whole to a new home. He became one of the two or three most senior persons in the County NatWest organization, while maintaining his role as head of Wood Mackenzie. Together these firms ought to be able to give Warburg and Barclays deZoete Wedd (BZW) a run for their money, or so Chiene must have thought until the Blue Arrow affair surfaced.

Blue Arrow was an employment and temporary-help agency. It was run by an aggressive, self-made man named Antony Berry, who decided in early 1987 to launch a surprise tender offer for an American counterpart, MAI, Inc., for $1.3 billion in cash. This was an extraordinary offer, not only because of the high price, but also because Blue Arrow was a much smaller company. The tender offer was opposed by MAI, but the sizzlingly high price won out in the end. Blue Arrow borrowed most of the money to complete the purchase from banks eager to finance leveraged takeovers of all types, but apparently it had agreed to fund out most of the loans by issuing shares in the ever-rising stock market. Thus it declared a rights offering for common stock in September 1987 valued at £837 million ($1.4 billion at the time), the largest such issue ever done, and certainly one of the highest ratios ever of new money to existing capital. As dramatic as the issue was, it appeared to go fairly smoothly, reassuring the skeptical yet again that in such equity market conditions, anything was possible.

The merchant banker handling Blue Arrow's acquisition and rights issue was County NatWest. The firm had gotten a leg up in the City by pulling off both of these spectacular deals. No longer should aggressive, dynamic companies feel that they had to knock on the doors of the long-established merchant banks, hat in hand, to get the kind of expertise they needed and wanted. They could get both at County NatWest, which had been strengthened and invigorated by Big Bang and now wanted to challenge the older firms for the business of the newer companies.

The trouble was, the market really wasn't as dumb as it seemed. Only about 50 percent of the rights issue was subscribed; the rest was left with the underwriters. To support the issue, County NatWest bought stock itself, in the end owning 9.5 percent of the company, a fact it had not disclosed as it was required to do. The Blue Arrow stock price, already sagging from the unsuccessful rights issue, was flattened by the market crash of October 19, 1987. By the end of the year County NatWest reported a £50 million loss on the Blue Arrow position alone, plus another £95 million in other losses, mostly attributed to the crash. Its parent, National Westminster Bank, reported profits down for the year by 30 percent, due to these losses and additions to loan-loss reserves.

County NatWest could, perhaps, be forgiven its poor earnings record since Big Bang. After all, the markets had been tough, competition tougher, and the marathon was still in its first couple of miles. Blue Arrow, however, was not only a further loss—it had been a scandal. The bank was alleged to have knowingly failed to disclose as required information concerning its holdings of Blue Arrow shares so as to mislead the market. Worse, County NatWest had involved its parent, National Westminster, Blue Arrow's broker, Phillips and Drew, and its parent, UBS, in the cover-up. Charles Villiers, County NatWest's chairman, and Jonathan Cohen, its chief executive, were dismissed in February 1988, only three months after the Wood Mackenzie acquisition. National Westminster appointed Terry Green, a deputy group chief executive of the parent, to become chief executive of County NatWest, but Green seemed unable to come to grips with the errant child and he was replaced in early 1989 by Harold Macdonald, a hard-boiled former treasurer of Shell who had most recently served as chief executive of Dome Petroleum in Canada. A few days later County NatWest announced plans to trim its sails by withdrawing as a market-maker in gilts and as a primary dealer in U.S. government securities in New York, the first such departure from the coveted position by a major foreign firm.

The other U.K. banks, though spared the scandal, had not fared a great deal better than National Westminster during 1987. Barclays, which had spent around £100 million in developing BZW into the major player that they wanted it to be, had suffered operating losses from market-making after the crash in the area of £50 million. Midland had done so poorly in its market-making operations that its new chief executive, Kit McMahon, a former deputy governor of the Bank of England who had taken over at Midland following its final disposition of the much-ailing Crocker Bank, shut them down within six months of their beginning. The bank recorded losses of about £46 million in its brief experience in the equities market-making business. Lloyds Bank did little better, though its commitment to the securities business following Big Bang was the least of the four major clearing banks. Lloyds had bet most heavily on the gilt market, from which it withdrew in the first half of 1987 after losses reported to be in the area of £20 million. Lloyds Merchant Bank, the bank's securities affiliate, had capital at the end of 1986 of £135 million and a staff of 800. By mid-November 1988, its capital had been reduced to £50 million and its staff to 370. Many other firms were feeling the heat too. Citicorp's London affiliate closed its gilt operations and substantially cut back on equity market-making. In December 1988, Morgan Grenfell announced it was withdrawing from all market-making activities in both gilts and equities and would save approximately 450 jobs.[3]

To be sure, 1987 was a terrible year in which to launch something as radical and as extensive as Big Bang. The bond markets were very troubled after the first quarter of the year; the stock markets continued to be very strong, but only until October 19, when the crash in New York collapsed markets all over the world. In the United Kingdom, the failure of the giant BP issue hung over the market for months. Firms everywhere had competed with each other in terms of the amount of commitment to the global securities business of the future, but these commitments were coming at the tail end of a five-year boom period in the markets. By the middle of the year, many firms were starting to cut back on their overheads, sensing correctly that there would be more difficult times ahead.

There were. Volume of trading on the stock exchange in 1988 fell 30 to 40 percent behind the level of the previous year, and competition for the business that was done intensified. Conditions in the government gilt-edged securities market were even worse, caused by overcapacity on one side and a sharply dwindling supply of new issues by the Treasury on the other. In early 1989, the Bank of England reported that market-makers in gilts had lost £190 million in the first two and a quarter years since Big Bang. The report noted that seven market-makers (all British firms) had withdrawn and that two had been added, both Japanese, but it was "too soon to conclude that either the structure of the market or the number of market-makers has yet stabilized." A financial-services consultant at McKinsey & Company was expecting in late 1988 that the number of U.K. government bond market-makers would fall into the mid-teens from twenty-four, and the number of equity market-makers would drop to between ten and fifteen from thirty-one, with half a dozen or so commanding relatively large market shares. Prospects for the future were considered gloomy by many City observers; one career counselor was forecasting a loss of as many as 50,000 of the City's 450,000 jobs over the next few years. Eddie George's comparison to the marathon footrace seemed apt. Many would start but not all would finish.[4]

Clients, of course, had done much better. Commission rates on stocks had about halved since October 27, 1986, and volume, and thus liquidity, had approximately doubled, until the crash set it back again somewhat. Many shares were bought and sold net, that is, without any commission, the transaction being done with the market-maker off his own books. In such cases, the only profit the broker makes is from the dealer spread, which can be high or low depending on how skillful the dealer has been in managing his position.

Institutions for the first time were being offered markets in large blocks of stocks by several firms who were competing hard for the business. Some firms would also offer to trade with them for whole portfolios of securities as the client changed money managers and needed to liquidate the portfolio of the departing manager and acquire the portfolio recommended by the incoming manager. Firms were prepared to make a one-price bid for the securities being sold, or a one-price offer for those being acquired, or a one-price offer for both. These kinds of transactions, increasingly common in New York, were unheard of in the United Kingdom previous to Big Bang.

Institutions were also being offered more and better research information in particular securities, on economic conditions, on portfolio structuring, and on global financial outlooks. Far more information than could be processed was being received by large and small institutions, much of which, however, was extremely useful to the institution's portfolio managers. The research was free, thus enabling the increasingly performance-minded institutions to pass much of their information costs on to their brokers. The receipt of useful research meant that commission orders would be sent to the firms responsible; otherwise, however, the portfolio manager would execute trades at the lowest cost possible. In a very short time, the City had upgraded its research and liquidity services, nearly to New York levels. After a period of several months, while institutions exploited the competitive situation immediately following Big Bang, they began to recognize that if they did not pay out just a bit more to their eager suppliers, there would be far fewer suppliers out there to supply them. Thereafter, they became a bit more generous, and indeed must have been grateful for the services of those brokers who would buy large blocks of stock from them in the difficult days following the market crash.

Corporate clients benefited too. Equity capital was raised on many more occasions than in the past, sometimes in very large amounts, as evidenced by the Blue Arrow rights offering. Convertible debentures and other forms of debt capital were also raised, both in the United Kingdom and in the Euromarkets, by merchant and investment bankers who were in frequent contact with them in London.

Merger-and-acquisition services also improved, as U.S. firms, supplementing their commitments to the U.K. capital markets with advisory services, began to compete aggressively with traditional merchant bankers, frequently being invited in "alongside" the traditional firm for a raid defense, or acquisition. Sometimes the traditional firm was displaced by one of the U.S. firms, or a combination U.S. and British team that had not worked for the company before.

The British government too was a beneficiary. Its government-securities market had been completely modernized, its equity markets too. London was more securely established than ever as the supreme financial center of Europe, something that will stand Britain well after 1992, when the European Common Market countries are scheduled to dismantle any remaining financial frontiers existing between member countries.

The government had also been able to turn Big Bang into an important political victory. It is highly doubtful that the enormous privatization programs that made possible the sale of tens of billions of pounds of equity securities to British institutions and individuals could have been possible if Mr. Parkinson and Mr. Goodison had not worked out their compromise in the hot summer of 1983.

The City had been greatly strengthened as a result of the three years of preparation for Big Bang. There was very little danger that, despite the crash and the hung-up BP deal (which the government refused to call off despite the huge losses that would be spread throughout the City), the Guinness and the Blue Arrow scandals, and the dropping out of the marathon of several of the starters, the City would not emerge from the early years following Big Bang stronger, more global, and more competitive than at any time in this century.

In the coming years, many further changes are inevitable. There remain a number of practices and procedures in the City that are uniquely British that will be tested by national and international competitive forces. The long-standing practice of raising new equity capital through rights issues at a discount to existing shareholders may be replaced in time by more efficient market-price issues as it has in the United States and in Japan. Institutional shareholders in the United Kingdom still resist vigorously any dilution of their preemptive rights, however, and for the time being they have been successful in fighting off changes.

Such resistance, claims Andrew Large, an Englishman who heads Swiss Bank Corporation's overseas investment-banking activities and who is chairman of Britain's Securities Association, "is a backward step. People forget that British companies are competing for funds,

218

and increasingly so on a global basis.''[5] Global shareholders tend to believe that companies are better off raising new money at the highest price they can, not by offering 10- to 15-percent discounts to existing shareholders.

There will also be changes in the procedures used in the United Kingdom, but no longer in the United States, Japan, or the Euromarket, in which new issues are priced first, then marketed during a two-week subscription period, then allocated to either investors or underwriters depending on how well the issue has been subscribed. Such procedures, last used in the United States in the days of Jacob Schiff, make accurate, market-based pricing almost impossible, and impose far too much risk on the underwriters, as the BP issue demonstrated all too clearly. Certainly for large global issues, a change to the system in which the underwriting does not commence until a marketing period has first been completed will be inevitable. There will no doubt be resistance from older firms in the City, but as Andrew Large suggests, if British firms want the benefits of the global markets they will have to make further adjustments to secure them. The infrastructure is in place, thanks to Big Bang, for global-market integration to occur in the United Kingdom. All that remains is a willingness to experiment with different ways of doing things.

With such experimentation will come preferences for the systems that work best, which in turn will become standardized, at least until they too are improved. As Gordon Pepper reminded us, once you unleash the forces of change, you can't turn them off halfway.

Too little time has passed since Big Bang to evaluate who the winners and losers are, though this has not stopped many observers from attempting to do just that. After Mayday in 1975, the first crisis that came along was in the back offices of U.S. firms, something no one had anticipated. It then took at least five years for the consolidations in the U.S. securities industry to take place, after which there was a considerable change in the rankings of investment-banking firms.

Big Bang, however, is the closest any national securities market has ever come to being driven by almost entirely free-market forces, for better or for worse. The most important consequence of this is

219

that the level of competition rises to eye-watering levels. With such competition must come emphasis on the part of all participants in the market on performance: performance in brokerage services, in market-making, in research, in money management, in the management of new issues, in the rendering of advice on various subjects, and in the ability of competitors to be creative and to offer new and better ideas. Those who do not perform well will be dropped by their clients, who themselves must measure up under closer scrutiny. It will not be possible to carry weak performers because of long-standing relationships. After all, in such a competitive market, there may be very little difference between the price of good and bad performance. Everyone will want the best performance, and will probably insist on it. The motto of the City may have to be reinscribed, from "My word is my bond" to "What have you done for me lately?"

As performance criteria become more important, there will be a greater emphasis on trading, turnover, and the volatility of relationships. No doubt, for example, it will be harder for portfolio managers to hang on to their clients, as has come to be the case in the United States. This will encourage bright, young, and confident money managers to break off on their own, into boutiques. The same may become true in other fields such as merger advice, where, as in the United States, such new firms can easily be set up.

Another change that seems inevitable is the conversion of U.K. companies to the Dennis Dammerman and John Browne school of modern corporate financial management in which multiple suppliers of investment-banking services are used. As this conversion takes place, the chief financial officer will begin to look at services offered in a somewhat different way. If he wants to raise debt capital, he will want to know all of the options, all of the capital-market possibilities that can result from tapping into global sources. He will want to know what the banker is willing to offer in the way of new ideas, or how willing he is to position an entire issue on his books for resale at his own risk.

If he wants to raise equity, he will want to know how the issue should best be structured to attract the highest price in global markets, and what the banker is prepared to contribute in the way of research and market-making support. Market-making services have become inseparable from corporate-finance services. And both of these must be global in order to be sure of offering the most competitive service. Those offering corporate-finance services in the United Kingdom will have to make the adjustment, whether they are U.K. firms or U.S. or Swiss or whatever.

In a relatively short time, nationality will be comparatively unimportant. Some banking firms will assimilate British ways; others, knowing that they can only hope to get some of the company's business, will stick to their specialty, financing in the United States or Japan, for example. The chief financial officer will see all of the firms that he thinks can offer useful capabilities. He will select those that have the best approach to whatever it is that he wants to do at the time.

Big Bang has already had great effect in many other countries. Similar deregulation has been tried in Australia and in Canada, and somewhat more cautiously in Germany, France, and Switzerland. But the world's accommodation to the principles and the results of Big Bang has just begun. Many important financial markets, especially those on the European continent and in Japan, have miles to go before they catch up with what has already taken place in London. Certainly the expected financial deregulation of the EEC in 1992 offers the next large-scale opportunity for those who succeed in operating in highly competitive, global markets. And after the EEC, we can have a short rest before undertaking the task of finding a place for ourselves in a fully financially deregulated Japan in the next century.

As profound as the experience of Big Bang has been and as challenging to its participants, it may be only a scratching of the surface compared to what lies ahead. Many new skills and capabilities will have to be developed, and in many ways, what it will take to succeed in a deregulated environment of the future may be quite different from what it has taken to succeed until now. The Big Bang has been our laboratory, our experiment, our learning opportunity. It has also been our combat zone and our trial by ordeal. It hasn't been easy.

Maybe we are all being hurled through the great void at the speed of light; well, if so, such things take getting used to.

Others may agree with a popular singer of the 1930s and 1940s who had become terminally ill at a relatively young age, but announced, upon hearing her first rock-and-roll number on the radio, ''Lord, thank you for calling me when you did.''

# 7

# Breaking with Socialism

In the last days of the Heath government in 1974, the lights went out in London. The coal miners had gone out on strike, other workers had supported them, coal deliveries ceased, power was rationed, and the country went on a three-day workweek. In the City, electricity was in short supply and was only available for a few hours at a time. It was dark and cold in our offices that winter. We were there, in them, during this particular labor "dispute," as it was called, but we didn't always have any clients to talk to.

The strike was finally settled when Heath caved in to the miners' demands, which were pretty substantial. He had valiantly tried to stand up to the unions, whose ever-increasing wage demands had long since awakened the gods of inflation. Heath, however, had picked the wrong union, the popular coal miners, whose harsh lives had inspired sympathy and guilt since the days of D. H. Lawrence, and the wrong time, in the winter when above-ground coal inventories were low. Heath's concessions to the miners caused his opponents in the Labour party to torment him and his colleagues in the Tory party to scorn him. He called an election, to seek the wisdom of the voters on the question of "who rules Britain?" and was voted out.

Heath was replaced by Harold Wilson, who had been defeated by Heath in 1970. Wilson had originally been elected in 1964, the second Labour prime minister since World War II, and the one, perhaps, who made the most difference. Under Wilson, the government had carried out its policy of turning Britain into a "welfare state" and had nationalized several industries with high employment such as coal, railways, steel, shipbuilding, aerospace, and parts of the British-owned automobile industry. In 1965, government expenditure, central and local, accounted for 37 percent of gross domestic product; by 1970 it had climbed to 41 percent; in 1975 it would reach 50 percent.

Heath's administration was largely spent trying to control inflation through price controls and battles with unions, which would either strike officially, walk off the job for a day or two, or "work to rule," meaning by the book, so as to slow everything down. Heath's efforts, however, were not successful. By June 1975, just after Heath's defeat at the polls, inflation in Britain reached 27 percent.[1]

The unions were extremely effective during this time. They not only got the high wage settlements they were looking for; they also got their wages indexed to rises in the cost of living, making themselves inflationproof, even if the rest of the economy was not. Mainly, the unions acted against nationalized industries, where their power was greatest and their opponents the weakest. Many of the union leaders described themselves as socialists; others were known to be members of the British communist party. A Gallup poll taken in 1974 found 52 percent thought that Jack Jones, leader of the Trades Union Congress, was the most powerful man in Britain. Only 34 percent thought the prime minister was.[2]

After the first oil crisis, many once-powerful industries in Europe were brought to their knees. High energy costs and the withering effects of high inflation threatened the existence of many of these industries. Their governments, for the most part, stepped in and kept them alive, to prevent a surge of large-scale unemployment, which always possesses the power to sweep those deemed to be responsible out of office.

Europe as a whole was a patchwork of socialism. From Britain to Germany, Sweden to Italy, Holland to Denmark, almost all subscribed to the notion that governments were there to protect their citizens against harsh economic reality. Social programs were extensive, taxation to pay for them was also. The distribution of wealth and income was becoming more uniform in Europe, and class distinctions were beginning to fade, or at least that was the aim. Only in France, during the presidency of Valéry Giscard d'Estaing, were politics not explicitly socialistic, and in France, few things are as they appear.

Europe was firmly in the grip of socialistic political philosophy. *Capitalism* was a dirty word, associated with greedy Americans, or Hong Kong Chinese, neither of whom were considered role models at the time. The wealthy were careful to be inconspicuous, to disguise or to conceal what they owned. Too often they watched legislation aimed at confiscating their property through wealth or inheritance taxes, or through nationalization, work its way through legislatures, often to fail only at the last minute. To escape the possibility of losing their wealth, many emigrated to more hospitable locations. Others smuggled funds out of their countries into Switzerland, where they were invested in anonymous, bearer-form Eurobonds and other instruments.

Corporate life in the 1970s was difficult. In most European countries, foreign-exchange controls were a continuing nuisance, local labor regulations and wage and benefit packages weakened productivity, and capital was expensive because interest rates were high and stock prices low. Despite the existence of the Common Market, shipments between countries reduced efficiency because of local regulations, differing languages, and awkward cross-national distribution networks. The general view of economic life came to be: "So what if our corporations aren't efficient!" The main thing was that the ordinary person got along better than in the old, dark days of the class system that lay in every European's past. This attitude came to be represented by the popular expression "I'm all right, Jack."

During the 1970s, much was heard about Eurocommunism, which

225

was often described as an evil and insidious boring-from-within that didn't look all that harmful because that was the way it was planned. In the longer run, those worried by the phenomenon would predict gloomily, Eurocommunism would turn all of the free-world economic institutions, its bastions of enlightened capitalism, into mush. Economic rule by the state would replace them, piloted by the grim little men from the KGB who had planned it all.

There was certainly a lot of open communist activity going on in Europe at the time. In Italy, communists had returned more than 30 percent of the vote in some elections, and coalition governments in other countries that included communists were becoming common. Americans felt they had to keep certain secrets from their NATO allies for fear that communist cabinet members would pass them on to unauthorized recipients. The French communist party was highly visible in elections before the end of the decade, so much so that it took special efforts to reassure U.S. investors to whom French government–guaranteed commercial paper was being sold in 1974. Communists, or at least very leftish socialists, were visible in all parts of Europe from Sweden to Spain. Many American companies hesitated before proceeding with investments in some of those countries for fear of losing them when the revolution came.

Americans, of course, were conditioned to fear and loathe communists in whatever form they appeared. Their reaction to a Europe, revived by American aid and commercial activity, slipping into the grasp of the enemy was predictable. In time, though, they consulted with their wiser, more adaptable European friends and colleagues, who advised them not to worry too much. "In Europe," they said, "communists were only important when they had no real chance of controlling things." The secret, Frenchmen and Italians would tell their American friends, was to let the communists get just a whiff of how difficult it was to govern one of these Eurodemocracies, and to taste the political consequences of failing to meet everyone's unrealistic expectations, and they would fade from the scene. Not very many Americans, however, were reassured by words of this kind.

Americans are not used to such a wide difference in political philosophy as exists in Europe. Republicans and Democrats may appear to represent a "real ideological choice," as George Bush in-

226

sisted recently, but by European standards the two seem to be branches of the same party. No Democrat has ever proposed, let alone carried out, a nationalization of American industry for reasons of social policy. None has ever proposed an annual tax on wealth, or that government subsidies be paid to protect employment in, say, the steel industry. We have only rarely seen our government interfere in corporate activity, such as the rescue of Chrysler, and of Lockheed before it, and then the rescues were justified on the grounds of protecting national security interests of the United States. No laissez-faire Republican has so abjured the powers of his office as to turn his back on affairs in distant lands or to leave the Federal Reserve alone to set whatever monetary policy it wanted or to fail to complain about Japanese trading practices.

In Europe nationalizations were not uncommon; nor were protective measures against imports, or monetary or fiscal policies that ignore the dictates of the free market in order to accomplish a specific social or political objective or to protect economic sovereignty.

But it all changed in the 1980s.

Humiliated by the flickering lights of the Heath government, the Conservative party decided to select Margaret Thatcher as its next party leader, over older and wiser men who smelled of status quo, men who could be expected to make all the same compromises as their predecessors and accomplish little. Thatcher, as we have all come to know, was different. She had a different ideology and believed that principles worth having were worth fighting for, no matter what. But she was only the leader of the opposition, not the prime minister. Harold Wilson held that office until he retired, passing the baton to James Callaghan in 1976.

This last Labour government, which shared its time in office with the hapless Jimmy Carter, had little hope of steering Britain through the continuing periods of ruinous inflation, strikes (which even Labour couldn't control), and more demands for social services and job protection. Severe financial difficulties followed, resulting in 1976 in once-mighty Britain having to stand, hat in hand, in the queue for IMF rescue loans like a Third World country.

The British economy's saving grace was the production of North

227

Sea oil, which had come on-stream just at the time of escalating oil prices. This fortunate situation, however, combined with a global financial rejection of Jimmy Carter, drove the pound up and the dollar down, despite a highly uninspiring economic performance in the United Kingdom. The exchange rate came to be too high to export anything but oil and caused massive imports. The country's trade position was severely distorted. Financial difficulties continued. There seemed to be no way out of Britain's economic difficulties. A malaise developed in the United Kingdom that was the equal of anything Mr. Carter had in mind when he referred to the declining spirits of Americans.

When Callaghan called the election he could postpone no longer, he lost. Britain's first woman prime minister took office in 1979 and changed the course of Europe at least for a decade, probably much longer.

The British, even at the beginning of the Thatcher era, tended to think of their Honourable Lady as the quintessential English nanny, something, they point out, most other countries don't have. Many British children are raised by nannies, who, like British teachers and bank managers, are universally seen as stern disciplinarians whose job is lovingly to take their charges by the ear and make them do the right thing. She was as tough as only a nanny can be, knowing in her heart that the steps she prescribed were in the best interests of her charges and that no matter how tearful, all protests had to be ignored.

Right from the beginning, castor oil was administered to the economy and the country. Disinflationary policies were enacted, interest rates raised, the fiscal budget choked. The unions were challenged, especially by Scottish-born Ian MacGregor, a case-hardened retired American industrialist Thatcher had hired to head British Steel, thinking perhaps that no British executive could be found willing to put 100,000 workers out of work within the first year or so on the job. She cut income taxes to give people an incentive to work harder, and she encouraged corporations to do what it took to restore profits.

The effect was visible right away. Companies perked up, then began extensive cost-reduction programs. Payrolls were reduced, factories closed. Industrial unemployment rose sharply, but, before long, so did corporate profits. A lot of political grief was blamed on Mrs. Thatcher, who stood fast. A few cabinet ministers who demonstrated their sympathies too openly were let go and returned to the back benches. Though her majority was small in 1979, she risked it often by sticking to her guns and telling everyone, in her perennial nanny's tone of voice, that "there is no alternative" course for Britain so "we must all stick it out as best we can."

I arrived in London to take over the Goldman Sachs office there shortly after her election. The delight of the British business and financial people was obvious, though most might have felt the same way even if Heath had got back in. There was no great love in the circles in which I traveled for Labour.

I was astonished to discover something about the life of the average British salaried person, as a result of a visit from a delegation of Goldman Sachs employees who had come to complain that we (and everyone else in our business) had three different ways of paying our people in London, and these resulted in vastly different incomes for employees doing the same job.

The problem was their respective tax rates. Americans who had been sent over from the United States were told on departure that the move would not cost them anything, that their salary would be adjusted for foreign-exchange and cost-of-living differences. Americans had to pay taxes in Britain, though about half their income then was excluded from such taxes, which in turn could be deducted from the U.S. taxes that the employees had to pay on the basis of their adjusted "global" incomes.

Non-British European employees also were able to deduct half of their salaries and most of their bonuses from British taxes. These employees did not have to pay any taxes in their home countries because their governments did not tax global income; therefore more than half of their income was tax-free. Though because of this advantage, they did not receive cost-of-living and foreign-exchange-rate adjustments, they still came out way ahead of their U.S. and U.K. colleagues on an after-tax basis.

British employees, of course, had to pay full taxes in the United Kingdom, which were at a very high marginal rate, and these rates took effect at comparatively low levels of income. Our British employees were being taxed heavily, and their U.S. and European colleagues were not.

Most British firms dealt with this problem by simply accepting it. British corporate and financial executives in 1980 earned very little, after tax, compared to their U.S., German, Swiss, and other counterparts. They did, however, receive a number of perks: cars and drivers, use of company apartment houses, first-class travel, and similar benefits were provided, along with low-interest-rate mortgages for houses. The British seemed to be satisfied with these arrangements, which assured low after-tax incomes, that must have been accepted in exchange for the prestige of being known as a so-and-so company executive, and presumably for a tacit understanding of job security.

Goldman Sachs's British employees were not at all happy with the arrangement. They had joined a foreign firm to make money and get ahead, not to fall back into second-class citizenship simply because they were Brits. We believed that the best employees, of whatever nationality, are attracted to opportunities to be advanced in accordance with their ability, and to be paid well for it. In cash, not in perks. We wanted our employees to be able to accumulate some wealth, but in the case of British employees it was difficult. We and other firms in the City at the time used a loophole in the system to pass more after-tax income to our people. We "leased" them at a nominal rate some of the things they would buy with money if they had it: their cars, their household furniture, even their suits. Naturally, they appeared to be extremely conspicuous consumers, something that neither their neighbors, their customers, nor we liked at all. Yet the only way we could get money that would not vanish in taxes to them was through this device. No one was happy with it, but it was the best we could do.

After Thatcher, though, our employees had their marginal rates reduced to levels comparable to U.S. rates. We continued to lease them their cars but scrapped the rest. Most of them paid their taxes and invested their savings in the rising U.K. or U.S. stock markets and were, at long last, able to behave as capitalists again.

Despite the salutary effect of her tax policies on the better-off portion of the population, the early years of Thatcher's economic programs were very rough. By the end of 1982, unemployment had reached about 3 million, or about 12 percent of the work force. Many of these were thought at the time "never to work again." The excess manning in the system had been reduced. Employment would not be increased again until significant amounts of growth had made it necessary to increase capacity. Young people coming out of high school at sixteen or seventeen were unable to find jobs. Many families had two generations out of work at once. These people, it was thought, might be beaten at the union level, or by bills their representatives in Parliament couldn't defeat, but just wait until the next election! In mid-1982, with a year or so to go before the next election, it appeared to many people that with the increasing horde of unemployed voting against her, the Iron Lady would be unhorsed.

Meanwhile, in 1981, the French had elected François Mitterrand to a seven-year term as president. Giscard d'Estaing had been turned out, despite a demonstrated record of exceptional competence behind the economic wheel. Though not a conservative in the same sense as Thatcher, Giscard was certainly no socialist. Mitterrand was decidedly one, and he shocked many people by appointing communists to secondary posts in his cabinet. "Better to have them inside where one could watch them" was supposedly the reason for the appointment, but no one could be sure. The socialists had a manifesto, on which they had campaigned for some time. The principal banks, insurance, and industrial companies would be nationalized. Thereafter, France would operate as a planned economy.

They weren't fooling, and everybody knew it. From the time of Mitterrand's election until he took office and could exercise his powers, huge amounts of foreign exchange left the country. Goldman Sachs did more business during those three weeks with French customers than we did during the rest of the year. Frenchmen were afraid that their investments might be confiscated through nationalization programs or that they would be subjected to witch-hunts and persecution by tax authorities trying to uncover their property.

There were no ambivalent views among those who possessed even modest amounts of wealth—if it remained in the country, it was likely to be diminished. Thank God, they must have thought, for numbered accounts in Switzerland. Soon after Mitterrand took over, he closed the borders to financial transactions to prevent a massive exodus of funds. He announced severe penalties for those caught smuggling wealth out. Then he went on to announce his nationalization program.

In the first phase of nationalization, announced within weeks after he took office, Mitterrand provided for the purchase by the government of all of the stock of the large manufacturing companies and those banks and insurance companies that were not already owned by the government. The three largest banks already were. Payment would be made in marketable French government bonds. The prices to be paid for each company were determined after some negotiation and advice taken from investment bankers. In all, companies representing a majority of the market value of the Paris Bourse were acquired in this way.

Shareholders on the whole could not claim that their property had been confiscated, as many had feared. Though they did not receive payment in cash for their securities, they received government notes, which could be sold in the market. No longer able to own the stocks that had been nationalized, many became content with the bonds they had received in exchange. Some invested in the remaining shares on the Bourse to rebalance their portfolios. New rules were passed encouraging investment in shares listed on the Bourse. The market in Paris, in both stocks and bonds, was surprisingly active just following such an uncapitalistic act as a massive nationalization. People shrugged and went back to work.

Life at the nationalized industries was something of a shock, however. The new government, a bit short of socialists in high places in business and finance, began appointing new people from other walks of life to the top jobs in the banks and manufacturing companies. It all appeared random and chaotic to outsiders, maybe even to the French.

The head of the Banque Nationale de Paris, Jacques Calvet, a former aide to Giscard d'Estaing, was suddenly sacked, only to reappear a few months later as number two (soon made number one) at Peugeot, the automobile manufacturer. The head of the Compagnie Générale de Electricité (CGE) was replaced by a man who had just recently been the French ambassador to Japan. I called on him soon after he had received his new appointment and asked for his initial impressions. He replied that he really had no idea why he had been selected for the job, he was quite unqualified for it, but he would try to do his best. As a diplomat he might be able to contribute more than was expected, because it would now require considerable political skills to be able to negotiate each company's annual budget and funds requirements with the government so they could continue to do business. As a politician, perhaps he was used to this kind of environment, but the new situation took some getting used to on the part of the rest of the people at CGE. The company was to be run one year at a time; capital budgeting and long-range planning became impossible.

Other French companies, now nationalized, were simply broken apart and recombined in ways that seemed best to the young technocrats of the administration. Of all of the restructurings that took place in the 1980s, the French ones were the crudest. If the government decided to do something, transfer a chemical division from one company to another, that was enough, it happened. Many such transfers occurred as a result. It will be years before all of the costs and effects of the French nationalizations of 1981 are known.

Meanwhile, in Britain, the Thatcher government was pressing on with its commitment to free-market policies. She received much support from a kindred spirit across the Atlantic, Ronald Reagan, who took office in January 1981 and commenced to undertake similar free-market experiments in the United States. Thatcher was given much advice and comfort by a British economist, Alan Walters, whom she had recruited back to the United Kingdom from the United States, where he had become a disciple of Milton Friedman, the Nobel Prize–winning monetary economist from the University of Chicago. Things were going according to plan, she would reassure her anxious countrymen; we will just have to give it a bit more time.

In those days, the consensus of informed opinion was that Thatcher's policies would have to be applied without interruption for about ten years before the positive effects of free-market economics could be felt and the permafrost of socialist thinking and attitudes that existed in the United Kingdom melted down. There was, of course,

233

some doubt that she would last that long, in view of the rising un-employment and the political objections to her rule.

One of these forms of objections came to life in the formation of the Social Democratic party, a new political party organized by four senior members of the Labour party who felt that their party had drifted too far left. Led initially by Roy Jenkins, a former cabinet member, its aim was to attract moderates from both the Labour and the Conservative parties and to forge a new center for British politics. They were less confrontational than Labour, more sympathetic to the plights of the unemployed than the Tories. They came across, during a time of distinct policy differentiation, however, as the mush in the middle. Nevertheless they attracted away some of Mrs. Thatcher's support.

What may have saved Mrs. Thatcher's political life, and kept her economic programs going, was a totally unexpected event that was motivated by an anxious Argentine general seeking to extend his shaky popular support. The Falkland Islands, located deep in the South Atlantic 250 miles off the Argentinian coast, had long been a source of conflict between the two countries. Britain had seized the islands a hundred years or so earlier, and the Argentines were always hoping to show they could get them back, not that either side had any use for them. They were remote, cold, wool-growing islands of about two thousand people that contributed little to the might and main of Great Britain.

Lieutenant General Leopoldo Galtieri, Argentina's president, sent Argentine troops to occupy the Falklands on April 2, 1982. The Brit-ish were furious. The foreign secretary, Lord Carrington, "carried the can" and resigned. Thatcher rallied the nation, banged together an ersatz invasion fleet consisting of an old aircraft carrier, some frigates, the ocean liner *Canberra,* which was commandeered (its passengers being dumped in Naples), and several leased container ships. The fleet embarked on the eight-thousand-mile journey to the Falklands at ten knots, so it wouldn't run out of fuel, and to allow time for the Argentines to withdraw voluntarily.

They didn't. The *Canberra* and her escorts arrived in late May, and a number of surprisingly intense battles ensued, during which four British ships were lost. Argentina's largest warship, the cruiser

*Juan Belgrano,* was sunk by a nuclear-powered submarine with the loss of more than three hundred. Finally, British commandos and Army troops landed on the Falklands, and the Argentines there surrendered. By June 14, it was all over. The lioness had roared and the sound of it, long silent, brought joy and pride to the hearts of all Englishmen.

That July my wife and I were visiting on the Isle of Wight, just off the southern coast of England across from Southampton, a principal port. As we were about to return to London, our friends urged us to stay another day. "The *Canberra* is returning tomorrow," our host explained, "and we are all going out to welcome the lads home." We stayed. The next day we boarded a small boat and sailed into the harbor to await the moment. There were nearly a thousand boats making up the greeting party, mostly private craft filled with patriotic Britons. *Canberra* appeared at last, stained, streaked, and weary. Manning her rails were a thousand or so of Her Majesty's finest troops, still in combat clothing: young faces, all waving proudly. Our boats all followed along in *Canberra*'s wake, singing, cheering, weeping. English hearts had not felt the sharp stab of patriotic pride for a very long time, perhaps not to this extent since El Alamein.

During July and August of every year, a series of musical concerts are held in Albert Hall called the Promenades, or Proms for short. The last night of the Proms is always devoted to patriotic music. In 1982, when the orchestra broke into perhaps the most stirring of all such music, Elgar's "Land of Hope and Glory" (we call it "Pomp and Circumstance," but theirs has a great blood-warming fanfare and words), everyone present joined in the singing. There was not a dry eye in the house or in any of the millions of houses in Britain that had tuned in to Albert Hall that night on television or radio. It was a rare emotional, wonderful feeling for the British. After so many years of being on the slide, their fortunes and influence diminishing relative to those of other countries, to have again the teeth-gritting, chest-filling surge of pride in one's country was a glorious thing. And Thatcher had given them that.

Thatcher's advisers immediately set to planning the next election, which was hers to call at any time before her five-year term was up. Some wanted to have the election right way, or in the autumn, to capitalize on the Falklands. She resisted, not wanting to appear to be taking unfair advantage of the event. The election was held in the spring instead, and she was returned with a majority of more than seventy seats, a landslide by British standards. Labour's returns were the lowest since 1918.

Patriotism is fine, of course, but what about the political effect of all those unemployed? Presumably most of them voted for the Labour candidate, Michael Foot, who ran an uninspiring campaign. Perhaps some voted for the Social Democrats, who had allied with the Liberal party and won a few seats. Surprising some, however, Thatcher's victory came because ordinary, working-class people voted for her. They believed that she was the strong leader that Britain needed and had lacked for so long. They didn't understand her economic policies, nor would they have agreed with them if they had, but they felt a firm hand on the tiller was what they needed most, and if "that bit" worked out, the rest would too. Ronald Reagan was elected for the same reason in both of his election campaigns.

There were, of course, other factors at work too. The so-called *underground economy,* in which British craftsmen, builders, and laborers worked for undeclared cash payments, went a long way toward providing means for the unemployed to survive and, in some cases, prosper. Most people had had it, too, with governmental promises and programs that always seemed to disappoint. Thatcher told them they could do better, and they wanted to believe her. Many submitted to her, as they might to a nanny, obediently but without understanding. In any case, she won the election by a very comfortable margin, despite economic policies that could only have been described as antipopulist.

Installed into her second term, Mrs. Thatcher set out to do some of the heavy work of her administration, the privatization of nationalized industry and the creation of a stake in these industries on the part of the common people.

Past Labour governments had nationalized various British industries, beginning with the coal industry in 1947, railways in 1948, and steel in 1951. The government also owned the public-service utilities and extensive oil and gas properties. Although steel had been denationalized by the Churchill government in 1953, the Conservatives had made little effort since to return the government's large industrial holdings to the private sector. Indeed, little opposition was offered to the renationalization of steel in 1967 or to the subsequent takeover of the aerospace and automobile industries.

Labour and other opposition parties opposed privatization, on the grounds that the country's crown jewels were being sold for a pittance, and that only the rich would benefit. There was some objection too on the grounds that, run privately, the industries might be unresponsive to full-employment policies of the government. However, as this government had no such policy, the issue seemed somewhat moot.

The real issues of privatization, however, were economic and financial, not political. How could industries that did not earn a profit expect to be sold for a decent price? How would the price be determined for these creatures anyway, so that it was not too high for individual investors to participate, and not so low that the *stags* (those who buy new issues and immediately sell them to capture the aftermarket profit) would make all the money, rather than the taxpayer, on whose behalf the government was acting? How should the new companies be capitalized, that is, how much debt should they have on their books as they began life as stand-alone enterprises? What should the regulatory policies be in certain industries, such as the telephone industry, so that both the public and the private interests are properly served? How, if at all, could the government attempt to prevent takeovers of these new companies, possibly by foreigners, should they be launched? How many companies should be sold, and over what period of time? In what sequence should they be offered for sale? Could the markets accommodate such large offerings as would be needed to get the government out entirely? Billions of pounds would be involved—was the City strong enough to handle the sale? Would foreign underwriters have to be involved also? If so, to what extent? And how could these offerings be structured so as to encourage wide-scale participation by ordinary citizens? No matter how it was to be done, the privatization program would be large, complex, back-breakingly hard to do, and controversial.

"All the more reason to get on with it," we can imagine Mrs.

Thatcher saying. The first step, as noted in Chapter 6, was to prepare the City for the effort, by settling the outstanding restraint-of-trade suit that the last Labour government had brought. Then to make up a schedule, then select advisers, then—

The warm-up to privatization was a sale of British Petroleum shares owned by the government. In November 1979, 80 million shares of BP, worth £290 million, were sold. A number of smaller issues followed, including one in which the government sold off all of its holdings in a radioactive chemicals company, Amersham, only to see the offering become twenty-two times oversubscribed and the share price shoot up in the immediate aftermarket. The government was much criticized for "giving away" valuable government property. The next issue would be different. When the shares of Britoil, part of the government's stake in the North Sea oil fields, came to market in a sale valued at £549 million, the largest issue so far, the government was very tough on the price of the shares to be sold, and the issue flopped.

Despite these early difficulties in pricing issues, the government continued to bring them to the market. In November 1984, the sale of £3.9 billion of British Telecom shares was completed, this time with a portion of the shares sold to investors overseas through a global underwriting syndicate similar to the one that was used in the $1-billion Texaco convertible-debenture issue of the year before. The market for privatization issues was proving to be very strong—there seemed to be no limit to what could be done.

This was because of the exceptional steps taken by the government and its bankers in designing and marketing the issues. In the British Telecom sale, great efforts were made to attract participation from individual investors. An advertising campaign was launched to promote the issue, featuring Busby, a British version of Big Bird from "Sesame Street," who shamelessly appeared on television and on billboards across the country to promote the sale of shares. Individuals who subscribed to the offering would, at the end of a year, receive "bonus" shares for their "loyalty" and some kind of discount on their phone bills. Investors were urged to subscribe in order to become shareholders in British industry, something comparatively few Britishers were and a great many would become during the four-year period when privatization sales were at their peak.

Like many investment bankers, Goldman Sachs had been soliciting the government to be included in the British Telecom global syndicate and to lead the U.S. tranche. The solicitation was not at all easy, because it was very difficult to know whom to be calling on. We already knew the senior people at Telecom, and had presented some ideas to them, but they were not the selling shareholder—the government was. In previous sales, a merchant bank was appointed to advise the government and another to advise the company. This was a necessary arrangement because there were a number of issues that had to be resolved by negotiation prior to the sale, such as the capital structure of the new company, the extent to which earnings in the future would be affected by regulatory factors, how voting arrangements were to work, and so on. Each side needed to be advised, to prevent either one from overwhelming the other. Neither the government nor the industries concerned had any experience in public offerings before; they had to know how each step might affect the market for the shares.

The selection of merchant banks to perform the advisory services, which carried more prestige than income, was done by *beauty contest,* a process in which a number of banks would be asked if they would care to be considered for the job(s), and if so, a series of meetings would be held with appropriate government officials from the British treasury and the ministries concerned with the company whose shares were to be sold. As each company came from a different ministry, the officials involved in the sales were different for each deal. The Treasury too varied the staff assignments on privatizations so it was never clear who the decision makers were until the last moment. At the meetings, the officials would give the bankers a chance to make a brief presentation on their ideas for the coming sale and the reason why the particular bank should be selected to be an adviser. These meetings, involving only British firms, were, I suppose, more informal than the ones in which the foreign firms were involved.

Once the U.K. merchant bank who would advise the govern-

239

ment was selected, this bank would play a key role in determining whether a global marketing effort was to be made, and, if so, how it would work and which foreign banks should be involved as leaders of their respective tranches.

In other words, when the beauty contest for the selection of the foreign advisers was held, the principal "judge" was to be one of our competitors in several areas of international banking. This competitor could ask any questions it wanted about our business, narrow the focus of selection so as to exclude our known collateral strengths from consideration, or reward or punish us for some prior encounter. There was never any evidence that the lead merchant bank misused its advantage over the foreign banks, but the situation was always a little unnerving for the beauty contestants.

In the British Telecom case, the beauty contest was very tightly structured. The government's adviser, Kleinwort Benson, sent each contestant a questionnaire, which posed the questions Kleinwort wanted addressed. These were sensible questions about how the issue should be structured, whether the U.K. underwriting risk could be shared with the U.S. underwriters, what pricing factors would operate in the United States, and how much demand for the shares could be expected. The pricing questions were a little premature as the issue was still many months off, but they were to give an idea of each firm's thinking. We sent back the questionnaires, along with all the information about the firm that had been asked for. We were then told that seven firms were to be interviewed, and we were asked to present ourselves at Kleinwort's offices at a certain time.

We had a large delegation, including two or three senior people from New York. We had prepared an oral presentation to supplement our written remarks. The oral one would emphasize our enthusiasm for the job and our qualifications for doing it based on our rankings in several product areas. We had rehearsed and sharpened our remarks to a bright competitive edge. Finally we were led into the presentation chamber.

There, sitting in a half-moon configuration in front of us, were the committee that had been assembled to hear us. There were two or three from the Treasury, two or three from the ministry responsible for Telecom, someone from the Ministry of Trade and Industry, and two or three from Kleinwort, one of whom was chairing the meeting.

The meeting began. We were told we had an hour and a half. "The committee has certain questions to put to you," said the chairman, "but would be willing to wait to ask them if you would like to make an opening statement first."

"We are delighted to be here today to present our best ideas on how a British Telecom issue in the United States ought to be handled," one of us replied, "and we are very enthusiastic about the prospects for such an issue, but we will be glad to defer our prepared remarks until we have addressed the committee's questions."

This was probably the only response we could provide—it would be death to fail to answer all of their questions completely, but we had no idea how long it would take to do so. In the event, our response was probably the wrong one.

The chairman then proceeded painstakingly to read out, one by one, the questions that we had previously responded to in writing. We then got to reply orally to those questions, our written answers to which were on the table in front of them. Occasionally, a Kleinwort man would ask for clarification on a point of detail. Occasionally, one of us would volunteer something. By the time the process of answering all of these questions that we had already answered was completed, most of our time was up. We then had to figure out how to compress our forty- or fifty-minute pitch into about ten minutes, without appearing to our audience to have fallen completely apart. We had passed notes back and forth between us and had reshuffled ourselves, some major parts being cut out altogether. What we were able to get across was very general and, if I do say so myself, unmemorable. Stripped of our air time, our presentation must have come across as pretty flat.

Immediately when our time was up we were politely thanked and shown out. No one on the other side had uttered a word other than the Kleinwort men.

"Well," said one of our partners from New York, who headed trading of equity securities, "one of you guys from over here is going to have to tell me how that went. Maybe we were great, but if this meeting had been held in New York, I'd have said we probably didn't knock 'em dead, because they were dead already.''

The meeting had been deliberately overcontrolled, perhaps to keep the unruly Americans from doing something inappropriate, or to keep the issues very narrowly defined so the committee would not see much difference between the players, or simply to enable the advisers and the committee to go through the motions of a beauty contest when the actual selection had already been made. In any event, Morgan Stanley was selected as the U.S. leader, the principal firm the U.K. government had used in the United States since the beginning of the century.

The issue, when it came, was a great success. The U.S. portion, however, had been a little disappointing as many of the shares sold there were resold for a quick profit in the aftermarket in the United Kingdom. For the next issue to involve a series of foreign tranches, the government was going to look harder at its alternatives for a better distribution within the United States. This event arrived with the planned sale of £5.4 billion of shares of British Gas six weeks after Big Bang had come into effect.

For this issue, the committee was interested in having the U.S. underwriters financially committed during the two-week subscription period in the United Kingdom and to have assurances that the shares sold in the United States would remain there and not come back to the United Kingdom in the immediate aftermarket. Goldman Sachs was able to satisfy the government on both points and was selected to lead the U.S. tranche for the British Gas issue, which turned out to be a great success.

Partly because of this, we were chosen again to lead the largest of all privatization issues in the United Kingdom, the sale of the government's remaining shares of British Petroleum, representing 31.5 percent of the company and valued in the market at approximately £7 billion. This was the issue that had the bad luck to have the stock market crash of October 19, 1987, occur after the commencement, but before the completion of, the issue's subscription period. As in the British Gas issue, the U.S. underwriters were committed to purchase their share of any unsubscribed shares at the end of the subscription period. Because the shares had dropped well below the subscription price following the crash, hardly any were subscribed for, which left the underwriters with some very substantial losses.

The British privatization programs, notwithstanding the BP issue, have been enormously successful. Through the end of 1988, they raised more than £22 billion for the Treasury, distributed billions of shares of premiere British companies to investors in the United Kingdom who had never owned stock before and to foreign investors who had never owned British companies before. Newly "liberated" businesses made their appearances as privately managed companies and immediately realized the benefits of no longer being nationalized industries. They also were an example for other countries that was widely and extensively followed.

Many other European countries adopted privatization in the wake of the British success. French companies, now nationalized, invented a certificate of participation that they sold in lieu of common stock. The German government sold shares in Veba, a large multi-industry company, and Volkswagen. The Austrian government sold several businesses, as did the Canadians and the Finns. The Spanish government sold shares in its telephone company; privatization even worked its way into Argentina and Chile and was being considered by the Israelis. Japan sold parts of its holdings in Nippon Telephone in the world's largest initial public offering of stock, which was valued at over $13 billion. Singapore and Malaysia sold some of their holdings in their respective airlines. The U.S. government, after several years of deliberation, finally sold its holdings of Conrail, the freight railroad company, in a $1.7-billion privatization offering to the public. Even Libya sold its holdings in Fiat through a large Euro-equity issue.

Despite the market decline after the October 1987 crash, privatization programs are continuing. Japan has sold two additional tranches of Nippon Telephone shares. In December 1988, £2 billion of shares of a rebuilt, streamlined, and competitive British Steel were sold in a successful global offering. Several further issues are scheduled in Britain and in other European countries for 1989.

The French government, having nationalized in 1982 and pursued socialist economic doctrine for a few years, was overcome by rising inflation, a huge trade deficit, and devaluations. Mitterrand's

243

government moved back to the center and imposed austerity measures. These resulted in a loss of the majority in the legislature by the socialists in March 1986 to a centrist-Gaullist conservative coalition led by Prime Minister Jacques Chirac and a period of his "cohabitation" with President Mitterrand in the principal offices of state. During this period, the legislature passed a denationalization program, which Mitterrand was unable to prevent. The program, which provided for the sale of seventy large banks and insurance and industrial corporations, was begun immediately. The first stock sale, of shares of St. Gobain, the large industrial concern, occurred in September 1986.

The French market quickly absorbed the shares, and trading in them, and in subsequent offerings, became active. Many foreign investors were attracted to the French market, which had been truncated by the earlier nationalization program. French regulations were relaxed to allow foreign brokers and bankers to operate in France and to become members of the Bourse. Within a year, twenty-two companies with market capitalizations of $17 billion were returned to the private sector.[3] The market crash in October 1987, of course, halted the flow of issues.

In the French presidential elections of 1988, Mitterrand was reelected and Chirac's effort to take control repelled. The legislative elections that followed reestablished the socialist majority and resulted in the selection of a socialist (though a moderate one) prime minister, Michel Rocard. There are no plans at present for the renationalization of those companies that have recently been denationalized, but there are also no plans for further privatizations either.

By the mid-1980s, the pendulum of political-economic philosophy in Europe had swung from left to right, from the planned economy and "welfare state" to the free market, from socialism to capitalism. The British experience has caught the imagination of other European countries, which have attempted to copy it. The French, while voting socialist, have acted as capitalists in the things that actually affect their lives. Socialist economics have been replaced in Germany and diluted in Scandinavia. Much change has occurred in the last ten years.

These changes are the result of many forces and events that were unforeseeable a decade ago. Certainly the world economy had some tough times following the 1973 oil-price rise. High inflation, high interest rates, high unemployment, and high levels of volatility in foreign-exchange rates put a great strain on government management of economies. Few did the job well, and those that performed best were those that had comparatively small public sectors in their economies. People came to understand that governments were more than partly responsible for the poor economic results that they had experienced.

At the same time, they came to realize that the plight of the unemployed was not exactly what it had seemed. When the unemployment was caused by layoffs necessary to match production with consumer demand, it was usually reversed in the next cycle, as many European economies have experienced in the last few years. Also, the "underground economy" seemed to do a better job of providing welfare for the unemployed than the state did, and many taxpayers came to resent the idea that they still had to pay unemployment benefits to those who were doing fine underground. Whatever the truth of these perceptions, the perceptions themselves have reduced the political effectiveness of socialist appeals for more government welfare programs and central economic control by the state. In due course, the free-market programs were seen to be more effective, and voters began to back away from their socialist heritage.

There was also the idea that the Japanese, and the Americans too, were increasing their share of world economic growth and wealth and that Europe was being left behind. Europe, as a collection of separate states none of which was big enough to make much difference, could not hope to prevent the Japanese and the Americans from increasing their share further. But if they could be united, economically at least . . . ?

Of course, the EEC was supposed to provide the vehicle for unification, but so far it really hadn't. Individual countries had chafed and strained in the harnesses that efforts to forge common regulations had created. Agricultural policies were constant sources of friction between the members.

One of Margaret Thatcher's more popular moves at the beginning of her term was to bash the EEC on the grounds that Britain's financial contribution was excessive. To everyone's surprise, she was able to secure substantial refunds from the EEC as a result of her efforts. In much of Europe, the perception of the EEC was of a large, inefficient, overfed bureaucracy that provided a forum for debates but was otherwise more the subject of jokes than of admiration. Its greatest achievements, no doubt, have been the heading off of problems more than the solution of those that could not be deflected.

After nearly a decade of deregulation, Big Bang, privatizations, political changes, and renewed efforts to become competitive in the free market, Europe had become more confident in the effects of deregulation and its own collective economic future. In 1985, the EEC, fortified by appointees from conservative governments, moved forward toward greater economic integration. This was to be accomplished not by an effort to force increased amounts of "harmonizing" regulation on its members, but instead by abolishing regulation altogether, except in such essential areas as safety standards for products and prudential rules for banks. There was to be no centralized regulatory body in Brussels overseeing all of this. There was to be nothing, just the free market.

The commission affirmed that, by the end of 1992, all trade and financial barriers restricting the flow of goods and services between different countries would be removed. All surviving tariffs would be removed, goods would be able to cross borders without becoming subject to local inspections, banks would be free to set up branches in other countries, securities would be bought and sold throughout Europe under a common set of regulations. The EEC hopes within the next year or two to have passed a total of three hundred directives intended to remove all physical, fiscal, and technical barriers to completely free trade between its members; more than one hundred of these directives have already been passed. Although there are plenty of doubts as to whether the EEC will achieve all of these measures on time, everyone takes them seriously. Plans have begun to be made throughout the continent to prepare for "1992," the name now commonly given to this next great step in European economic and financial integration.[4]

• • •

If it all works out as it is supposed to, 1992 will be to Big Bang what the hydrogen bomb was to the atomic one, a manyfold increase in its explosive power.

Europe is a huge market, weakened competitively by its lack of singularity. The population of the twelve countries that now make up the EEC is about 325 million, as compared to 240 million in the United States. The combined gross national product (GNP) of the twelve countries was about $4.3 trillion, as compared to the United States' $4.7 trillion, in 1987. The growth rate of the EEC and the United States has been comparable over the past decade.

After a period of further, extensive deregulation of commercial and financial barriers, the efficiency of the market should improve, perhaps considerably. The EEC itself estimates that improved efficiencies from the removal of customs formalities and barriers affecting production, the greater competition that will result, and the increased energy that will go into the exploitation of economic opportunities will add $230 billion to the combined GNP of the EEC. Such a boost would add considerably to the political commitment to European economic unification and greater opportunities for harmonization of different national economic policies. Even without a common currency, a common central bank, or even a common language, the 1992 proposals are the stuff of big changes to come, of watersheds that could affect European economic life for several generations.[5]

None of this has been lost on large corporations and banks, who have begun to restructure their businesses to accommodate the changes in the European market that they see resulting from 1992. Many believe that the EEC will become the world's largest consumer market, displacing the United States for this distinction. Companies will be free to operate as they wish from inside the EEC, but may find it difficult to cross into it from the outside. This has alarmed many U.S. and Japanese companies that have not developed manufacturing facilities inside the EEC. Over the next few years, a substantial increase in the rate of investment in the EEC countries is expected, much of it coming from Japan and the United States. Some of the irresistible appeal of the United States as a place in which to do

247

business and to make direct investments will be lost to the EEC, creating at least potentially the danger that funds flowing into the United States from abroad will be diverted to Europe, where they will be used to enhance further the competitive potential of the EEC.

The U.S. government is concerned about 1992. There are many areas where the reforms could operate to its disadvantage, creating potentially serious trade frictions. Instead of helping to integrate the EEC into the global economy by increasing its competitiveness, it may have the opposite effect. If intra-EEC trade is to be encouraged, does that mean that extra-EEC trade is to be discouraged, at least during the "transitional" period? If both the EEC and the United States, and Japan for that matter, argue that financial activities may be permitted within their barriers by foreigners, but only if the laws of the foreigner's country permit reciprocal activity, does that mean that the United States and Japan would have to give up their banking and securities regulations? Does it mean that in order to play in the other's market, reciprocity considerations will militate toward a great convergence of financial regulation, and perhaps of laws pertaining to trade, finance, and corporate governance?

It is certainly too early to tell. There are a great many problems yet to be solved, many difficult bridges to cross. The 1992 reforms may not happen at all, or perhaps more likely, not on schedule. We have already learned, however, how powerful the influences of the global economy operating outside one's borders can be. We have already seen with our experiences in the Eurobond market, with Big Bang, and with the efforts further to deregulate financial markets in Japan and many other countries that domestic regulations cannot long resist the need to conform themselves to the practices and standards of active international markets.

Meanwhile, those most likely to be affected, the banks and industrial corporations seeking to continue doing business in the EEC, have moved ahead to prepare themselves for 1992. Many cross-border mergers, acquisitions, joint ventures, and the sale of divisions to other corporations have occurred since 1985. The French electronics giant, Compagnie Générale d'Electricité (CGE), has acquired ITT's European operations. One of CGE's rivals, Thomson, S.A., has bought not only Telefunken Electronic G.m.b.H. of West Germany, but also the consumer-electronics division of General Electric (US). Philips and Siemens have formed a $1 billion joint-research venture in semiconductors. Whirlpool Corporation of the United States acquired a majority stake in Philips's appliance business, which the two companies will now operate jointly to produce major household appliances.

Deutsche Bank has bought Bank of America's subsidiary and branches in Italy. Other European banks have acquired foreign branches of troubled American banks that have had to divest them. The Compagnie du Midi, a French insurance company, has bought Equity and Law, a U.K. insurer. A dozen or so acquisitions in the brokerage community have also occurred inside the EEC.

The long-dreaded hostile takeover has also appeared within its walls. In late 1987, Carlo Benedetti, an Italian version of Carl Icahn, began to accumulate shares in Société Générale de Belgique, the impregnable Belgian holding company for the country's largest bank, the Société Générale de Banque, and other industrial holdings. If there was ever a "Belgium Inc.," this company is it. The idea of an unfriendly effort to take over the company was thought to be absurd. Benedetti knew, however, that Belgian regulations regarding the accumulation of shares and the disclosure thereof allowed him to acquire an 18.6-percent interest secretly. He formed alliances, maneuvered, schemed, and plotted. His actions struck panic into the hearts of the Belgians, forcing them to take illegal steps to protect the company, steps that were later overturned. The episode pushed the price of Société Générale's shares through the roof. In the end the Belgians won, but it was a Pyrrhic victory at best. The largest shareholder group was a French-Belgian consortium led by the French Cie. Financière de Suez. The Belgian chief executive was replaced, and Benedetti, who retained a 16-percent stake, was made vice chairman of the company and allowed to appoint four of the company's directors (one of whom is Peter Cohen, chief executive of Shearson Lehman Hutton). Certainly Benedetti is still very much on the scene.

One consequence of Benedetti's strike was a general realization throughout Europe that hostile, that is, contested, takeovers were viable and would increase. These, of course, would have to be done in a Continental way, dissimilar from practices in London and New York.

There are plenty of potential raiders like Benedetti who know their way around the Continent. The recognition that the board of Société Générale had resorted to extreme measures to entrench themselves behind formidable Belgian barricades led to criticism of the board and a consensus that raiders had a right to make fair offers to share-holders, who should be allowed to determine the outcome of these struggles for themselves.

Following the Benedetti raid, and the deregulatory thrust of the movement toward 1992, the Amsterdam Stock Exchange relaxed the rules that were intended to prevent hostile takeovers of Dutch companies. Other countries, however, have not gone this route at all, still reflecting concern about retaining national ownership of important properties. Once efforts to acquire companies begin, however, even in the more conservative countries, global influences come into play and the outcome of these struggles cannot be taken for granted.

Securities regulations, including those protecting investors from fraudulent activity, are at present a patchwork. This is one of the areas in which a series of directives will be necessary both to dereg-ulate out of the old dissimilar regulations and to reregulate new ones. As difficult an undertaking as this may seem, it is probably no more complex a task than the acceptance of standardized regulations re-garding bank-capital adequacy, which were proposed by the Bank for International Settlements (BIS) after the work of its Basel committee and adopted during 1988. These standards were also adopted by the United States and Japan, indicating again how major regulatory changes have to encompass these countries and Europe to be workable.

There are bound to be more changes in banking regulations in the United States, Switzerland, and Japan as a result of 1992. Rec-iprocity is a big issue with the regulators. Freedom for a non-EEC company to compete in the EEC may be denied if EEC companies are not granted comparable access in the home country of the non-EEC company. Further, regulations that restrict activities in home markets such that they preclude competition on a global basis will have to be changed; otherwise home-market players will lose out. A sobering thought is that after 1992, banking regulations for all of Europe will be far less restrictive than those within the United States. The restrictions on interstate banking in the United States, scheduled for expiration in 1991 (or thereabouts), and the Glass-Steagall prohibitions on universal banking in the United States will have to go if the United States has any hope of harmonizing its banking and financial businesses with those in Europe. After a generation of unsuccessful efforts to get these laws changed, the U.S. banking community may have to thank a group of farsighted, free-market bureaucrats in Europe for the changes that will inevitably have to occur in the United States. All of this is equally true for Japan, whose banking and financial regulatory system is similar to the United States'.

So much has changed in the five years since Messrs. Parkinson and Goodison met. We have had Big Bang, and soon 1992 will arrive, both breathtaking events. We've also had the greatest period of corporate restructuring in the United States in its history, with powerful free-market forces, arising out of a period of financial deregulation, creating the liquidity and the techniques for massive takeovers and redistributions of corporate assets. Giant transactions have occurred, such as GE's purchase of RCA, that have generated requirements for billions of dollars of financing, for the sale of unwanted divisions, and for the reshaping of American industry into a leaner, more competitive posture.

*Restructuring* has become the most coveted, most criticized, and most profitable activity of investment and commercial banks. It has created demand for most of the banking industry's most important, most challenging, and most competitive activities such as merger-and-acquisition advice, leveraged buyouts, junk bonds, bridge financing, global finance, mortgage-backed finance, and market-making in the securities that result from the transactions. These services have displaced the traditional banking and investment-banking activities of deposit taking and lending, and of the buying and selling of plain vanilla stocks and bonds. The banking industry in the United States has been substantially restructured by its involvement in the greater restructuring of industry over the past twenty years or so. Similar restructuring of both European industry and banking can be expected to follow the many deregulatory changes of the past few years.

Nineteen ninety-two will almost certainly result in a huge wave of corporate restructuring crashing all over Europe. Much of this restructuring activity will be beneficial to European industry, which seeks reinvigoration similar to that which has occurred in the United States and the United Kingdom. Many of the methods and techniques used to acquire and refinance companies in the United States and the United Kingdom will, no doubt, be applied in the process, though in a different, European context. These actions surely will not always be applauded; indeed, they may be staunchly resisted, even at the regulatory level, for a time. But the influences and the forces of the larger global market will be very powerful and hard to withstand.

Financial-services providers, universal banks, and niche players alike will have to reorganize their businesses as well, to develop new market-oriented skills, and to learn how to market these skills in areas in which they do not have indigenous networks. This will take time, a lot of time, and a similar quantity of talent, most of which will have to be developed from within. Inevitably, the competitive advantages will flow to those banks that master the management and training side of the task. This is a large task for everyone, and at the outset few financial institutions have special advantages over others; each is at a relatively early stage in the globalization of its business, but its success in the next century will depend greatly on how well the firm manages the competitive globalization process over the next decade.

In 1985 fifty-four-year-old Mikhail Gorbachev became general secretary of the Soviet Union. His appearance on the scene was a surprise to everyone, as not long after his taking office extraordinary changes began to occur in that ancient, ossified, and hostile land. He apparently realized that his country had nothing but a bleak future ahead of it unless it could modernize its economy. To do so meant escaping the arms race by befriending the West. Neither the West nor many of his own colleagues trusted him at first. After four years in office, many more came to do so, including the most conservative, Russian-fearing president the United States has had in the postwar period.

As a result, after struggling with much difficulty to get the Europeans to accept intermediate range Pershing missiles in 1983, we are now abandoning them as an investment in the possibility of a new era of peace and prosperity in which all countries, including the

Soviet Union, will participate. None of this was imaginable a few years ago.

Equally, none (or very little) of the economic and financial restructuring that is occurring today was imagined before 1983.

Restructuring, *perestroika* to the Russians, is revitalizing industry in the United States and in Europe. It is also being applied to save the world's two most socialistic societies from economic ruin in the next century. In the Soviet Union, the old ways of the planned economy are being abandoned in the hope that a free marketplace can create the worker's paradise that socialism has not been able to. In China, experimentation has reached the point where ideology itself appears to have been abandoned. How else can one explain the rebirth of the Shanghai stock exchange in the world's largest communist nation?

It is ironic that the word that symbolizes the possibilities for streamlining and reinvigorating capitalist industry should be adopted by Mr. Gorbachev, the world's chief communist, for whom it means the same thing. These indeed are early days in the process, but the retreat from socialism appears to be taking place on many fronts throughout the world: in the Western democracies, in the communist world, and increasingly in the developing countries. For all the retreat signifies the end of an era of excessive nationalistic regulation and economic self-containment. What it will lead to is hard to predict, but the opportunity for the winds of free-market capitalism to blow from the Atlantic to the Urals, and possibly beyond, is one the world has never experienced before.

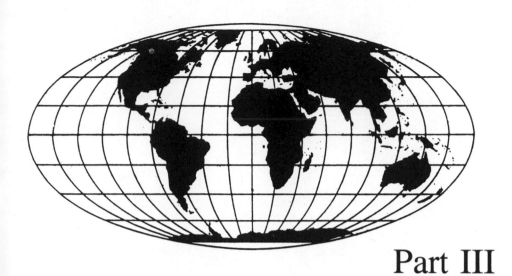

Part III

# The Floating World of Japanese Finance

# 8

# Pacific Gridlock

**R**ecently retired U.S. ambassador to Japan Michael Mansfield, a former majority leader of the Senate and someone generally acknowledged as one of our wisest old men of politics and affairs, has often described the relationship between the United States and Japan as the "world's most important bilateral relationship, bar none."

The United States has done, and still does, a lot for Japan, and Japan does a lot for it in return. Certainly our broad political relationship is almost totally harmonious. Japan supports the United States internationally, or at least does not oppose it, and it has granted political hegemony in the Pacific to the United States, despite its own rising power and influence in the region. The two countries are of great importance to each other in both national security and economic terms. We are very close allies. It is almost impossible to imagine Japan acting forcefully and independently in the world again as it did in the 1930s. Our combined strength in the Pacific represents a solid anchorage for democratic principles and an example of peaceful and prosperous governance for the rest of Asia. It is a fascinating and extraordinary alliance, perhaps made more binding because it has united former enemies. It is of invaluable importance to the people on both sides and their future generations. But it is under constant attack, not by outsiders, but by continual, increasingly bitter clashes between the two principals themselves over a crucial issue—trade.

The Japanese-American trade relationship is based on a number of deep-rooted economic, social, and political fundamentals that are extremely difficult to change. On the surface, the relationship can be affected by actual or threatened political actions and by changes in financial markets. For example, toward the end of 1985, the U.S. Congress was fairly far along in its consideration of overtly anti-Japanese protectionist legislation when the dollar-yen exchange rate reversed course and the yen strengthened by 30 percent against the dollar in a little more than three months. Apparently this was enough to persuade the Congress to set aside their trade bill for the time being.

The basic fundamentals that underlie the Japanese-American trade relationship continue, however, only partly affected by periodic changes in policies or in the value of the yen.

Japan's population may be half that of the United States, but more significantly, its people live in an area smaller than the state of California. It has the world's most homogeneous population, being even today almost 99 and $^{44}/_{100}$ percent pure Japanese. It is one of the oldest continuing civilizations in the world. In 1600, for example, before any European colony had been established in North America, the Japanese, who then numbered 10 to 15 million, were about to conclude a civil war and enter into a period of 250 years of peace and internal development. Its arts and letters were highly advanced by the year 1000. A fierce warrior class protected the country from invasion or occupation until the end of World War II. It has since forsworn war as an instrument of national policy.

Japanese concepts of duty, honor, and obligation have always been extremely high. Their moral and ethical codes, which place great emphasis on group loyalty and service, are differently directed from the Judeo-Christian ones that we take for granted. Japanese also have extremely high standards of quality, particularly for things that they personally use. Since World War II, Japan has become principally a manufacturing nation that has had to export goods in order to prosper.

Thus it has lived for at least twenty years under a dark cloud of trade frictions that, like the threat of nuclear war, tends to be discounted as people get used to it. The Japanese have succeeded in overtaking the United States in GNP per capita, but they are not rich. The rigid structure of Japanese agricultural and commercial society protects huge inefficiencies, once needed as a safety net under the economy, that drain away the value of the high incomes and savings of the Japanese. They are very proud of their achievements, but the country they most wish to be like is the United States.

America, by contrast, is a large, heterogeneous, loose society of rugged and not-so-rugged individuals. Its skilled workers are members of the prosperous middle class. Its people are free to choose how they want to live and to consume what they like. They care about government, but are difficult to govern. America's social and economic success has diluted its true grit as a manufacturing nation. It has been unable, or unwilling, to produce high-quality manufactured consumer goods at competitive prices—so it has imported these from Japan and elsewhere. We take our wealth for granted, are careless about investing it for the future, and have become the world's most ardent consumerists. As a result, America has run an increasing trade deficit with Japan for a long time.

Trade frictions between us periodically reach hysterical levels in the United States when all sorts of responsible people call for drastic action. The government sends trade delegations to Japan to threaten harsh reprisals if the Japanese don't reduce competition in the affected areas, remove "nontariff" barriers that restrict imports of our goods into Japan, and stimulate their domestic economy so their people will export less and import more.

There is an equivalent hysteria that develops in Japan, as people there resent America's bullying interference in their affairs, the unwillingness of Americans to understand what the Japanese think of as the "special circumstances" of the Japanese economy (such as its dependence on imported raw materials) that make it necessary for it to export so much to the United States, and the constant charges of unfair trade practices.

This, in a nutshell, has been our basic trade relationship with Japan. It hasn't changed very much for nearly twenty years. But as the trade deficit has increased, so have the heat, the noise, and the tensions.

A new factor, however, has emerged and changed the relationship considerably. Japan has become an international financial investor on a scale equal to that of its standing as an exporter.

A form of Japanese art that has been popular in the West for many years is the traditional woodblock print of the eighteenth and nineteenth centuries. Japanese refer to the lovely, delicate, dreamlike prints as *ukiyo-e,* "pictures of the floating world." The term has several meanings: an ephemeral world of transition and change; an illusory world of fantasy and make-believe in which things are not as they seem; and also perhaps, a notion of retreat from the hard and Spartan to the soft and hedonistic. The term applies well to today's discovery of the Japanese as international financiers.

A decade of low domestic growth, high export earnings, persistently high household savings, and progressive deregulation of financial institutions has created a Japan that is bursting with cash—and has no place to put it. Cash can move fast, if permitted to do so, and their cash has recently moved into the American economy, replenishing depleted savings, helping to fund the government's deficit, propping up bond and stock markets, and financing new manufacturing investments in this country.

The flow of Japanese savings into U.S. dollar investments has, at times, more than compensated for the outflow caused by the trade deficit, as big as it is. For the years 1982 through 1985, the investment cash inflow offset the outflow of dollars spent on Japanese goods, thus preventing the yen from rising to a level that would, more or less, stabilize the trade balance. The yen was substantially undervalued during these years, and the trade deficit increased by 1985 to more than five times what it was in 1980. In 1985, the inevitable correction occurred, and the yen strengthened back to and indeed

260

well beyond its 1980 level, substantially braking the growth in the trade deficit, which nonetheless has become exceptionally large.

Such a large deficit puts American industry under severe competitive pressure, heightening the ever-present trade frictions. However, the problem has now changed. It involves not just the Japanese addiction to American markets for their exports; it now also involves our dependence on access to Japanese savings—these dependencies have created a "gridlock" that now characterizes our relationship.

Our financial dependence on Japan limits our freedom to force changes in the trading relationship, and vice versa. Thus the gridlock is one that neither of us can escape without damaging ourselves as much as the other. It is frustrating, complex, and seemingly unchangeable, but probably for the most part it is more beneficial than harmful.

The gridlock results from a rapid acceleration of the U.S.-Japanese trade deficit during the years of fast-increasing budget deficits in the United States. Trading with the United States in those years was very easy for the Japanese, who were unwilling, or unable, to restrain themselves from allowing the trade imbalance to become acute. Their financial surpluses were redeployed by the market back to the United States, where they helped to finance the growing financial shortages. The equation was, on the whole, in balance. But the balance was only struck "on the whole," as required by economic law, not in its parts. Several U.S. industries lost market share to the Japanese; U.S. financial and investment circles became uncomfortably dependent on Japanese capital. The heat and fury grew, but so did the interdependent linkages between the two countries, hardening the gridlock.

Japan, as everyone knows, is an extraordinary economic nation. But just how extraordinary, I believe, is missed by most people. Japan entered the modern world only in the late 1860s after an American naval officer, Commodore Matthew Perry, paid an unwelcome but well-intended visit to the islands, in 1853, which for the previous two centuries had lived in a kind of splendid, feudal isolation from the rest of the world. Perry's visit sparked political changes that resulted in the overthrow of the Shogunate and the restoration of the Emperor Meiji, who led the country to open relations with the rest of the world and who tried to develop Japan along Western lines. Following Perry's visit, an American consul general, the eccentric but indefatigable Townsend Harris, was allowed into the country in 1855 to attempt to negotiate Japan's first commercial treaty with a foreign country. After some extremely baffling and frustrating years, he managed to conclude agreements in 1858. Unfortunately, he then left, and Washington, distracted by the events leading to the Civil War, neither replaced him nor put the treaty to much use. European countries, led by Britain, which clearly saw the economic potential of Japan, followed Harris's lead and took over the guiding and teaching that modernization required. One can only speculate how later events might have unfolded if the United States had more fully developed its early special relationship with Japan.

The Japanese modeled themselves on the great European powers and in a relatively short time developed substantial manufacturing and trading capabilities as well as comparable political ambitions. Japan began to think of itself as a major power that ought to be able to participate along with the others in the exploitation of China that intensified after the Boxer Rebellion of 1900. Japan wanted special trading relationships and colonies, too—even then, it was mindful of its dependence on imported raw materials.

Fewer than forty years after the Emperor Meiji was restored, Japan was able to stand up militarily against one of the great powers of the time. It defeated the entire Russian grand fleet in the stunning and rightly famous battle of Tsushima Straits in 1905, financed in part by the efforts of Jacob Schiff. This event was, in its own time, no less surprising to the rest of the world than if, in our own, Argentina had been able utterly to destroy the British naval forces that had come to the Falkland Islands to teach it a lesson.

Events evolved into the fateful developments of the 1930s and 1940s, which are well known to all. Postwar Japan was physically and financially ruined. The yen, which had been valued at about 2 to the dollar before the war, was reconstituted at 360. Everyone who had anything before the war had lost it. But the Japanese did have the MacArthur governance, which was exceptionally enlightened in many ways. Numerous reforms were made, following American practices, but the principal reorganization of the country was left to the Japanese, who were fortunate to have some exceptional leaders to accomplish it.

The period from the end of the war until the outbreak of the Korean conflict was spent trying to remove the rubble left from the war and put new structures in place. The giant prewar conglomerates, called *zaibatsu,* were broken up and new government and financial institutions were set up to marshal savings into productive investments. Some but not a great deal of economic aid was provided by the United States. The Korean War, however, provided a boom for the Japanese economy that helped it get back on its feet. Clearly the Japanese motivation to rebuild was enormous, and much of the strong societal bonds, cooperativeness, and work ethic that we see today can be traced to the harmonious adaptations the Japanese were forced to make for the first time to survive the early postwar period.

By the 1960s Japanese economic strategy had become one of export or die. Without foreign-exchange reserves and needing to import raw materials, they had no choice but to manufacture items for foreign markets, and in doing so, develop the domestic economy. Ruthless centralized controls by the government were brought into play whenever the balance of payments went soft. Credit would be curtailed, a highly meaningful action in a country with huge debt-to-equity ratios, and the economy would cool off, reducing imports. It was rough on those who were displaced, but larger companies employed people for life and didn't lay men off. Throughout the 1960s, however, intelligent economic planning, firm controls, hard work, and determination paid off. Japan's real growth rate for 1961 through 1970 averaged 11 percent. Its recovery indeed was a "miracle." In those years, however, Japan had two advantages it would not have after 1971. Its exchange rate was still fixed at the early postwar level of 360 yen to the dollar, which became a form of subsidy that made Japanese exports very price-competitive; and the relative prices of oil and other imported raw materials were low.

About this time, in late 1969, I made my first of more than fifty trips to Japan. I was accompanying two senior partners of Goldman Sachs—Charles E. Saltzman and Henry H. Fowler—as aide-de-camp. The

263

purpose of our trip was to visit top officials of government agencies, banks, and corporations to assess the business potential for the firm in Japan.

At the beginning of the year, Mr. Fowler had joined Goldman Sachs as a general partner after completing several years as U.S. secretary of the Treasury. (I knew Mr. Fowler, whom we called "Joe," quite well as he had been my father-in-law for seven years. He obviously thought sufficiently well of my employment at Goldman Sachs in 1966 to join the firm himself in January 1969, in a gesture of what Joe has always called "reverse nepotism.") Charlie Saltzman, a West Pointer who had been a Rhodes scholar, a general in the army, and under secretary of state under George Marshall, had joined the firm in the 1950s. Both Fowler and Saltzman were dedicated internationalists who insisted that the firm do more than it then was doing to develop its business in Europe and Japan.

Unrelated to their efforts, Goldman Sachs the preceding year had taken on as a trainee in corporate finance a young Japanese from Nikko Securities named Yuji Shirakawa (now a managing director of the firm in Tokyo). Yuji sat near my desk at New York headquarters and we became friends. I invited him to my home for Thanksgiving 1968, when my in-laws were also present. Yuji distinguished himself on that occasion by inquiring of my esteemed father-in-law where he worked. "In Washington," Joe replied modestly.

"Well, what do you do there?" continued brave Yuji.

"I'm secretary of the Treasury."

"Ah, so."

One day Yuji received a message from his head office, to which he had reported weekly about his education at the firm. "Goldman Sachs must become involved in Japanese business and you must persuade them to do so immediately." Yuji was distraught. Many of the major Wall Street investment banks had been active in Japan for years: Kuhn Loeb had financed the 1905 Japanese war with the Russians; antecedents of Smith Barney had been active in Tokyo since the 1920s; First Boston and Dillon Read had floated the first bond issues for the government after World War II; European bankers had been pushing hard since the mid-1960s for Eurobond business. By comparison, our experience was very modest. In 1961 we had managed an $8-million

264

issue of common stock by Honda Motor Company, which Nikko had introduced to us. The issue had been successful but difficult, and the firm had not followed up to gain additional business in Japan, which was scarce anyway because of the interest equalization tax that had been imposed by the United States in 1963. This tax was one of the United States' first postwar capital controls; it shifted most of Japanese financing business from the United States to the Eurobond market. Yuji felt that it would be impertinent for him to suggest to the august partners of Goldman Sachs that they do more than they were to develop business in Japan. So he came to me. Would I make the point to the senior people at Goldman Sachs so he wouldn't have to? My activities were entirely related to the domestic market at the time, and I had no business submitting recommendations about Japan to the partners, but knowing of and sharing the growing interest in international activities, I said I would if Yuji would give me a list of all the deals that had been done in the past several years by our competitors. He did, and I passed it on pretty much without comment. A few weeks later I was asked by one of the partners if, because of my "keen interest in Japan," I would go there and develop a plan for Goldman Sachs to become a major factor in the business of managing overseas financing for Japanese companies, spending, however, "not more than a quarter of your time" on the project as I was needed in the domestic business. I had no credentials for the job at all, but despite reservations, I agreed to do it because it was so exotic. I asked Joe Fowler and Charlie Saltzman to help with senior-level contacts, which they agreed to do as long as it was understood that making things happen in Japan was to be my job, not theirs. They knew how hard this would be, especially when the firm's commitment to Japan was to be limited to one-quarter of the time of a single third-year associate plus a little door opening on their part.

Our first trip was as fascinating as we expected. Thanks to Joe Fowler's exceptional reputation, we saw many very prominent government and business leaders. Among the more interesting personalities we met then was Takeo Fukuda, then minister of finance (later prime minister), with whom Joe had a close relationship. He often tells the story of a luncheon that he gave for Fukuda in Washington while Treasury secretary on the occasion of a visit by several Japanese cabinet ministers. The purpose of the trip had been for the Japanese to get to know their counterparts. During the day, Joe had complained to Fukuda that Japan had not delivered on agreed balance of payments "offset" measures to compensate for U.S. military expenditures in Japan. Fukuda's reply, always in Japanese through an interpreter, was evasive. Finally, at the end of the lunch, gifts were exchanged with pleasantries. The gift to Joe was an elaborately wrapped series of boxes within boxes. The final box, however, proved to be empty, much to everyone's apparent surprise. Nonplussed, Fukuda leaned over and said to Joe, in English, "Just like Japanese offset." Joe never did learn whether some young Ministry of Finance official was invited to commit *hara-kiri* for such a blunder as presenting an empty box, or whether the message was intended: no matter how pleasantly offered and well-wrapped the package, in the final analysis, there is no offset!

Our conversations with Fukuda on this trip were instructive. He repeated a refrain we were to hear often: Japan is a poor country, having to import all its raw materials and to struggle to survive in the world where so many unexpected things can happen. Its financial system had to be very rigid and could not support free-capital markets like those in the United States. Most people of Fukuda's generation, and their children, had known great hardship in their lives, and they were extremely conservative and fatalistic. They enjoyed their current prosperity but did not entirely believe in it. Something would probably come along and take it all away. Therefore they had to be prepared for the worst.

Everything in Japan then depended on how its balance of payments was running. They had then only $3 billion of foreign-exchange reserves and had to use harsh government measures to protect them. (Japan's foreign-exchange reserves at the end of 1986 were $30 billion.) This poor-mouthing of their own situation has been a standard feature of Japanese negotiations with the United States for a long time after Mr. Fukuda enlightened us with the original arguments, though they are less persuasive now than they were in 1969.

Not long after this trip, the Nixon administration sent a tough-talking trade delegation to Japan complaining about the imbalance of exports, especially in the area of textiles. This mission was not especially successful. It occurred, however, during a period of increasing tension and turmoil in the international monetary system. Cracks in the Bretton Woods Agreement of 1944, which provided for the global system of fixed exchange rates that had been in place since the war, were already visible. The dollar was under a lot of pressure. On August 15, 1971, President Nixon decided to end the convertibility of dollars into gold and, perhaps more surprising to the Japanese, to impose a 10-percent surcharge on imports into the United States. The Japanese called this event the "Nixon Shock." It was greatly objected to, but it did appear to give bargaining power to Nixon, who used it to bring about the Smithsonian agreement in December 1971, which provided for a general realignment of currencies. This dam, like others we have discussed, could not hold either, and the pressure on even these new alignments became so great that the Smithsonian accord was swept away by 1973, when floating exchange rates were adopted.

Just as the Japanese began to accept the new exchange rate system that was imposed on them by the U.S. decision to float the dollar, and, in some quarters, to start feeling like a fairly prosperous nation again, the "oil shock" of 1973 occurred, slamming them back into high inflation, high interest rates, and the need for great sacrifices. It is hard to feel secure when you import 90 percent of your oil and the price suddenly quintuples. The Japanese worked extremely hard and stabilized their economy fairly quickly, but they were reminded once again that for a country with a high import bill, exports were essential.

In the decade following the introduction of floating exchange rates and the oil shock, the Japanese growth rate declined to 5.3 percent, or about half the rate of the preceding ten years. A rising level of exports made up a significant part of this growth: the domestic sector of the Japanese economy only grew at about 3 percent, hardly a miraculous pace. Capital expenditures were not required at the same rate as in the 1960s. Consumers were cautious and preferred to save rather than spend their money. Their anxieties had been heightened by the recognition that, in addition to great earthquakes, fires, and typhoons, they were vulnerable to things like oil shocks. Most Japanese are conservative with their money anyway. They live in a country in which land and housing prices are sky-high, social welfare and pension systems are much less well developed than they are in the United States, and credit is generally harder to come by for those who want it. They save because they feel they have to, whether they want to or not. The household savings rate in Japan is about 18 percent versus less than 5 percent for the United States.

This propensity on the part of the consumer to save is a problem for the Japanese government when it wants to take fiscal steps to stimulate the economy. It is difficult, through government spending, to induce consumer demand that will then pull through the economy. The traditional practice in Japan is for the government to try to increase domestic growth by increasing spending for public works and other projects; this helps some but doesn't provide nearly the same benefits of the multiplier effect that we are used to. Over a period of time, inefficient efforts by the government to stimulate the domestic economy through spending led to the accumulation of large fiscal deficits without a commensurate increase in domestic growth. Indeed, the Japanese government deficit in 1985 was 25 percent of the total budget, larger than the U.S. deficit, which was 21 percent of its budget then. Both countries have since reduced these percentages somewhat. The Japanese *cumulative deficit* (national debt), in relative terms, however, is significantly larger than ours. The Japanese worry about the size of their deficit, just as we do, and most politicians and Ministry of Finance officials feel that the deficit should not get any bigger and indeed ought to be eliminated by 1990. Feeling as they do, it is not surprising that the U.S. government encounters resistance to the suggestion it periodically makes that the Japanese try to stimulate domestic growth through further government fiscal actions.

Despite Japan's development of a strong export bias in the years following the war, Japan's exports today do not constitute that large a portion of its total GNP—only about 10 percent in 1987, which is almost the same as the 11 percent for 1970, but still a considerably lower percentage than for, say, Germany, France, Britain, and Italy, each of which exports more than 24 percent of its GNP. The United States exported only 7.5 percent of its GNP in 1984, but this rose to 10 percent in 1987 as a result of the weaker dollar. The problem of course is that almost all Japanese exports are manufactured goods. In a country with an export bias and little else to export but manufactured goods, it is not surprising that this concentration developed. In managing their exports, the Japanese have shown unusual sophistication and intelligence. Their basic policy has been to secure export markets by manufacturing high-volume, high-value-added goods, such as consumer products, at as low a cost and with the highest quality that they could manage.

One of the first tasks the Japanese undertook in the 1950s and 1960s was to learn how to manufacture things efficiently, through good process engineering, quality control, and management of production costs. To make consumer goods, companies do not have to concentrate on pure science and R&D, so theirs didn't. Their engineering energies were spent on the manufacturing process itself, and on product design. The technology for this was not hard to come by or to improve if one spent the time on it. It was not hard for a country with a high literacy rate, a good technical-education system, and a history of superior engineering skills going back at least to 1905, to master the science of manufacturing.

Quality control was an early concern of the Japanese, as the Japanese consumer was and is unusually quality-minded, and a high manufacturing reject rate was costly and wasteful of precious raw materials. Ironically, the Japanese learned their modern quality-control procedures from an American engineer, W. Edwards Deming, of New York University, whom they hired as a consultant after the war. With a highly disciplined and dedicated work force, they were able to apply and perfect these early lessons very successfully.

Japanese companies became skillful at managing production costs by employing some unique ideas. They would seek economies of scale by manufacturing larger quantities of goods than they knew they could sell in Japan. They would estimate worldwide demand for each item, and the share of the world market they would target for themselves, and then turn the ensuing output over to their domestic and overseas sales departments or to the giant trading companies whose job it was to distribute goods. As long as they could sell them, on the whole, for more than cost, they were ahead. The emphasis was on building up sales volume—through the stimulation of primary demand for their products and through seeking an increasing share of market. If they did this well, profits would take care of themselves. Building up sales volume and market penetration were more important than current profits. Companies grew, many new jobs were created, the greater public was served, and prestige enhanced. These remain important values to the Japanese company, but were especially so in the postwar recovery period.

I gained a very useful appreciation of how Japanese manufacturers regard market share for their products from a conversation with Ryuzaburo Kaku, president of Canon, Inc., for whom Goldman Sachs was doing an issue in the United States in 1979. We were discussing the company's pricing for a new 35mm camera that Canon had introduced into world markets. The camera was an excellent one, reflecting many new technical qualities, and it was selling well. However, its retail price was set quite low, and I asked Mr. Kaku when they planned to raise the price to increase their profits. Demand for the camera was so strong that dealers were often out of stock. Kaku's response was that they had no plans to raise the price. By keeping it low, they would increase their world market share and keep competitors out. They would just have to work harder at increasing production and at lowering their manufacturing costs, both of which were things they were good at and could control. The strategy worked; they secured the leading market share for that type of camera and the profits, after a while, grew to be considerable.

On the whole, the strategy of worldwide market development has worked very well for the Japanese. However, more than a good strategy was needed to produce the results that were obtained. Japanese industry also had some help from the government and from their financial system. Most of this help has been much more indirect than the theory of "Japan, Inc." would imply, but it was certainly useful. Japan's centralized economic system is very effective. There is a national industrial planning effort organized by the government to keep Japan's manufactured goods moving up the technology curve. Representatives of industry participate in this planning process, the results of which are well known and reasonably consistent. Which industries are to be fed with resources and which cut back are identified in advance. For years, restrictions (now much liberalized) were placed

270

on foreign competitors' manufacturing in Japan. Vigorous competition among Japanese companies was encouraged. Social-welfare programs provided by the government were few, so employees looked to their companies for their long-term welfare and, in exchange for long-term employment, were willing to accept comparatively low wages and extend their loyalties to employers.

It was not, however, only the large labor-intensive companies who prospered in the 1950s and 1960s. There were a number of start-up companies led by highly entrepreneurial individuals who exposed themselves to great risk and hardship in the early days of their companies' history, but who succeeded brilliantly. Many Americans are familiar with such postwar stars as Sony, Honda Motor, Kyocera, and others. Two of my favorites are Ito Yokado and Wacoal.

Masatoshi Ito put together Japan's most successful retailing company in 1948. In 1961 he constructed the first of many large "superstores" that are located in the suburbs of Tokyo. The stores sold food, soft goods, and appliances and other "white" goods. Mr. Ito admired American retailing companies, and though he speaks no English, he studied them very carefully, visiting hundreds of stores in the United States. He became an expert in U.S. retailing, merchandising, and methods used to control costs, inventories, and suppliers. In 1973 he acquired the Japanese franchise for 7-Eleven Stores from the Southland Corporation. By the end of 1987, 7-Eleven Japan had three thousand stores in Japan, generating revenues of $3.6 billion. Mr. Ito also acquired the Japanese franchise for Denny's restaurants and built a chain of low-cost, family-style suburban restaurants in Japan. Mr. Ito, now a very wealthy man in his early sixties, continues to devote his time and considerable energy to new retailing ideas.

In 1945, Koichi Tsukamoto was a young soldier in Burma. His unit was attacked and wiped out, though he managed to escape. Somehow he made his way back to Japan. When the war ended, Tsukamoto believed he had been granted a "second life." He decided to found a company that made ladies' undergarments, though at the time, Japanese women were too poor to afford Western-style clothing and the undergarments these required. Most Japanese women didn't wear Western clothing and never had, but in any case, were too slender to need support garments, so the manufacturing of brassieres and girdles seemed a fairly pointless exercise. Tsukamoto ignored such facts. He knew that Japan too had been granted a "second life" in which it would adapt fully to Western ways and fashions. Japanese women would assume Western habits of dress, but to be absolutely sure they looked right in the new styles, they would insist on all the correct undergarments. He prepared a fifty-year plan for his company. Its goal was to become the world's largest and most respected manufacturer of high-quality ladies' undergarments, with its products being sold in the most elegant stores in New York, Paris, and London. Today, Tsukamoto, a vigorous, handsome man in his late sixties, with about ten years to go on his fifty-year plan, can take comfort in knowing that he has achieved all the original objectives he set out for Wacoal in 1950.

The manufacturers and retailers were not alone in contributing to Japan's miraculous recovery. Much help was also provided by its remarkable financial system. In the early days after the war, and for some time thereafter, Japanese industry was short of capital. This was dealt with in several ingenious ways. First, the banking system was set up under very tight controls by the Bank of Japan and the Ministry of Finance. The banks, both commercial banks and the long-term credit banks, were designed to be highly leveraged themselves and to extend as much credit to manufacturing companies as they could. Everything was very highly geared, but bank supervision was extremely tight. The larger companies tended to belong to groups, some based on the old *zaibatsu,* in which cross-shareholdings and supplier-customer relationships were encouraged to provide a group economic support system. If a member company got into trouble, the group, together with its bank, would be called in to rescue it, with the government watching closely from the sidelines. Because there was some assurance that the company would survive no matter what, larger enterprises were free to take greater risks with new products and capital investment than their American counterparts and their stockholders might have thought prudent. The ability to take these risks has greatly enhanced the competitiveness of Japanese manufacturers over the past twenty years.

The Japanese also rebuilt their capital markets in the 1950s. A corporate bond market, which required all debt to be secured by mortgages on plant and equipment, developed to some extent, but not as much as the unsecured debenture market, which enabled the powerful long-term banks to obtain funding for medium- and long-term industrial loans, which they made to their corporate clients.

The stock market rejuvenated very quickly. The principal stockholders of companies after the war were other group members and allied financial institutions. Their holdings were never traded, but added to periodically as companies issued new shares through rights offerings. Most of the trading in stocks was done by individuals, who paid no capital gains taxes and who had no other place to invest small amounts of money in speculative investments. The retail-brokerage industry developed rapidly to supply the infrastructure for share dealings.

Throughout the 1960s and 1970s, Japanese companies also issued large quantities of securities abroad, predominantly in the Eurobond markets, and foreign investors became large holders of Japanese stocks. The growth in the Japanese economy and the influx of foreign funds into the stock market, together with increased investment by Japanese individuals and corporations, pushed stock prices steadily upward over quite a long period. The Nikkei stock average was more than four times, at the end of 1987 (eight times in U.S. dollar terms), what it was at the beginning of 1974. The American Dow Jones at the same time was only a little more than three times what it was at the beginning of 1974.

For most of the past decade, Japanese interest rates have been relatively low, reflecting the low inflation rate, its trade balance, and government regulation. For companies operating with large amounts of leverage, low interest rates, and high price-earnings (P/E) ratios, the cost of capital is very low. For example, in the United States a company with A-rated debt outstanding might have 40 percent of its capital structure in the form of borrowings, for which it has been paying about 6 percent after-tax on average for several years. Sixty percent of the company's capital is represented by equity, which has traded in the market at an average P/E ratio of, say, 12. One simple indication of a company's cost-of-equity capital is obtained by taking the inverse of the P/E ratio (or earnings as a return on investment as measured by current market value), or 8 percent in this case. In this example, the weighted debt and equity after tax cost of capital for the American company would be 7.4 percent. Using the same method to calculate the cost of capital of a comparable Japanese company, which would instead have a debt-to-total-capital ratio of, say, 60 percent and an average P/E ratio of about 36, you get a weighted cost of capital of about 3 percent.

This lower cost of capital makes a difference, a large one, in the comparative manufacturing costs for Japanese companies as contrasted with their Western competitors. The Japanese use this advantage well, as they constantly invest in new capital equipment, which they depreciate rapidly, for both expansion and for improving their manufacturing efficiency. As a result, Japanese plants and equipment are among the youngest in the world and their productivity among the highest. Further, their propensity to invest in plant improvements encourages them to lead the way into totally new manufacturing technologies, like robotics, which can produce quantum leaps in productivity improvements for the future.

An impressive demonstration of Japanese manufacturing improvements in the electronics industry can be seen in a display in the waiting room at the offices of Sanyo Electric Company in Osaka. Here, lined up on several long shelves, are television chassis manufactured by Sanyo in each of the last fifteen or so years. The change is very noticeable. The size (and the cost to manufacture) of the chassis has shrunk appreciably, year by year, to its present petite form. Each annual model represented a substantial change and, therefore, investment in new tooling and machinery. Annual investment in each year's totally new product had become routine. Sanyo, of course, as a large Japanese electronics firm, makes thousands of products, most of which are subject to this process of annual renewal.

At the Smithsonian conference in December 1971, the yen was realigned when the dollar was; it went from 360 yen to the dollar to 303. In 1973, along with all other currencies, the yen began to float. The market would set the exchange rate. From then on, imbalances in trading or capital flows would be adjusted by changing exchange rates that would reflect market forces. Capital and foreign-exchange

274

controls in place around the world would no longer be necessary. They began to be dismantled. Even in Japan, which had the most severe controls of any developed nation, it became permissible for capital to flow into the country and out of it at the same time. A two-way street for financial transactions was allowed. Not only could Japanese corporations borrow in the Eurobond market if they wanted to, and were authorized to do so, but also foreigners, when authorized, could finance in Japan.

About this time I made my first long journey into the illusory "floating world" of Japanese finance, in which very little is as it seems. The occasion was a financing for Ford Motor Company, a long-standing and important client of Goldman Sachs. The episode began when my friend Yuji, now returned to Tokyo, told me in October 1972 that things were changing in Japan and that it was now possible to do a financing for a major corporation as a placement made directly with large Japanese institutional investors. Some European governments had borrowed in this way, but not yet a major U.S. corporation. "How about Ford?" said Yuji. "It is well known and admired in Japan and would be an ideal borrower to take to the Ministry of Finance [MOF]," he continued, revealing why he had tracked me down. Yuji was trying to kill two birds with one stone. The MOF would appreciate Nikko for bringing in a major multinational corporation to demonstrate the new openness of the Japanese capital markets; if Yuji could persuade the MOF to grant suitably attractive terms to Ford, then Yuji, through me, could gain Ford as a client for Nikko, an acquisition of enormous prestige for Nikko and for Yuji. What made this chicken-and-egg thing possible was the fact that Yuji knew me. I could deliver the message to Ford and persuade it to go along with Yuji. Simple.

"What do you have in mind?" I asked.

"Why not offer Ford $60 million of convertible debentures to be placed with Japanese banks. The terms would be very competitive."

This didn't make sense. Existing regulations limited placements of this type to $30 million, and although low-interest-rate debentures convertible into common stock of Ford might be attractive to long-

275

term equity investors like insurance companies, they were not the sort of security that a bank would ever buy for its own account. Even if the banks would buy it for some reason, the extremely conservative banking bureau of the MOF would never allow such a long-term investment to be made if it had to be financed with short-term Euro-dollar deposits. Banks should not borrow short to invest long, especially with foreigners. I made these points to Yuji.

"Not important," he said. "My friend at the MOF wants this deal to go ahead," now revealing that he already had discussed a possible Ford deal with the MOF. "It will make the MOF look good for opening the market. Trust me." Timing was everything. The MOF had changed an important policy, signifying a more liberal approach to capital-market regulation, and the MOF official in charge (Yuji's "friend," whom he had probably known for all of about six weeks) now wanted some international beauty to take advantage of it. The MOF was probably fishing for a deal, with Yuji and with all the other securities firms.

With grave reservations, I mentioned the matter to John Sagan, Ford's treasurer. Sagan thought that the Japanese policy was sensible and that Ford ought to try to tap into Japanese money sources, but he was clearly skeptical of the deal. I indicated to Yuji that we would need very aggressive terms to "induce" Ford to come to the market in Japan, much better terms than they could get in the United States or in Europe. Also, he would have to come to Dearborn himself, with a senior colleague, to explain why Ford could rely on Nikko's extraordinary offer. He and his senior colleague came. They convinced Sagan and his people that they should give it a try. The Ford board of directors was asked to approve the deal in principle. It did.

We set off to Japan to prepare the papers. Sagan was to follow later for a high-level protocol visit. This was Nikko's first opportunity to lead an issue for such a world heavyweight. They went nuts. Nothing was too much for Sagan: a lavish dinner at Shinkiraku, Tokyo's most famous and luxurious geisha house, a weekend trip to Kyoto for sightseeing, invitations to play golf, and so on. At one point, I was called by a young Nikko aide who asked me what sort of decor Mr. Sagan had in his home. Naturally, I had no idea, but cautiously asked why he wanted to know. "We are planning to give him a suitable gift while he is in Japan, and we thought we would give him something memorable for his home to demonstrate the close relationship between Ford and Nikko Securities.''

"What are you going to give him," I asked, "a sofa?"

A pause. "Yes. Well, we were thinking of a small piece of furniture, but not a sofa. However, in view of your recommendation, I am sure we can change to a sofa. Will he want to take it with him or have it sent?" It took days to persuade them that I had been kidding and that, in any event, a smaller gift, like a Japanese camera, would be better.

Just before Sagan arrived to bless the transaction, my colleague, Don Gant, the Goldman Sachs partner responsible for the Ford account, and I went to dinner with Yuji and his friend, the MOF official responsible for the deal. The official acted as host, indicating where we should sit and then ordering an incredibly sumptuous meal of Japanese and Western delicacies (sushi and steak) with a suitable accompaniment of whiskey, beer, wines, saké, and brandy. Yuji, who was paying for the dinner, sat still and let our "host" enjoy himself. We hardly discussed the Ford deal at all. The simple fact that we were dining together and having a good time was enough to confirm that we should just sit back and leave the driving to them, the MOF guy and Yuji. Later we were taken by our "host" to a giant Japanese nightclub where we enjoyed the floor show until the evening's end, a further good sign.

Subsequently, I returned to New York to finish up the legalities, which included having two separate $30-million transactions at the same terms occur on successive days to comply with the MOF's rule limiting individual issues to $30 million. Where there's a will, there's a way.

Don, meanwhile, stayed in Japan to look after the marketing of the deal. As far as Don could see, nothing was happening. He pressed. Finally a day or two before the commitment date, he insisted on a full status report.

"How many of the debentures have been placed?" he asked. There was squirming and the sound of air being sucked in over the teeth.

"It is very difficult to say" was the gist of the reply.

277

Don called me on the telephone. I was feeling good. My stock was rising. Doing a highly visible financing for Ford in Japan would be a real feather in my cap.

"How's it going out there?" I asked Don.

"Pretty good," said Don, "except these bastards haven't sold one goddamn bond."

I am sure I turned white. In a few hours, I was to report to Gus Levy, our formidable senior partner, and John Sagan on the status of the deal. If this deal didn't go through, I would be dead. I called Yuji, who laughed nervously and said I shouldn't worry because "It was now in the hands of fate." I went to my last resort: a telex from Joe Fowler to Mr. Fukuda's successor as Minister of Finance, the heads of Nikko Securities, and the banks who were supposed to be investing in the deal. The telex was very strong, implying that the Japanese had acted in bad faith and had hurt and embarrassed our client and ourselves in luring us into a transaction they couldn't deliver.

The telex was dispatched about 5:00 P.M. New York time (7:00 A.M. the next day in Tokyo). I had to attend an unrelated dinner that night and was surprised at the restaurant at about 8:00 P.M. by a call from Tokyo. It was Yuji's boss, Yasuo Kanzaki. "Whatever you do, do not send any more telexes. This one has been like dropping a 'stinking bomb' down the chimney at the MOF. All of our senior people have been ordered to the MOF to explain your reckless action." He said Yuji would call back later. By the next morning, Don Gant had reported that the deal had been fully subscribed and we could go ahead. Then Yuji called to say that the telex had really not been necessary as the deal was beginning to come together. Our MOF friend, after explaining what was happening to the Minister of Finance, had taken certain matters into his own hands and that tended to "clarify" the situation.

The deal was to be agreed upon immediately, but it would not be finally signed until an appropriate ceremony could be arranged in Tokyo. That would be on December 23. Sagan would attend. So would I. I left New York on the morning of December 21 on the eighteen-hour flight that refueled in Anchorage. On arrival at the airport on the afternoon of December 22, I was met by a very long-faced Yuji. "We have problems," he said.

The Japanese lawyer representing the underwriters (us) was refusing to grant his required legal opinion on a technicality. No one else, including the MOF, had a problem with this technicality. Only our lawyer. We went to my small hotel room to discuss it, Yuji, the lawyer, three other guys, and myself. I never understood exactly what the technicality was or why it should interfere with the closing. I ordered sandwiches and beer and we resolved to battle it out all night. Our lawyer was having a wonderful time. So, it seemed, were the others. I was not. I had not been to bed for more than thirty hours, my room was a mess, and most of their conversation was now in Japanese. At 2:00 A.M. I told them all to get out and to solve the problem somewhere else. The next morning the closing occurred without a hitch. There was no mention of the technicality or its remedy, if any. After the picture-taking and the little speeches, I went back to the airport and departed the floating world on the return flight to New York, where I arrived on the evening of December 23.

Yuji later explained that the reason the deal had not been signed up when it was supposed to have been was that the banks, who wanted to be in the deal only for the prestige of a tie to Ford and for doing the first deal of this type, could not agree with each other on how much each should take and who should appear to be in the lead. They were all hanging back to let someone else make the first move, which they could then top if they wanted to do so. They would have come together in time, Yuji said.

The investment itself made no immediate economic sense for the banks, which had to fund at something like 6 percent an investment with an interest rate of 4.75 percent that was convertible into Ford's common stock at $81.90 per share. The stock price was then $70 or so. After the rise in oil prices, Ford's fortunes declined for several years. It was not until the early 1980s that Ford's stock, after adjusting for splits, reached $81.90. The fifteen-year debentures matured in 1987. Nevertheless, several of the banks, particularly Sumitomo Bank, took a much longer term view of the investment. They used it as a way to get much closer to Ford, which, as we have seen, later became a major investor in Mazda Motors, an important Sumitomo client.

After the oil shock of 1973, Japan's growth rate slowed, but its export industries took up the slack. It began to run balance-of-payments surpluses, so the yen continued to strengthen. At the end of 1980, it stood at 202.

There was a growing Japanese trade surplus with the United States, but several factors seemed to keep it in proportion. The magnitude was manageable—in 1980, for example, the Japanese trade surplus with the United States was $9 billion; Japanese exports were concentrated on particular industries, but the concentration moved from textiles to steel to electronics to automobiles in the period from 1970 to 1980 and was then heading toward semiconductors and computers. There were plenty of complaints from the affected industries, but mostly these were dealt with by temporary quota systems or by administrative tribunals that were set up to adjudicate dumping and other unfair trade charges. Also, Japan was a big buyer of commodities from the United States, which helped to offset the surplus, although it was stunned by a Nixon embargo on soybeans and other grain exports in 1973 (the "soybean" shock) and afterward began to look elsewhere for more reliable suppliers. Mainly, however, the U.S. trade deficit with Japan, though objectionable, didn't matter that much because the value of the yen was reflecting it and overall stabilizing influences were in play; U.S. surpluses with the developing world offset the deficit with Japan. The worldwide trade deficit of the United States in 1980 was $36.4 billion.

All of this began to change in 1981, when a new president, with a strong popular mandate and a different set of economic policies, came into office. In the ensuing eight years, these policies have had many effects—some salubrious, some not—but nothing has affected the magnitude of the numbers that define the Japanese-American economic and financial relationship as much as Reaganomics has. Whatever trendline the relationship between our two countries was on in 1980, the new policies were like a shot of adrenaline to it. In some respects, the relationship has become distorted, but it has also resulted in the entanglement that I think of as the "gridlock." The adrenaline charge has now been spent, and market forces (especially in the foreign-exchange market) have counteracted its effects, but the adrenaline accelerated the emergence on the international stage of Japan as an investor on a massive scale and this has forged a permanent change in the nature of our relationship.

When President Reagan came into office, he had three principal goals: to reduce taxes so as to liberate the entrepreneurial spirit of America; to reduce the role of government as an expensive impediment to economic progress and personal freedom; and to reverse the decline that the United States had fallen into in international affairs, in particular, strengthening its national defenses vis-à-vis the Soviet Union. For the most part, Mr. Reagan's popularity reflected the appeal of these fairly basic policies. To implement these policies, however, he required dramatic action: a large cut in federal spending; a big increase in defense spending; and a large tax reduction despite the fact that the budget would be thrown into the largest peacetime deficit ever. The new deficit provided a lot of stimulus to the economy, which reversed the slump of 1982 and grew during 1983 through 1985 at an average annual rate of 4.3 percent, far more than during any recent three-year period. Inflation, however, didn't rise; in fact, it declined sharply. This was partly because of the exceptionally high real-interest rates in the economy that resulted from the deficit. The financial markets refused to provide long-term money to a system with such a high deficit except at interest rates that reflected a substantial premium over the rate of inflation that the market rather cynically expected the deficit to produce in the long run.

Such high interest rates, coupled with strong domestic growth and lowering inflation, appealed greatly to the foreign-exchange market, which on the whole endorsed President Reagan's three policies too. The dollar strengthened, foreign investment poured into the United States, and financing the deficit wasn't too hard with so much money coming in from abroad. Corporate profits improved; Detroit turned around; business really picked up. The securities markets bottomed out in 1982 and began a five-year bull market. This helped to suck in even more foreign capital and boosted the dollar further. Reversing directions sharply, the yen *fell* from 202 in 1980 to 235 in 1982, to

281

a low of 260 before the end of 1985. The Japanese trade surplus with the United States exploded: the cheaper yen and the growing U.S. economy made exporting to the United States extremely easy. It was also necessary; Japanese domestic growth was down to about 2 percent in 1984, and other markets for Japanese exports were doing even worse. The Japanese surplus with the United States became $19 billion in 1982; $34 billion in 1984; and $50 billion in 1985.

Another change also took place: Japan's great surplus of savings began to flow into the United States, ultimately in such a large quantity as virtually to offset the pressure in the foreign-exchange markets caused by the extraordinary trade surplus. For years, Japanese domestic savings had been accumulating. Now the foreign-exchange proceeds of the surging export business were added to them. Japan did not need to retain the foreign-exchange earnings as reserves; it already had an excess. If the Japanese were to promote private sector recycling of the funds back to the United States, the yen would remain at its comparatively low level, ensuring that exports would continue, and this would prevent the Japanese domestic economy from sinking into a recession.

A recession in Japan would be very difficult socially and politically. There was no further room to increase Japan's fiscal deficit, already overextended in the opinion of many; besides, the more evidence of a slowdown in Japan, the more the cautious Japanese would increase their savings to provide a cushion against the inevitable layoff or wage reduction that was bound to hit the family someplace. True, Americans complained that the export surplus could not be tolerated forever and that Congress would retaliate with protectionist measures if Japan didn't do something. But what could the Japanese do? Open up domestic markets for foreign goods? They were already open, but the foreigners didn't understand that in the slowly growing Japanese domestic economy, consumers weren't buying much, and what they did buy was sold on the basis of intense competition between strong, efficient Japanese companies with long-established, loyal customers. A foreign company might have no more success in the Japanese consumer products market than would, say, a well-known Japanese company like House Food Industrial in competing against Procter & Gamble, Beatrice, or Philip Morris in the United States.

Besides, the quantities involved are enormous. Americans export airplanes, military equipment, computers, and agricultural commodities to Japan. In 1985 these exports totaled about $15 billion. Japan had already bought as much of these as it could, and there was still a $50-billion difference. That's about $1,000 per Japanese family, far more than they could be expected to consume in additional purchases of anything, not to mention grapefruit and hamburgers. Opening up Japanese markets for more U.S. exports of consumer products simply will not amount to more than a drop in the bucket in relation to the overall problem.

The Americans were still not satisfied; they wanted some important concessions. "Well," they said, "if you can't do much to open up your commercial markets, how about your financial markets?"

"All right," said the Japanese, "if that's what you want."

Financial deregulation, which had been occurring in Japan for the past ten years, accelerated. All sorts of new foreign securities were created that could be sold to Japanese institutional and individual investors. *Samurai bonds*—yen-denominated issues in Tokyo by foreign governments and a few corporations—were joined by *Sushi bonds*—nonyen issues in Europe by Japanese companies sold back into Japan—and *Shogun bonds*—nonyen issues sold in Japan by foreigners. They also were encouraged to purchase U.S. Treasuries and government agency securities, corporate and other Eurobonds, and to increase their rather modest holdings of U.S. equities. Japanese life insurance and trust companies were urged to learn how to manage foreign portfolios and to take advantage of the rule that permitted up to 10 percent of net assets to be invested abroad. They could form joint ventures with foreign portfolio managers to manage these investments if they wanted to.

The capital outflow from Japan that resulted was incredible. The pressure to deregulate had uncorked a deluge. In 1980, the Japanese net capital account showed an inflow of $2.3 billion; in 1983 it was an outflow of $18 billion; by 1987 the outflow had reached $137 billion.

All of these activities sent the floating world into overtime. Many puzzling situations similar to our Ford transactions were experienced by foreign investment bankers used to a strict economic rationale for everything. At Goldman Sachs we had many colorful adventures in assisting in the first Samurai bond issue for a U.S. corporation, Sears Roebuck and Company, in 1979; in several zero-coupon and other Eurobond placements in Japan by U.S. corporations such as General Electric from 1982 to 1985; and in connection with the sale of equity securities of Rockefeller Properties, Inc. in 1985. Knowing what Japanese investors were doing, and why, became extremely important. No major financing could be structured without consideration of how the Japanese would fit in. Japan had become the world's newest and hottest source of capital since OPEC. International securities firms flocked to Japan to open representative offices, branches, trading desks, and finally, in 1986, to become members of the Tokyo Stock Exchange. By the end of 1985, the Japanese owned almost $30 billion of U.S. government securities, and their appetite or lack of it was crucial to each new Treasury auction. The investment flow out of Japan was predominantly, but not exclusively, limited to portfolio investment. Direct investment, especially in the United States, also surged as Japanese banks increased their U.S. assets, new manufacturing facilities were built, and marketing and distribution investments in the United States were expanded. Japanese companies headquartered in Tokyo found, in an increasing number of cases, such as those of the large trading companies, that the largest part of their business on a consolidated basis was centered, if not actually located, in the United States.

During 1985, however, the rocket began to level off and slowly begin its descent to Earth. The dollar turned and began a decline that accelerated rapidly. From its high of 260 to the yen, the dollar fell to 125 by late 1987, from which level it subsequently bounced back a bit. A 50-percent change in the value of the currencies had occurred in the very short time of two years. The rate of increase in the trade deficit, which had reached $50 billion in 1985, slowed dramatically. In 1986 the deficit grew only marginally, to $60 billion. In 1987 it declined to $52.1 billion and in 1988 to $47.5 billion, though by early 1989 Japanese exports had begun to increase again. The deficit was still very large, however, and it had not responded all that much to the extraordinary adjustment in the exchange rate, but at least it was a change in the right direction. Trade shifts, as all economists who can describe the "J curve" know quite well, take some time to overcome the momentum of existing customer orders and price levels. Before long, some economists were predicting, the trade deficit should show a further $10- to $20-billion improvement because of the residual exchange-rate change.

The Japanese are aware of the J curve too and have been adjusting their businesses accordingly. They have increased direct manufacturing investments in the United States; they have increased their purchase of component parts from low-cost sources in Asia and elsewhere; and they have pushed much of the foreign-exchange adjustment burden onto their suppliers. They have tried to squeeze more cost out of their manufacturing operations, and they have begun to do the one thing no one ever thought they would, to reorient more of their production to their own domestic market.

For years observers of the Japanese have wondered how Japan's consumers could be motivated to buy more of the goods U.S. and Japanese companies make. They had plenty of funds with which to do so, as a result of their higher wages and savings, and at present exchange rates, large price differences between U.S. and Japanese goods ought to be able to stimulate a lot of new consumer demand for U.S. imports. But even if the Japanese preferred Japanese goods, no matter what the price, a booming consumer market would lower the amount of goods available for export. All that has to be done, these observers say, is for the Japanese to arrange for 5- or 6-percent domestic growth, and the problems of trade imbalances will start to fade away.

This idea has been near the heart of U.S.-Japanese trade negotiations for many years. The United States usually asserts the position and then demands that the Japanese take immediate steps to implement such a plan. If they don't, Congress will retaliate, the negotiators say. The Japanese reply that they too think it would be a good idea, but Japan is different and it is very difficult to motivate the conservative, savings-minded Japanese. Besides, they might add, where would they put the new items they had purchased. Japanese houses are so small. . . .

They also sometimes add how difficult it would be to raise funds for government stimulation efforts in view of the Diet's concerns about the government's fiscal deficit.

They most likely would also add that even if they could stimulate growth a little, the benefits would not necessarily show up in consumer spending. Goods are very expensive in Japan, because of old-fashioned distribution methods, the need to preserve harmony with suppliers, competitors, and middlemen, and so on. So in short, forget it.

Every so often, when the temperature is high, and the flack is flying, the Japanese will come up with an ad hoc mission of high-level businessmen to tour the world to find ways to promote imports, or appoint a prominent individual to head a study effort on some kind of major direction shift by the economy as a whole. One of these efforts some years ago was headed by Prime Minister Tanaka himself, perhaps one of the most powerful postwar prime ministers (until he lost his office because of his involvement in the Lockheed bribes scandal, a predecessor by a few years of the recent Recruit bribes scandal). Tanaka recommended that Japan do something to break up the overcrowded, overexpensive Tokyo-Osaka corridor and make it possible for growth to occur elsewhere in the country. Even he was ignored.

Normally, Japanese special trade missions or special study efforts are treated the same as our blue ribbon presidential commissions, which are appointed to make it look as if something is happening, but whose reports are usually buried and forgotten even before they are received. The process may seem wasteful, but it is part of the way we do things in democracies.

In 1984, Prime Minister Nakasone, a Tanaka disciple who like his master was a powerful leader in Japan, at least as these things are judged by Westerners, in order to convince his friend Ron [Reagan] that he meant business, appointed a high-level group to make a special study of the future of the Japanese economy. This group was headed by a distinguished civil servant, Haruo Maekawa, until recently governor of the Bank of Japan. It was to study how the Japanese, still living in the Tokyo-Osaka corridor, could this time free themselves from their excessive savings habits and consume more so the domestic economy could grow faster.

Maekawa and his colleagues studied the problem and made their report. It had lots of hard-hitting stuff in it, almost all of which was eliminated during the political and ministerial review of the report. Nakasone was furious at the deletions and ordered the report rewritten. It was still unable to escape emasculation, however. Published at last, it received much praise in Washington, but very little in Japan, where everyone knew it was either a pro forma exercise only and thus could be disregarded, or an unapproved effort by Maekawa and Nakasone to rock the boat, something their political masters would not allow to happen.[1] A year after the report, which offered a very general outline for a five-year plan to develop the domestic economy, nothing had been done. Maekawa, understanding these things better than we do, told a group in New York that he was not discouraged at the lack of action so far, because he was sure that in the end his recommendations would be followed—he just was unable to say when.

The free-fall collapse of the dollar in 1986 had a major effect on the appetite for dollar investment on the part of Japanese private-sector investors. During the previous years, these had not only been reinvesting the trade surplus, but much of their savings as well. The liberalization of capital markets had let loose a huge stream of Japanese investment funds, which flowed first into Treasuries and high-grade Eurobonds, and gradually, but to a much smaller extent, into U.S. stocks. However, these investors were losing large amounts as the dollar dropped in value.

Under Japanese accounting systems at the time, the unrealized losses on their Treasury positions did not have to be accounted for unless the bonds were sold, and then they could offset domestic gains. In the meantime, their high nominal coupon rates could continue to boost investment income. The institutions had a cushion against foreign-exchange losses for a while, and they continued to buy dollar securities, but were selling some also. They had become short-term traders and were not continuing to place the large orders for Treasuries and other dollar securities that they had in days when the dollar was rising. By the end of 1987, they were running out of patience.

Net investments in overseas bonds fell in 1987 to $73 billion, from $93 billion in 1986. In 1988, such investments fell again to about $50 billion.

The trade deficit still had to be balanced, however, and the difference had to be made up by the Bank of Japan, by increasing its dollar reserves. The private sector was leaving the market, because of the drastic decline in the dollar, to the public sector.

This was particularly the case for Japanese institutional purchases of U.S. government securities. Some investors, however, were still attracted by U.S. real estate and direct investments, and to a growing degree by the stock market. October 19, 1987, brought a stop to the latter, however. By 1988, with the dollar trading in the 130s through most of the year, investors reappeared in all sectors of the market, but not with the wild abandon that characterized their earlier activities.

The major change since the massive adjustment in the yen-dollar exchange rate, the "dollar shock," is even more remarkable because no one believed it could happen. The Japanese domestic economy started to grow. In yen terms, domestic growth was rising, and exports were falling. Maekawa's report was read after all. The long-awaited shift, the one that was always ruled out for a dozen reasons, had begun.

In 1983, domestic growth was 2 percent and exports 1.5 percent, for a net total growth of 3.5 percent. In 1985, domestic growth was 3.5 percent, exports .5 percent, for a total of 4 percent. In 1988 domestic growth was about 6.5 percent, and exports minus 1.5 percent, for a net growth overall of 5 percent.[2] Was this the result of a consensus having been formed in Japan that exports were getting to be too much trouble, too hard to make any money on? Had threats and pressure from the United States made the Japanese finally fear the possibility of protectionism? Had the consumer, after saving all he would ever need, decided to hell with austerity and thrift and gone out to blow his bonus on a wild spending spree?

For whatever reason, the shift is welcome in the United States as is the 130-ish exchange rate, which should help us export more products to Japan. We are also glad to see the investment frenzy of a few years ago taper off to a healthier level, and we should be glad to see Japanese funds spreading out into a variety of different types of investments, avoiding concentrations in any one. Things have calmed a bit. Representative Gephardt's harsh words, so critical of Japan, that supported his cry for protectionist legislation did not bring him the Democratic party's nomination. Neither Michael Dukakis nor George Bush had much to say about the Japanese trade problem during their campaigns for the presidency in 1988. We are in a comparatively quiet period.

In the long run, however, we must expect frictions to reappear. America cannot prevent a huge quantity of imports coming in from Japan. Japan cannot afford to give up the markets in the United States that have been created for its products. Investment flows will continue, no doubt exceeding from time to time the trade flows. On some occasions, the investments will be welcomed, on others not. But until the U.S. savings rate increases, and the fiscal and trade deficits are reduced, the United States is going to be an importer of capital, and the world's largest source of free investment capital for the foreseeable future is Japan. This money, like the Saudis' in the middle 1970s, has to go somewhere. Japan needs to be able to invest it in a secure, profitable economy, so it needs the United States to make investments in.

Neither the Japanese nor the United States can significantly change either the trade or the financial situation over the short run. The order of magnitude is just too big. To cut the recycling of money to strengthen the yen and thereby to accelerate the adjustment of the trade imbalance would create immediately painful consequences for both of us.

Over the long run, however, the problem of excessive Japanese imports and investments into the United States will be influenced by four basic factors.

*First,* absolute manufacturing costs in Japan will be faced with increases that will make it harder to sell products in the United States at a relative price advantage. There is a shortage of expensive land and a shortage of labor that will deter Japan from indefinitely increasing manufacturing output to supply the whole world's markets. New and expensive social programs are also being introduced. American industry, on the other hand, has had considerable success recently in lowering its manufacturing costs.

*Second*, the Japanese domestic economy will probably undergo more deregulation, which ought to loosen up the domestic market further, permitting greater growth. There is plenty of room for improvements in the commercial sector like those that have coursed through the financial sector. Social changes are also taking place, and perhaps we can expect some further loosening up on the part of the consumer. Credit is now more available to assist consumption. New opportunities for entrepreneurial activities, especially in the service sectors of the economy, are also being developed, which help to redirect consumer activity. Like Koichi Tsukamoto of Wacoal, now may be the time for tomorrow's entrepreneurs to ignore the obvious and bet on the future, one in which increasingly prosperous Japanese consumers become big spenders, shameless users of leisure time, and just a little bit inefficient.

*Third*, Japanese companies are increasing their manufacturing activities in the United States, which, as discussed in Chapter 3, improves the trade equation and also provides jobs and a transfer of sophisticated manufacturing technology to American workers. Much of the Japanese consumer-electronics product line is now manufactured in the United States; all of the major automobile makers have committed themselves to large factories in the United States, and these factories have attracted others to be built by Japanese companies that supply tires and other auto parts to them. Most major Japanese manufacturers now realize that they have to do this or they will lose their position in the U.S. market, either because of protectionist action or because they won't be able to sell products here at the same price advantage that they have enjoyed in the past. It is a difficult task for them to erect a large manufacturing plant in the United States, and it is not hard to see why it has taken so long for Japanese companies to undertake the task, especially when the short-run profits inherent in the export boom of the past five years were so attractive.

*Fourth*, it is likely that the more the Japanese become manufacturers and investors abroad, the more they are going to have to fit into the global scheme of things. They are going to have to be less Japanese and more Western.

•    •    •

In a very short time, the Japanese financial industry has experienced the thrills, chills, and spills of global competition. They know that they have to find the lowest cost of funding and the highest returns for their services in order to succeed in the free market beyond their shores. They do business with customers and suppliers from all over the world and offer banking and securities products that were unheard of in Japan only a few years ago. They embarked into the global marketplace because they had to. If they did not keep up with their competitors abroad, their customers would leave them. They had a lot to learn before they could hope to be successful. At first, they were very out of place. The head office controlled every little thing. Now they are more confident, have employed more foreigners in senior positions, have undertaken business without the approval of the head office, and have competed with the best of their foreign competitors in their home countries. They are not shy wallflowers, though they still have much to learn, but no one can deny that they have come an enormously long way in the fifteen or so years that they have been in the game.

When industry takes the same tentative step to become a free-market player beyond its own borders, the same experience will greet it. Japan, more than any other major industrial nation, manufactures for export. With price advantages, or capital advantages, or technology, or work-force culture, any country can for a time succeed in manufacturing goods for others. Britain did. The United States did. No country, perhaps, has been as successful at this over the short run as the Japanese, but as we are all aware, Japan's great success makes it the model for Korea, Taiwan, Malaysia, Singapore, and ultimately, China. It is too early to feel sorry for Japanese manufacturers for the drubbing they are likely to receive from their fellow Asians, but it is not too early to contemplate it. Surely the Japanese have. It takes a lot to pull the Japanese out of their factories in Nagoya, and Sapporo, and Hiroshima, knowing how difficult life afterward will be, but it's happening.

As industry gets settled abroad, it will lose some of its Japaneseness, as surely Bridgestone will do having acquired Firestone. No longer will "obligations" to parts suppliers dominate the decision to purchase their products; economic factors will. General Electric used to have a traditional banker that did all of its business, but no more. Welch and Dammerman know that if financial relationships are not state-of-the-art competitive, then the company isn't either. Those companies from the United States and the United Kingdom that have been overseas manufacturers for years know that traditional relationships have to compete with the rest when the company goes abroad. The Japanese will too.

It is clear that the United States benefits by these longer run trends affecting Japanese industry. There are several things we can do, however, to accelerate them through policy actions. First, we can stop wasting time and energy on threatening the Japanese to "open their markets to our goods, or else." There is no way that such statements can produce more than lip-service responses. Instead, we should press on the Japanese to increase manufacturing investments in the United States and to accelerate the pace of commercial deregulation in Japan. Mainly, we have to have a clear, focused understanding with the Japanese that these steps will help, that they are what we want, and that they must cooperate with us on seeing that such things work. If our two governments work at it by focusing on realistic objectives, there is a lot they can do to improve the situation.

Naturally, we can expect to make a lot more progress in restoring our relationships with the Japanese, and with the Europeans, to a healthier equilibrium if we make some necessary repairs to our own economy. Perhaps with a new administration it will be possible to address the basic problem of the fiscal deficit, and the longer term forces shaping our overall performance in a competitive, global economy.

According to Rimmer de Vries, chief economist at Morgan Guaranty Trust, the United States has been lagging behind Japan and the four principal manufacturing countries of Europe in gross fixed capital formation, as a percentage of GNP, since 1967. Unless increased investment is made, the United States will fall behind its competitors. The dismally low savings rate in the United States cannot go on forever without seriously undermining our ability to protect our relative standard of living. Most policymakers know this but don't act on it.

De Vries also points out that what investments we do make are two-thirds financed from foreign funds. This is obviously too high a dependency, but to fix it requires fixing the savings rate and the fiscal deficit, the latter of which continues to be substantially more out of line than our present circumstances can tolerate. In February 1988, the Congressional Budget Office forecasted a deficit in 1993 in the area of $120 billion if current policies are continued, a figure de Vries agrees with. He proposes a unilateral policy of flexible taxes and spending to reduce the deficit and to increase savings. He adds that "U.S. fiscal action should not be contingent on successful negotiation of a complex, broad-ranging, and coordinated international program for the better management of the global economy," something he doubts is obtainable. De Vries goes on to add, however, "If the U.S. is seen to forge ahead with fiscal correction . . . U.S. trade partners will see their self-interest in basing their growth more firmly on their own domestic potential—as Japan, for one, has done during the past twelve months."[3]

Mr. de Vries's remarks, made to a blue-ribbon commission studying the national economy, must have been received by his audiences in much the same way as Mr. Maekawa's. The foreigners, in agreement, praised it highly; the locals, not wanting to hear it, either ignored it or disparaged it. Certainly much of de Vries's message has been heard by President Bush and his economic people before this. They have always replied that there was nothing wrong that postponing or reducing government spending further couldn't cure.

Japanese, however, may build up their expectations for an American housecleaning based on the influential words of the distinguished Mr. de Vries. Americans hearing of such expectations would be astonished to think that anyone would take such a statement seriously, however well-respected its author. No one in the United States expects a long overdue policy change to be made as a result of a single statement from an expert.

But will the policy be changed? Maekawa had confidence in his prescription, just not the timing. Surely de Vries could say as much.

Let's hope that the symmetry of the two statements will be complete, and that de Vries's recommendations will be adopted too, as surprising as that may seem to those who heard the remarks when they were made.

# 9

# Samurai
# Finance

A year or so after our first trip to Japan, in 1969, Goldman Sachs developed some business there. Our client was Mitsui & Co., one of the largest and most distinguished of the Japanese "trading" companies, or *Sogo Shosha*. Mitsui & Co., whose origins go back to the seventeenth century, is the principal descendant of the important Mitsui *zaibatsu*. After the war, it became the lead company of the Mitsui Group, then comprising twenty or so companies that had previously belonged to the *zaibatsu* or were allied closely with the group. By Japanese standards, Mitsui was an extremely prestigious company for a newcomer to do business with, a real catch.

Before making our trip to Japan, Joe Fowler had contacted Tatsuro Goto, who was then president of Mitsui's subsidiary in the United States, to ask for an introduction to his colleagues in Tokyo. Goto, whom Joe had known only slightly before this, became a close friend. He was delighted to arrange a luncheon during our visit at which we could meet members of the Mitsui board. We did and all went well. When we returned to the United States, Joe and I called on Goto to tell him how the event had gone and to thank him for his assistance. During the next six months or so, I maintained contact with Goto and his staff in New York, proposing various financing schemes for their consideration.

Mitsui had been a client of Smith Barney for a decade or more at this time, and Mitsui felt considerable loyalty to the firm. There was never a question of our outshining Smith Barney during the various solicitations that went on, although that was what I was attempting to do. If Mitsui was to decide to do business with Goldman Sachs, it would have to be in addition to the business it was doing with Smith Barney, and the two firms would have to work "harmoniously" with the company in the process.

Goto decided that Mitsui would issue some *commercial paper* (short-term promissory notes sold in the money market) in the United States through Goldman Sachs, a service not offered by Smith Barney, and then the company could see how things worked out before advancing the relationship further. I was delighted with this competitive victory, which I ascribed to my painstaking efforts to explain every detail imaginable about commercial paper to at least a dozen people on the Mitsui staff.

The reality was that Mitsui had decided to see whether a relationship with the influential Joe Fowler, and his firm Goldman Sachs, could be developed. In doing so, it would pick a service offered by the firm that was not offered by any of its existing banking relationships and therefore would not upset the existing harmony too much, but would put the others on notice that Mitsui was quite capable of creating new and potentially powerful relationships. It would also teach Mitsui something new about U.S. financial markets for future use. They would give "this commercial paper, or whatever it was called," a try, but only in a small amount until they could see how it worked.

My next job was to present the program to our management committee, which had to approve all new clients of the firm. I was sure that the committee would be impressed with our capturing this prestigious piece of business so soon after our Japanese efforts had

296

begun. "It's like landing some business from IBM only six months after being assigned to the account," I told my wife, modestly.

Confidently I had sent a memorandum to the management committee setting forth the details of the transaction and the pedigree of the client. I went to the meeting prepared to answer questions about the issue, which involved a new approach in the commercial-paper market and for which this would be the first foreign issuer.

"What the hell kind of business is this?" one of the members asked. "It's got almost no equity—more than ninety percent of its assets are financed with debt—current assets are less than current liabilities, earnings as a percentage of sales are less than half of one percent, and the company doesn't make anything?"

"Besides that," added another member of the committee, "this outfit seems to take positions in the commodities markets for its own account, with the intention of selling the commodities to its clients later on." He volunteered an unprintable comment about the wisdom of financing a speculator in commodities in the conservative, blue-chip commercial-paper market.

Another member interjected, before I could respond to any of this, "Also, this company seems to lend its own capital to its customers to facilitate deals, like a finance company, and some of these customers make Mitsui look like AT&T.

"And I don't seem to find any consolidated financial statements. The company has a lot of subsidiaries, but their balance sheets and earnings statements aren't here."

I spend the next couple of hours trying to answer these questions.

"The Japanese system of corporate finance is very different from ours," I began. "As capital is in very short supply, there is a lot of leverage, provided mainly by Japanese banks and suppliers. They know that Japanese companies in their system of mutual support have access to extraordinary reserves if need be. Mitsui was the brains of the Mitsui *zaibatsu* before the war and is again the brains of the newly constituted Mitsui Group. More than a third of its stock is owned by members of the group, and Mitsui owns stock in each of the group members too. A large part of Mitsui's assets are receivables from group companies; many of its liabilities are provided by

Mitsui Bank, Mitsui Life Insurance, and other Mitsui group creditors. Other Japanese banks have sizable loans outstanding to the company too, and all of these loans are reviewed closely and periodically by the Ministry of Finance and the Bank of Japan, which regulate banks in Japan. Their safety net is carefully woven to permit large amounts of leverage, which is the only way they can finance the extraordinary growth that Japan has been experiencing.

"There have been no defaults on Japanese debt to foreigners, either before the war or since," I continued, finding my stride, "and every overseas financing, including this one, has to be approved by the Ministry of Finance, which doesn't guarantee anything, but does put the transaction into a basket they are known to watch very carefully for fear of having the rest of their credit from overseas banks cut off.

"Mitsui's business is unlike anything that exists in the United States or Europe, though you might compare it to a large grain trader like Cargill. Its job is to act as the eyes and ears of the group, to find resources, finance, and markets for it, so the main body of the group, that is, the manufacturing companies in it, can focus all of its attention on what it does best, namely, making things. Mitsui will find the raw materials its companies need from sources all over the world, and it will market the companies' products in Japan through its ancient distribution channels, and to foreigners. Mitsui maintains more than one hundred offices overseas to make this sourcing and marketing possible. And it provides relatively cheap financing for the flow of goods for its group companies. Banks finance some of this too, but mainly they lend for plant and equipment expenditures.

"Mitsui does purchase and sell commodities and other raw materials for their clients as principal, but these purchases are almost always covered by customer orders. There is very little speculation that goes on, though Mitsui is a service-providing company, and occasionally it will have to do something for Nippon Steel or somebody that might involve some position risk.

"The company's profits are not as bad as they look. Its profit-and-loss statement records gross sales of all goods and commodities; then below it, you see the 'cost of goods sold,' which when netted out gives you the 'gross trading profit.' Their net profits are about fifteen percent of gross trading profits, which compares favorably to banks and retailing companies that are in somewhat similar businesses.

"Finally," I continued, coming up for air, "we don't have consolidated financial statements because Japanese companies are not required to prepare them (they were in 1977). We have the statements of the U.S. subsidiary and the parent, which represent the bulk of their consolidated assets. The figures would be a little hard to understand, though, because the company has many assets recorded at nominal cost that are worth many times that amount today, such as land, buildings, and shares in group member companies. It also conducts substantial intercorporate transactions to save taxes. Most U.S. bankers, who have been lending to Mitsui and other trading companies since the war, believe that consolidation would increase earnings, perhaps fifteen to twenty percent. Anyway, Mitsui is in the process of having consolidated financials prepared by one of the Big Eight U.S. accounting firms."

The management committee voted—four to two—to take on the assignment, but mainly because they wanted to support our fledgling Japanese effort and back up Joe Fowler. All, however, were experienced bankers, several of whom had seen the 1930s, and they were very cautious about being drawn into another round of "damnfool" foreign financing. These people, hardheaded bankers who had nevertheless found ways to finance dozens of small, promising, but weak American companies, had no way of relating to the Japanese financial structure that then existed, and to a very large extent still does.

Japanese manufacturing companies work on the basis of what I think of as the enlightened anthill principle. This is, in essence, a highly specialized division of labor that makes everyone an expert in a tiny part of the whole enterprise. The enlightened part is that the anthill learns as it develops and changes as it goes. It changes the tasks that the experts perform, but not the structure of the anthill itself.

The great Japanese manufacturers—Toyota, Hitachi, and Nippon Steel—are the aristocracy of corporate Japan. They exist to perform their high purpose of manufacturing goods that can be consumed and exported to keep the Japanese system in circulation. They are charged with the task of upgrading technology to lower manufacturing costs and to create new high-quality products. They are not to be distracted from their sacred tasks by having to worry about such sordid things as finding raw materials and export markets, or arranging finance. They are surrounded by vassals who will happily perform these services for them, and do so very competitively. Trading companies, banks, foreign banks, suppliers, joint-venture partners, even customers are eager for their business, and they compete with each other for it. Before about 1970, the vassal relationships were fairly well fixed, emanating from the *zaibatsu* mainly, but also reflecting competitive forays to create new relationships. After 1970, the relationships became much more fluid and competitive, and some companies set up their own overseas operations to reduce their dependence on trading companies, whose sourcing and marketing skills were concentrated in commodities.

The manufacturers and their principal vassals created a safety net for themselves, which involved cross-holdings of shares, mutual extension of credit facilities, joint investments, and a form of common planning. When one of the group stumbled, the others helped it recover. The rescue of Mazda Motors by the Sumitomo Group, led by Sumitomo Bank in the early 1970s, is a good example of the safety net in operation.

Because of the safety net, corporations were comfortable in taking more risk than their overseas counterparts might. Balance-sheet risk, for sure, but also other business risks of various types. Mazda's trouble was a large commitment to the Wankel rotary engine, which at the time was supposed to be about to revolutionize automobile manufacturing. Mazda made a good engine and was probably the only car company in the world that had bet its future on the rotary engine. The oil shock of 1973 consigned the fuel-inefficient rotary engine to the scrap heap, and, but for the safety net, Mazda would have gone with it. There are many examples of the safety net in action, but thousands more where, because in place, it was never needed. With such a system, who wouldn't take advantage of it, and increase corporate risk exposure, which usually would show up in overexposure to a technology, a new manufacturing method or facility, or in finished goods inventories.

300

The job of the industrial aristocracy was to make things, as many as possible. The more that are made, the lower the unit cost, and the more they could afford moving up into a new model. As long as sales get made somewhere, the shipping docks are cleared for more outgoing orders. Volume was the key, the more volume, the more expansion, prestige, employment, everything. Every now and then, a new item would be made that captured a large volume of orders. Hand-held calculators were such an item in the 1970s. Electronics was not yet a settled business, and many new competitors began to make calculators. Canon, a camera maker, was one. Canon saw the long-term potential of marrying optical and electronic technologies and plunged into the calculator business to establish itself and gain some good experience in manufacturing and selling calculators. It made too many. The market became saturated, prices plunged, and Canon took a bath. The safety net, though not officially used, was helpful.

Yamaha is another company that exposed itself to product risks that U.S. companies might not take. Starting as a manufacturer of pianos, it found its precision-forging operations applicable to manufacturing motorcycle chassis and therefore went into the motorcycle business, which was mainly a highly competitive export market. Undeterred, Yamaha roared into the business and lost money on it for a good while.

Another motorcycle maker, Honda, must have horrified its bankers when it asked for money to design and manufacture a "four-wheeled" vehicle, but it got it. Without such a safety net, American businesses are much more careful about which high-wire tricks they perform.

The safety net, the relationships, and the competition for the business that goes on beyond the relationships makes raising finance relatively easy for Japanese companies. Financial officers until quite recently have not been market-oriented, or even aggressive. They are mainly trained, if at all, in accounting. Their main function is to maintain the many relationships that their company has, keeping them properly balanced and nurtured. Most of them know very little about modern corporate finance, which no doubt explains why Japanese companies have the lowest cost of capital of any in the world.

301

Finance officers receive a lot of visitors, for a half an hour and a cup of green tea. Usually a guest is received in one of several small conference rooms, though sometimes only at a tiny meeting area crammed in between the Xerox machine and the coatrack. Foreigners usually get a meeting room, a grander one if the visitor is deemed to be senior enough. Wherever the location of the meeting, all are about the same.

I'm not sure what they say to Japanese visitors, but all meetings with foreign bankers begin with "When did you arrive?" The finance officer is there because the banker has asked for an appointment; it is up to the banker to make conversation and say whatever he is going to say. Usually the banker has nothing special to offer—he is just maintaining contact, trying to pick up leads for something the company is planning to do. These meetings are often very uneventful, but they serve their purpose anyway.

Sometimes the finance officer will ask a question about interest rates or some competitor's financing. Sometimes he will drop a hint of something he's interested in, waiting for the banker to dig out what it is, if he can. Usually he is interested in the latest gossip from the market, the latest deals or new ideas. If he has some business he is willing or able to let the banker compete for, he will let him know, elliptically to be sure. After this, the banker is supposed to come up with his "proposal."

The proposal, which seems to be limited to one per customer, is submitted in a two- or three-page memorandum, where it is reviewed along with all other proposals that the company receives for the same matter. Once a banker makes a proposal, it is difficult if not impossible to change it, except perhaps for the interest rates, which may be changed by the market.

If, for example, the company man said, "We might be thinking about raising finance for a new factory in Nagoya," an ordinary banker accustomed to Western ways would say, "How much do you plan on raising and when do you need it?" He might go on to ask whether the company would like to issue bonds, or stock, or convertible debentures, and if so, would it be thinking about the foreign or the domestic markets. Sensible questions like these are useful in a discussion about corporate finance with a U.S. company. They are not with a Japanese company.

What a foreign banker operating in Japan should know from this conversation is that a foreign financing is being considered and has reached a fairly advanced stage. Otherwise, the matter would never have been raised with the foreigner. Proposals are being accepted from a limited number of bankers, including himself, and time is of the essence. The financing will aim at raising whatever amount of money the market will comfortably bear. The objective will be to have the lowest nominal interest rate possible, regardless of other factors. If a foreign-exchange risk arises, it will either be ignored, which the company can do if the issue involves bonds convertible into the company's common stock, or swapped into yen. If the financing involves the issuance of new shares, through convertible debentures or warrants, the company won't mind: dilution of shareholder's equity through the issuance of new shares is not closely followed in Japan.

The winning proposal will be aggressive, but it will also maintain harmony with the other banks and securities firms that the company uses. The selection will be "very difficult." Once it is made, however, the company will leave everything to the banker and hold him to delivering what he promised. The interest rate, the timing of the issue, and the date of the closing dinner are all penciled in. There can be no further changes except under extraordinary conditions and after having exhibited every possible samurailike effort to prevent them.

In 1979, Goldman Sachs was selected as lead manager, with Nomura Securities as co-manager, for an $80-million convertible debenture offering to be made by Canon in the United States. The process through which we were chosen was lengthy and never really clear to us. Many months went by after the first handkerchief had been dropped. We had made a proposal for a U.S. issue and periodically updated it. We had had many meetings with the financial staff to explain details. We had given them time schedules, checklists, a list of documents to be prepared, accounting requirements, and everything else they asked for. My colleagues in Japan and I had met several times with Ryuzaburo Kaku, Canon's president, to talk about the outlook for the U.S. market, the yen-dollar rate, and many other general subjects.

For every meeting, we prepared ourselves according to "suggestions" that we were receiving unofficially from Nomura, who had been told by the company of our approach to it. Nomura was not the "traditional" securities firm that Canon used—that was Yamaichi— but Nomura was trying to get some business going with Canon, and Canon was apparently receptive. An overseas offering, according to the evolving Japanese financial protocol of the times, allowed for the selection of a nontraditional securities firm without the charge of disloyalty. Canon was growing, would need additional finance in the future, and thought it should add Nomura to its stable of relationships, to create competition for Yamaichi and generally keep everyone on his toes.

Yamaichi was also competing for this business, and it had aligned with Merrill Lynch. Canon liked Merrill Lynch well enough, but the bigger question in its mind was whether to pick Yamaichi, after having let Nomura run it a merry chase, or to pick Nomura. Merrill was lined up with Yamaichi, so, by default, Goldman Sachs would be lined up with Nomura. Naturally, none of this was to be mentioned to us; each bank was acting in its own right.

Yamaichi, I believe, got going first. It had proposed, and Merrill had also proposed, separately but in cahoots, a "private placement" in the United States, in which the issue would not be registered with the SEC, saving Canon a lot of time, effort, and cost, but the issue could not then be marketed beyond a relatively small number of investors, and no trading market in the issue would exist afterward.

Nomura got word to us of this proposal and suggested that we prepare an alternative proposal for a registered public offering. We did so. We explained why a larger public offering would be better for Canon than a private placement. We explained our qualifications for handling the issue, the research and trading support that we would provide in the aftermarket, and we minimized the time, effort, and cost of the registration with the SEC, forecasted interest rates, stock-market behavior, and the effects of the midterm congressional elections of 1978 on the market. We also played heavily on the fact that

Canon's new, high-visibility advertising campaign in the United States would offer excellent support for a public offering where investors in their products would be encouraged to become investors in their securities.

Canon had already decided that it wanted to issue convertible debentures. The company did not have all that much debt by Japanese standards and would have been well advised to issue nonconvertible debt, at an interest rate at the time of perhaps 12 percent, or roughly 6 percent after tax.

Its shares were trading at a modest P/E ratio at the time, something like 14 times the coming year's earnings per share (based on U.S., not Japanese, accounting). If it sold stock to raise new money, it could do so at the market price. Such an issue would cost it the equivalent of about 7 percent after taxes, but it would create dilution in earnings per share that shareholders, if they were Americans, would not like to see. American companies only raise equity when they have run out of further borrowing capacity, or when stock prices are so high as to be irresistible.

Canon did not want to do a share issue. It was planning one of these for the Japanese market next year; it didn't want new shares issued abroad coming back to Japan at inconvenient times; and perhaps other reasons. It did not consider a convertible debenture to be the same thing as an equity issue—it was a low-cost way of raising debt. The coupon rate would be 6.75 or 7 percent instead of 12 percent. The conversion rate would be set at a price near the present market price so the value of the conversion feature would be high in the market and therefore the coupon could be low. Conversion into shares would occur in a year or two, most likely, and then the interest cost would disappear. The effect of the announcement of the creation of new shares resulting from conversion in the future was of no consequence to the Japanese market, where investors didn't care about such things. So they planned to finance at what they regarded as about 4 percent after tax, a low rate by any standard during the high interest rates of the time. As the dilution resulting from conversion seemed to make no difference to the market, the equity component of the issue seemed to have no cost at all, which is pretty cheap.

Canon had a choice also between the U.S. and the Eurobond

market, and further between borrowing in dollars or in Swiss francs or deutsche marks. The latter markets were available only periodically and not always in large size. Over the years, however, Canon used these markets to issue additional convertible debentures at coupons below 5 percent, spreading its business around. Canon, and most other Japanese companies, preferred borrowing in dollars, as usually the dollar rate at which they would pay their interest declined after the issue, thereby lowering the interest rate, in yen terms, further.

In the end, Canon selected Goldman Sachs and Nomura to handle the issue. A disappointed Merrill Lynch and Yamaichi requested of Canon that they be appointed co-managers, somewhat contrary to protocol. The losers are just supposed to creep away, but when U.S. firms are involved they rarely do. In this case, the losers made a great appeal to Canon, who asked us to include them. We said four co-managers was too many and resisted. They kept asking, and we kept resisting. Finally, Canon accepted our position and dismissed the others. The deal went on to be a big success, but we paid for our resistance in the end. The next year Canon did an issue in Europe led by Merrill and Yamaichi and neither we nor Nomura ever heard about it until it had happened. Naturally, we were not included as co-managers. Over the years the dual relationships were maintained, one issue being done by Nomura and Goldman Sachs and the next by Yamaichi and Merrill Lynch.

Despite their occasional use of capital markets, most Japanese companies relied for many years on bank financing for their principal funds requirements. In recent years, many Japanese companies have become unusually flush with cash, but until then they found banks very easy to use. They paid their clients a lot of attention, and, over the years, provided much cheaper financing than did the capital markets. This was because short-term interest rates were usually lower than long-term rates. The Japanese companies knew this and preferred the cheaper rates. Their balance sheets, like Mitsui's, used to be very heavy with short-term liabilities, which often equaled or exceeded short-term assets. In the United States, a healthy "current ratio" of current assets to current liabilities is 2 to 1. In Japan there is usually an understanding that the short-term loans are really long-term loans, that is, technically longer than one year, if the company wants them to be. In other words, the banks will continue to roll over the loans as long as the company wants them to, and they won't even think about calling the loan if they don't like what is going on at the company.

This may be another example of what Tokuyuki Ono, of Sumitomo Bank, means when he says that in Japan the clients are the "kings" and the banks are the "slaves." U.S. banks would not agree to letting a U.S. client build up its short-term liabilities to such an extent just to avoid paying long-term rates. This may also be another example of how the principles of corporate finance as we understand them in the United States are ignored in Japan.

The company's relationship with its bank, however, is meant to be a two-way street. Loans are not repaid if doing so would be inconvenient to the bank; instead, large deposit balances are maintained with the bank. Often this results in a situation in which the company's own deposits are financing its borrowings from the banks. Usually such a situation is convenient for the bank because it is thereby able to keep up its total assets, by which banks tend to be compared with one another. At other times it is convenient for the company, which can squeeze a very low lending rate from the bank, which it can use to demonstrate its financial prestige, or to aid it in rate negotiations with other banks. Gradually, however, regulatory pressure on the banks to discontinue such matched-funding loans, and the emergence in Japan of a commercial-paper market, has altered these arrangements somewhat.

The Japanese simply play by other rules when it comes to corporate finance. They have safety nets; they can borrow more on their capital than companies anywhere else (at least this was true until the leveraged buyout was invented in the United States); they don't worry about anything except how low the interest rate is; and they can borrow all the long-term money they want at short-term rates. The net result, of course, is that their system permits them to borrow a great deal at relatively low rates if they want to, or need to.

They are also frequent issuers of equity securities. During the years of rapid growth previous to the 1973 oil shock, companies issued new shares every year or two. These were offered to existing shareholders at the stock's "par" value, a price well below the market price. This practice was necessary when capital markets were not well developed, but it was inefficient and caused substantial dilution. Afterward, as growth slowed and much more of corporate financing was provided by internally generated funds, the need for new equity financing dropped. Many companies continued to issue shares and convertible debentures, however. The practice of issuing at par value was discontinued in favor of issues priced at the market. With slower corporate growth, more liquidity accumulated in the financial system. Some of this excess liquidity went into the stock market, and prices started to rise dramatically by 1980.

New share issues continued to be made by Japanese companies both in the Tokyo market and in overseas markets. For those companies who needed it, equity capital was available at extremely high P/E ratios, by U.S. standards, and such high prices pulled their cost of capital down further.

Some academic studies have been made in the past several years that have attempted to determine whether Japanese companies enjoy a cost-of-capital advantage over U.S. companies. These studies have been inconclusive until recent years, when Japanese prices reached such high levels that they had to show advantages for Japanese companies, no matter how the studies were conducted.

There are many things that make these studies difficult to do, such as limitations on the availability of data, the size of the sample of companies used, the number of years covered, and whether companies are compared on an industry-by-industry basis. The studies cannot really be accurate unless a number of accounting adjustments are made to earnings and investment data to put companies from both countries on a comparable basis. All of these adjustments have not been made for the purpose of studies done to date.

Also, it is especially difficult to adjust for the fact that the Japanese stock market does not appear to correct market prices for the issuance of new shares through the conversion of debentures or the exercise of warrants, or to care that dividend yields are extremely low. Further, the technical ways we use to evaluate whether or not a stock is doing well in the United States (for example, growth in earnings per share) are not used to the same extent in Japan. The market's inefficiencies (if that's what they are) appear to work for the benefit of the Japanese company in most cases, by masking factors that ought to cause a price adjustment.

My own conclusions are that before 1980 the "average" U.S. company and the "average" Japanese company (after accounting adjustments) had about the same cost of capital, which was lower than that of European companies. The averages, of course, cover up exceptional advantages being experienced by particular companies. Then, from about 1980 to 1984, Japanese companies appear to have had a modest (but important) 1 to 2 percent cost-of-capital advantage (for example, 5 percent versus 6 or 7 percent for U.S. companies).

Comparing P/E ratios with U.S. companies is an inexact process at best. One very knowledgeable observer, Paul Aron, while vice chairman of Daiwa Securities America, studied Japanese P/E ratio adjustments that should be applied to make them comparable to U.S. ratios. Some accounting adjustments are necessary, and some valuation ought to be assigned to an individual company's holdings of hidden assets, such as appreciated land, office buildings, and shares in other companies. When these adjustments were made for the 1981–84 period, Aron concluded that Japanese P/E ratios, which then appeared to be about 15 to 20 percent higher than U.S. ratios, would after his adjustments be about the same as the P/E ratios of U.S. companies.[1]

After 1984, however, the explosion in Japanese stock prices, and hence P/E ratios, has put the issue of comparative cost of capital out of doubt.

The Japanese today enjoy the lowest cost of capital of any country. This is an important comparative advantage for Japanese companies. It provides them with a continued incentive to invest in new equipment and technologies. It eases the burden of direct investment overseas. It enables them to streamline their cash flows, to conserve borrowing power for future needs.

How can this be? Japanese internal growth has been slow, until 1987. The yen is now the world's strongest currency. Interest rates in yen are in the 3- to 4-percent range. These are conditions that, by American standards, ought to justify an average P/E ratio of maybe 10 or 12, not the 35 to 40 that prevailed during the summer of 1988.

The Japanese, of course, currently see a high rate of domestic growth, a very low inflation rate, and a strong currency to spend abroad and feel somewhat differently about their P/E ratios.

It is typical of what happens in the floating world.

In the real world (that is, in the United States), institutional investors, largely pension funds and investment companies, dominate trading and establish price levels. Investments in stocks have to compete with investments in bonds, particularly U.S. Treasury securities. Selection of investments is made on the basis of careful analysis and research of macroeconomic data and information disclosed by corporations. Profits, and growth in profits, are the most important considerations, but dividends are also important; investors look at total return, which is the sum of dividends and capital gains.

In Japan, the stock market is driven by a mixture of institutional investors all buying and rarely selling shares and individual investors trading, that is, buying and selling them, in a market in which the outstanding tradable supply, or *float*, in stocks is perhaps no more than one-third of total shares outstanding. Institutions and group member companies own most of the stock listed on the exchanges, but they never sell it. What trading occurs is the result of big Japanese mutual funds adjusting their portfolios, a few foreign institutions, which account for about 30 percent of all trading (thanks mainly to their involvement in Japanese Eurowarrant issues), and individual investors buying and selling small amounts of stock, on which they were not required to pay any capital gains taxes until the tax law changes were effected in 1989. These investors are somewhat indifferent to dividends, on which they would have to pay regular Japanese, that is, high, income taxes. They would rather the company retain the funds to invest in something that will make the stock price go up. Institutions are passive about dividends—most would like to have them, but they are accustomed not to and don't complain too much. Most institutions have held the stock for so long that even the relatively small dividends look large as a return on their original investments.

Japanese investors are serviced by brokerage firms, of which

four are dominant. These—Nomura, Daiwa, Nikko, and Yamaichi—
are called the "big four." Together they account for about 35 percent
of all commissions paid in Japan. (The four largest brokers in the
United States, for example, account for about 25 percent of commissions.) The Japanese big four firms have approximately one hundred
branch offices and three thousand to four thousand salesmen each.
Forty percent or so of the salesmen are women who go door-to-door
to sell investment products to housewives who manage the household
funds in Japan. Other salesmen contact hubby at the office and corporate and institutional investors.

Brokerage commissions in Japan are fixed. Fixed commissions,
high volume, and soaring stock prices, together with heavy new issue
activity in equity and related securities, are what explain the extraordinary profits of the big four firms. In 1986, Nomura's pretax profits
grew by 86 percent to $2.7 billion, the largest profit of any financial
institution or bank in the country, and the second most profitable such
institution in the world, just behind Citicorp. Nomura's net capital
was then in excess of $10 billion, exceeding the capital of the top
three U.S. firms combined.[2] Profits declined by about 25 percent in
1987 as a result of the impact of the crash on Japanese brokers' foreign offices, and cuts in Japanese fixed commission rates.

Japanese investors reputedly do not scrutinize annual reports or
research their investments much. It is frequently argued that they rely
on their ability to know what's going on in the market from minute
to minute in making their investment decisions. They invest according to a kind of mass psychology "greater fool theory." One may be
a fool to buy a stock at a particular price, but since the crowd is
running that way, the investor believes, there will be another, "greater"
fool to buy the stock from him at an even higher price. Japanese fund
managers are beginning to change, however, into analytical and
research-oriented investors, and no doubt will continue to grow in
investment sophistication.

The Japanese, being a highly group-oriented people, are especially susceptible to the herd instinct in market situations. Accordingly, usually only a handful of stocks dominate trading on the exchanges on a day-to-day basis. During the market rally in the summer

311

of 1986, for example, the most active ten stocks accounted for as much as 50 percent of the total trading volume on the Tokyo Stock Exchange, although this ratio is usually much lower.

Investors get their investment information from tips, market letters, networking, and securities firms whose salesmen are given a quota of stocks to push during the week. This kind of market, where investors are constantly poised to lunge in the next fashionable direction, can lead to manipulation. Rumors can spread that the next hot stock is so-and-so, some early buying will support the rumor, the price will start to run, and the manipulator will start to sell. In Japan, this is called "ramping," and it is supposed to go on all the time, though it is always difficult to prove. Similar actions occurred in the late 1920s in the United States when "pools" were formed to manipulate the market in the same way. Premeditated manipulation is illegal today in both countries.

Stock prices in Japan tend to be high, for some of these reasons, and because of the increasing amount of "excess liquidity" that has flowed into the market since 1980. Excess liquidity is the difference between the growth in the money supply and that of real GNP. In Japan, during times of comparatively slow economic growth, corporations disgorge funds, but high savings activity continues. Funds available for investment rise, both at the corporate and at the individual level. The average growth rate for the Japanese GNP for the decade subsequent to the oil shock was 5.3 percent; from 1980 to 1987 it was 3.9 percent. Excess liquidity in 1980 was minus 3 percent, and only .2 percent in 1981, but thereafter it grew rapidly. From 1982 to 1987 it averaged 3.2 percent, but in each of 1986 and 1987 it exceeded 5.5 percent.[3]

In 1980 the average P/E ratio of the Nikkei 225 stocks, based on parent-company historic accounts, was about 20. In mid-1987 this average P/E ratio reached 80. A year later, after rapid and substantial earnings growth, the average P/E ratio had declined somewhat.

Excess liquidity is the principal explanation for the price escalation—all that money chasing the few shares that were in circulation and gobbling up most of the new issues of shares. The liquidity, however, was highly focused on the stock market as a result of deregulation occurring in Japan at the time.

312

Two things were happening: Japanese pension funds were being funded, at a comparatively high rate to catch up with past underfunding, and corporations had been allowed to invest their excess cash in tax-advantaged funds, called *Tokkin*. Between them new sources of institutional funds were flowing, rapidly, into the market. Investment trusts rose ninefold from 1980 to the end of 1988, when they represented 50 trillion yen ($400 billion); *Tokkin* and other corporate-fund trusts more than tripled just from the beginning of 1986 to midyear 1988, to 30 trillion yen[4]—recently, following the economic pickup in Japan during 1988, they have started to decline.

This extraordinary influx of funds helped push the Nikkei index upward. From the beginning of 1985 until the end of 1988, the index very nearly tripled.

The market value of Japanese stocks went to extraordinary levels. By midyear 1988, Sumitomo Bank had reached a market value of about $65 billion; Toyota, $55 billion; Nomura Securities, $49 billion; Fujitsu, $24 billion; and Bridgestone Tire, $7 billion. By comparison, Citicorp's market value at the end of July 1988 was $10 billion; Ford's was $14 billion; Merrill Lynch's, $3 billion; Hewlitt Packard, $12 billion; and Goodyear, $3 billion.

Japanese companies' comparative advantage in their cost of capital had increased considerably. Further, the market value of their stock gave them a powerful competitive weapon.

Already Japanese companies had learned a lot about financial arbitrage. Observing John Browne and his colleagues at BP, Japanese banks and securities firms began to take advantage of the opportunities to finance, swap, and reinvest excess funds in order to earn 20 or 30 basis points. They began to draw their clients into the game too.

The process of taking advantage of "high-tech" financial transactions in order to earn a return is called *zaitech* in Japan. During a time of slow growth and a rising yen, many companies found *zaitech* a useful and necessary way to offset a decline in operating earnings. Many old-timers, however, thought the practice was inappropriate for manufacturing companies, and there was something of a stigma attached to it initially. Most companies, nevertheless, learned how to invest their considerable excess funds more aggressively, though qui-

313

etly. Investing in a *Tokkin* was not considered quite so bad; because someone else, someone who really knew how to manage money, was handling the investments, not the manufacturer.

After a while, however, opportunities became too attractive to pass up, even if it meant practicing *zaitech* publicly. During 1987, Japanese companies found that they could issue deep-discount bonds with extremely low coupons (for example, as low as 1 percent) in the Euromarket if these bonds had Japanese stock purchase warrants attached. The warrants would be stripped from the bonds and sold to Japanese and to foreign investors wanting to participate in the stock market without committing much money. In the rising market, these warrants would be worth a lot to investors, but the market would ignore them in terms of adjusting the stock price to reflect their issuance. The bonds were sold to Japanese banks, who would bundle them together and swap them into floating-rate assets at a profitable spread. This business became so attractive that more than one hundred such issues, totaling more than $19 billion of bonds with equity purchase warrant deals, were floated in Europe by Japanese companies in 1987, about $30 billion in 1988, and $13 billion in the first quarter of 1989 alone.

For the issuers these deals were gold mines. They could borrow dollars at very low coupons and reinvest the proceeds at a spread of 5 or 6 percent. The warrants were more or less free from their point of view. *Zaitech* earnings became so important in Japan that separate tables were published to record those companies who were doing it and the extent of the incremental earnings from such activities.

Despite the fact that few Japanese needed to raise equity capital during the period from 1986 to 1988, hundreds did. By selling stock at such high prices, the companies were locking in the cost-of-capital advantages they knew they could only enjoy if they acted to realize them. The proceeds were invested initially in reducing debt (itself very profitable at the cost of capital represented by the stock sales), or in making various other *zaitech* investments.

Some used their newfound competitive weapons more aggressively. Sony moved to acquire CBS Records for $2 billion. Bridgestone overwhelmed a French-Italian group to purchase Firestone for

314

$2.6 billion. Sumitomo Bank, which had made a $500-million investment in Goldman Sachs in 1986, financed the whole thing with a sale of stock in 1987 when the bank's P/E ratio was around 85. This equated to an incremental cost of capital of about 1.2 percent; for it the bank was able to acquire a 12.5-percent interest in the profits of one of Wall Street's most profitable firms.

Looking ahead, four things stand out. As Japan begins to enjoy a resumption of domestic growth, its excess liquidity will decline. This should cool the markets somewhat, but Japan's exceptional savings position, its trade surplus, and its pension-fund activity will certainly contribute a lot of money to the equity markets for the foreseeable future. The basic structure of the market, in which float is limited, individual investors pursue stocks on a helter-skelter basis, and proceeds from *zaitech* issues get rotated back into the market, should continue, and "floating world" valuations will continue to be placed on all popular Japanese stocks.

This being the case, the competitive weapon of a high stock price will be available to many Japanese companies long enough for them to learn how to use it effectively. And once they have learned how to use it, they will do so.

Nevertheless, the cost-of-capital advantage in the long run is probably a wasting asset; it will disappear as the Tokyo market becomes more institutional and converges with other markets into a kind of global pool in which prices are set according to rational rules and procedures, and arbitrages between markets are common. But in the long run, as John Maynard Keynes has said, "We are all dead."

"By the time this pool thing happens," one knowledgeable observer of Japanese markets comments, "we may all be."

In the meantime, cost-of-capital advantages are of no real value to the Japanese unless they use them. They are likely to do so, as there are many uses for low-cost funds.

In coming years, many Japanese manufacturing companies, like Mazda, will have to finance the investments that they will be required to make to globalize their businesses. Financial operators will do the same. But foreign investments, however strategic, may not be large

315

enough to absorb all of the cost-of-capital advantages that at least some Japanese companies will enjoy relative to their domestic and international competitors.

What will the rest be used for? Will Japanese companies expand into totally new business areas, as many have done over the years, from pianos to motorcycles, for example? Will some try to carry out plans for developing Japan's living and working space outside the crowded Tokyo-Osaka corridor? Will more opportunities for investment in domestic Japan force changes in the archaic agricultural and retailing systems? Will Mr. Ito be able to see the day when 70 or so percent of Japanese retailing is done through chain stores instead of the approximately 20 percent currently? Will all this money that Japanese consumers are supposed to have saved, while having so little to do with it, end up in an explosion of leisure-time activities, a softening of the Japanese competitive spirit? Will the "floating world" begin to become a reality in Japan, instead of a fantasy?

These are questions for twenty-first-century observers to answer. For the time being, we must face a Japan armed with powerful financial weapons. Surely they will use them on their competitors in the West. We will have to develop our ability to survive against these weapons, perhaps by employing one of the ancient Japanese martial arts, jujitsu, in which an opponent's strength is deflected or manipulated so as to be used against him. Perhaps this can be done in connection with the continuing restructuring of the U.S. economy and the coming restructuring of the European, both of which can put low-cost Japanese capital to work. Maybe, too, the Japanese will restructure their own economy, using mergers, acquisitions, divestitures, and refinancing of companies to great advantage at home.

When they do, maybe they will adopt sophisticated Western corporate-finance methods to accomplish the restructuring. Maybe then the competitive advantages of samurai finance will begin to fade.

# 10

# Honorable
# Partners

In August 1973, after Goldman Sachs and I had gotten to know Mitsui & Co. pretty well, and indeed, had done some other business with them, I was astonished to read in the newspaper one summer's morning on the beach that Mitsui had agreed to purchase 45 percent and Nippon Steel 5 percent of AMAX, Inc.'s aluminum business for $125 million. There were, reportedly, no investment bankers involved, even though this was the largest Japanese acquisition of shares in a foreign company, and would continue to be so for many years, and one of a very small number of international transactions of any type to exceed $100 million. Having no bankers involved meant that Mitsui had not selected someone else to advise it, which was good; but also that they had not thought enough of our capabilities and potential to be of assistance that they had wanted to take us into their confidence, which was not.

It turned out that Tatsuro Goto had gotten together with his friend

Ian MacGregor, who was chief executive of AMAX previous to heading British Steel for Mrs. Thatcher, and worked the deal out "unofficially." AMAX is a metals company, and one of the world's leading producers of molybdenum, an important ingredient in steel-making. AMAX had been a supplier to Japan for some time and had had many occasions to work with Mitsui, which was a major importer of raw materials and exporter of finished products for the Japanese steel industry. AMAX was diversifying into other metal businesses at the time, which would be a highly capital-intensive affair. Its aluminum business, called Alumax, was the fourth or fifth largest company in the industry in the United States. It was growing, consuming capital, and producing a great deal of metal that was shipped to Japan.

The opportunity to take Mitsui on as a partner offered AMAX the chance to take some money out of the aluminum business, solidify a long-term relationship with an important purchaser of aluminum, and strengthen a relationship with one of Japan's most important companies, one that was involved with all of AMAX's businesses and with whom other partnerships might be possible. Many projects in the metals business were syndicated, so this deal was not so different from what AMAX did every day. As a result of the transaction, AMAX would dispose of 50 percent of Alumax, and could avoid consolidating it (and its large and growing amounts of debt) in its consolidated financial statements. AMAX would continue to control the management of Alumax.

From Goto's point of view, the Alumax deal was just what Mitsui needed. Japan was running out of aluminum capacity, and it was too expensive to create more there—aluminum-making required vast amounts of electric power, which was tightly regulated and expensive in Japan. Industry would prefer to purchase aluminum in countries in which the cost of production was low and then bring it back to Japan. Mitsui had been looking for sources for aluminum production around the world. The United States and Australia were the most reliable areas and were currently producing a surplus. Indonesia, Brazil, Canada, and other locations were also being studied, but the United States was the best place to get aluminum from and would be for the foreseeable future.

Mitsui and the other larger Japanese trading companies were at the time looking to invest in their long-term sources of raw materials, to assure availability in the future, and as a hedge against price escalation, something that they were already beginning to be concerned about. Japan, as every schoolchild was told, is totally dependent on foreign sources of raw materials for its manufacturing industries. It must import oil, ores, coal, and special materials in order to survive. Why not, the trading companies came to realize in the early 1970s, purchase ownership of some of the sources? Mitsui had already entered into joint ventures in Australia, in Brazil, and in Iran for these purposes. They were already investors in an iron-ore deal in Australia with AMAX, whom they respected and related well to.

Although this was a large deal, Mitsui should have no real problem financing it. As a courtesy (one assumes) the transaction had to be shown to Mitsui's and AMAX's good customer, Nippon Steel Company, then the world's largest manufacturer of steel, who decided to take a 5-percent interest, perhaps as a small hedge against the displacement of steel by aluminum. The two Japanese companies would borrow from Japanese banks, with a guarantee of the Japanese Export-Import Bank (which financed raw-materials investments of this type), at partly subsidized rates. The Alumax shares would pay a decent dividend, which should cover the debt service in the early years.

The deal was a natural: two sides that knew and respected each other, each side getting something of strategic value, the finances working out. What would one need investment bankers for, presumably the AMAX side suggested, when we can work everything out harmoniously together?

One thing might have been to have some independent advice on the purchase price. Mitsui had absolutely no way of assessing the market value of the Alumax shares being offered it, no way of determining who else might be willing to purchase them, no sense of what their purchase of the shares would mean to AMAX; in short, they had no idea of how much they should offer to pay for them.

Goto is a very experienced businessman, a tough man in the give-and-take of commodities trading. He is also a charming and extremely likeable person, with very polished social and political skills, not at all like your ordinary commodities trader. His own subordi-

319

nates would say of him, "He's very tough, makes everyone work very hard; he is very cheap, he always flies tourist class to show us he wants to save money; and he is very loyal, once he likes you he will like you forever."

Goto was also a bit in awe of his American corporate counterparts and of the power of their companies. He was also extremely sensitive, as are all Japanese of his generation (those educated before World War II), to any anti-Japanese sentiments, of whatever kind, and would always maneuver carefully to avoid any potentially embarrassing situation. Haggling with a dear friend over the price of something that he wanted, and could afford, might turn their harmonious relationship into one of conflict. He was much less prepared than some of his counterparts to handle any sort of conflict in dealing with U.S. companies on U.S. soil, dependent as he was on the goodwill of his customers and suppliers. If an investment banker came in and told him the transaction was only worth $100 million, what would he have to do?

Perhaps Goto and others like him were willing to pay a little more to make the deal go through more easily. He knew these people quite well and was loyal to them. He believed that they would try to get a good deal for themselves, but not to cheat Mitsui.

The transaction was completed and Alumax flourished, despite the problems of the oil shock, declining prices, and escalating environmental problems. After a few years, however, the difficult times in the metals industry began to tell on AMAX's performance. In 1978 it became the target for an unfriendly takeover by Standard Oil Company of California (SOCAL), which had acquired 20 percent of AMAX in a friendly transaction in 1975. After several months, SOCAL agreed to suspend its efforts to take over AMAX, but it held on to the substantial minority interest in the company that it had acquired. AMAX tried to find someone else more acceptable to AMAX to come forward and offer to purchase SOCAL's shares. Naturally, AMAX thought of Mitsui.

Goldman Sachs, along with another firm, was working on the problem of defending AMAX against a further attack from SOCAL or any other unfriendly party. I was asked to join the team with respect to finding a Japanese partner, especially Mitsui. I was reminded by the AMAX financial staff how easily the Alumax deal had gone through and how close the relationship with Mitsui was. The price that would have to be paid to entice SOCAL to sell its AMAX shares was very high. The market had dropped since SOCAL's purchase, which reflected a premium over the market price at the time. Price could easily be a big obstacle in this case. The 20-percent block of AMAX owned by SOCAL was then worth about $400 million, valued at the price SOCAL was offering for the rest of AMAX's shares.

Everyone on our team believed that Mitsui was the best candidate for purchasing the shares, indeed possibly the only one that would have been acceptable to AMAX, as a friendly partner who had the resources. I was very doubtful.

Investing in Alumax was one thing. It tied directly into the strategic sourcing of aluminum, which was an important matter for Japan's economy at the time. It was conceivable that, if Ian MacGregor (who had left AMAX by this time) asked his friend Tatsuro Goto if Mitsui would like to buy more of Alumax, Goto might persuade his colleagues in Tokyo to do so. But investing in AMAX itself as a "white knight," at an extremely high price, would be a very different matter. Most of AMAX's production in 1978 was of metals and ores that Japan did not need at the time, or had already made supplier arrangements for. Getting involved in a "fight" in the United States would be a very high-profile thing to do and would surely complicate Mitsui's existing relationship with SOCAL, a major supplier of petroleum to Japan. It could also create "anti-Japanese feelings."

I further explained to my colleagues that making such an investment fitted into no pre-established policy groove, as the purchase of Alumax did, and to get approval from the Mitsui board would require a relatively long period of study by all the departments involved and the building of a consensus within the company. Unless this was something Goto had talked about already with his colleagues, I doubted that he would have any interest.

We went to Japan—whence Goto had returned a few years before to become deputy chairman of Mitsui & Co.—to ask him. He received us cordially, but made it sufficiently clear during this one meeting that neither he nor the company would like to hear any more about it. His was a totally flat, though very polite, turndown.

321

The experience of both the acquisition of Alumax and the refusal to acquire AMAX shares, the first being a matter I was not involved in, the second being a transaction that never happened, taught me two, possibly three, general lessons concerning Japanese behavior relative to mergers and acquisitions. These lessons would be proven out over and over again.

First, no one can sell a Japanese company something it has not already made up its mind that it wants. Second, once they have decided to do something, a process that will take all kinds of factors into account (many of which U.S. companies would not think were relevant at all), nothing can stop them from completing the mission. And, third, but related to the second, price is relatively unimportant once the decision to go ahead has been made. The Japanese company can expect, in its eagerness, to pay a higher price than an American company might, but so what? It won't matter in the long run. In the short run, they want everybody to be happy.

Japanese firms are ambivalent about mergers and acquisitions. Under some circumstances, mergers can be accepted as unavoidable or strategically necessary, but otherwise, Japanese businessmen will tell you, mergers or acquisitions are dreadful things, to be avoided almost at all costs as a sign of the complete failure of the company being sold and of massive headaches for the surviving company. Each company in Japan has its own strong identity, its own culture. Its employees are extremely loyal to the company and, as in feudal days, expect to have difficulty serving another master. To integrate two corporate cultures is always difficult, no matter how similar they may appear to be on the surface.

To acquire another company by the aggressive means common in New York and London is considered to be destructive of the balance and the harmony between the company, its employees, customers, and suppliers, and its shareholder-group members. To do such a thing only to make money would subject such a company to great public scorn, which few, if any, companies could tolerate. Besides, because of cross-shareholdings and group affiliations, the unwillingness of bankers to become involved in such unfriendly transactions, and the possibility of difficulties from government regulators, the

322

completion of a hostile takeover would be extremely difficult. There have been a few occasions when unwanted takeover attempts have been tried, threatened, or rumored, but none has ever been completed under ordinary circumstances for a public company in Japan.

One such attempt occurred when one of Japan's most notable corporate mavericks, Tamaki Takahashi, president of the family-controlled Minebea Corporation (formerly Nippon Miniature Bearing Corporation), which had earlier acquired a controlling interest in a U.S. bearings company, attempted a takeover of a Japanese precision-equipment maker, Sankyo Seiki. Takahashi was able to acquire secretly a 19-percent interest in the company through purchases of convertible bonds in the Euromarket. Once this position was achieved, Takahashi did not launch any kind of tender offer; instead, he resorted to behind-the-scenes pressure and threats, hoping to squeeze the company into submission. While this was going on, however, a British financier was able to assemble a 23-percent interest in Minebea, which he sold to Trafalgar Holdings Ltd., a U.S. company controlled by Charles Knapp, who was also chief executive of American Savings and Loan, soon to become the largest bankruptcy ever in the U.S. savings-and-loan industry. Knapp hoped to persuade Takahashi to reacquire the position in a "greenmail" transaction. Takahashi refused to go along, and Knapp attempted to put together a hostile takeover. He found, to his surprise, that not a single Japanese bank or securities firm would associate itself with the effort. Takahashi, meanwhile, conveniently merged Minebea with a Japanese kimono-manufacturing company and issued new shares to institutional holders, thereby diluting Knapp's holdings. Knapp thereafter withdrew. Minebea, however, continues to hold its Sankyo Seiki shares, but has not succeeded in taking the company over.[1]

In another much publicized situation, a group of corporate raiders led by Kozo Kotani managed to take control of the board of directors of a large aerial survey company, Kokusai Kogyo, which had a market capitalization of about $1.65 billion, at the annual shareholders meeting. In this case, however, the president of Kokusai was outmaneuvered by his own father, the founder of the company, who had pledged his shares to the raiders in order to force the son out of his position. Not a typical case at all.

323

Strangely, Kotani's successful efforts with Kokusai Kogyo led to another bizarre Japanese takeover episode. Prior to the Kokusai Kogyo matter, Kotani had been stalking an automobile headlight manufacturer, a major supplier of Toyota, with a market capitalization of $6 billion, Koito Manufacturing Co. Kotani hoped to secure a sufficiently troublesome position in Koito to induce it or Toyota to buy him out in a form of Japanese "greenmail." Strapped for cash after "winning" at Kokusai Kogyo, Kotani sold his position in January 1988 to Kotaro Watanabe, an uneducated, self-made billionaire who had made his money selling used cars and dealing in real estate. Watanabe had acquired some Koito shares in December 1987—with Kotani's shares Watanabe soon amassed a block of 25 percent of the company. However, his efforts to get Koito or Toyota to take him out at a profit were unsuccessful. He decided to shop the block elsewhere, and ultimately came upon Boone Pickens. After several meetings Watanabe persuaded Pickens to buy a 20-percent interest in Koito at a price of Yen 3,050 per share. By mid-April the stock had risen to Yen 4,900. The matter became a big issue; the institutional community ceased trading in Koito to drive the stock down. The stock did drop a bit before Pickens tried his usual approach of announcing that he was a major shareholder with lots of ideas about how Koito's business could be improved and that he would like to be elected to the board. All of this was a shock to the Japanese. Pickens went to Japan, had an icy meeting with Koito's president who rejected his requests, and has since sat back to watch how things might turn out. With over a billion dollars tied up in the investment, it is doubtful he can wait for too long. The Japanese, who have rallied to support each other, can wait forever.[2]

Acquisitions, to be successful in Japan, have to be voluntary, must be thought to be necessary, and have to be very carefully and patiently implemented. They have occurred in Japan more often than many people think. In 1971, two large banks, Dai Ichi Bank and Nippon Kangyo Bank, were combined at the suggestion of the government, which was trying to "rationalize" the banking industry in Japan. The transaction created the Dai Ichi Kangyo Bank, Japan's largest. For years it has been necessary to make staffing assignments, promotions, and retirements fit into a carefully orchestrated formula

324

for treating former employees of each bank evenly and fairly. In Tokyo, it is generally thought, more than a decade after the merger of the two banks, that the continuing requirement for balancing personnel assignments has been a substantial cost to it in terms of expense and the quality and timeliness of decision making.

Before the Dai Ichi Kangyo merger, there were a number of other rationalization moves promoted by the government, the largest of which was the creation of Nippon Steel Corporation from the Fuji and Iwata steel companies in 1968. Since Dai Ichi Kangyo, however, the government appears to have refrained from promoting strategic rationalizations through mergers.

Many smaller scale acquisitions have occurred in Japan because the companies involved wanted to create strategic rationalizations of their own. Sales of privately owned businesses in Japan go on much as they do in other countries. In some industries, in particular the "superstore" segment of the retailing industry, mergers have proliferated. Superstores are retail chain stores located in suburban areas that sell food, clothing, and various appliances and hardware. They are to crowded, land-poor Japan the rough equivalent of an entire shopping center in one building. Masatoshi Ito was one of the early pioneers of the superstore chain, though, unlike almost all of its competitors, Ito Yokado did not expand through the acquisition of smaller, regional retailing companies.

Many of the giant trading companies, if not all, have acquired smaller traders along the way to becoming as large as they are. The appeal has been simply to join forces with a winning team on its way to a bright future.

Very few of these transactions involving voluntary acquisitions of companies in the same field have involved public securities transactions such as cash tender offers or exchanges of shares of the acquiring company for those of selling shareholders. Friendly acquisitions, when they are done, usually don't involve publicly owned companies as sellers. More typical of Japanese transactions is the situation in which company A and company B decide to formalize their friendly, perhaps mutually dependent, relationship by purchasing a stake in each other's shares. This is only done with consent and usually involves back-room negotiations about where the shares are to come from. Sometimes a company will ask an institutional investor to sell some of its relationship holdings to the new friend; sometimes they involve the issuance of new shares directly to the company; occasionally they are purchased in the market. In recent years, high levels of corporate liquidity have made it possible for many companies to increase their cross-holdings in and by others.

Another form of merger that is common in Japan is the rescue merger, usually by one group member of another. Perhaps because of the safety net that most Japanese companies enjoy, a number of companies overextend themselves and become threatened with bankruptcy. When this occurs, the group, led by its bank, attempts to restore the company to solvency by restructuring it. Sometimes the damage done to the company is so severe that a restructuring won't do, at which time a merger is arranged. These have usually involved Japanese companies with little international visibility, although the merger of Ataka & Company into C. Itoh, a large trading company, occurred after Ataka had become hopelessly entangled in a large overseas oil refinery that collapsed.

In 1986 the seventh largest savings-and-loan institution in Japan, the Heiwa Sogo Bank, was found to be insolvent because of fraud. The Ministry of Finance drew up a list of potential merger partners. It would select one and then persuade the lucky bank that it had to acquire Heiwa Sogo, which of course was a mess. Somewhat to everyone's surprise, Sumitomo Bank, emulating the style of Citibank, with which it was often compared, stepped forward and volunteered to take over Heiwa Sogo, which was fine with the Ministry of Finance. Sumitomo knew that a rescue merger would be arranged and that Heiwa Sogo, which had a large customer and deposit base and about one hundred branches, offered some attractive strategic assets.

Sumitomo knew that if Heiwa Sogo had been healthy it would never have been available through acquisition. It also knew that most other banks would want nothing to do with the many problems of acquiring another company as troubled as Heiwa Sogo. It put two and two together and volunteered opportunistically for the job. However, taking the initiative in this case was controversial, even within the Sumitomo Bank. Later, as integration of Heiwa Sogo into Sumitomo proved to have its share of difficulties, the acquisition was considered to be a major factor in the unprecedented early retirement of Sumitomo Bank's president Koh Komatsu in 1987.

Despite their conservatism at home, however, Japanese companies are becoming quite familiar with foreign acquisitions. Approximately $12 billion of investments in controlling or minority interests in foreign corporations or real estate have been made by Japanese corporations, most in the last two years. Although it is quite usual in Japan for two sets of standards that govern corporate behavior to exist, one for Japanese domestic matters and another for foreign, most Japanese acquisition activity abroad has kept close to standards that would be acceptable also in Japan.

Acquisitions tend to involve companies the Japanese company already knows and with which it feels it has had friendly relations. Most acquisitions have been small in relation to the size of the Japanese company. Often the Japanese company prefers to invest only in a minority stake in the company so as to be able to rely on a foreign partner and/or local management to take care of all aspects of running the business for it.

Acquisitions have been confined to businesses the Japanese company already conducts and needs to expand internationally. They are not motivated by financial factors so much as by industrial factors. The acquiring companies recognize that they must adapt themselves to a more global environment, manufacture abroad, and become more international and less Japanese in their overseas activities. Nevertheless, many companies prefer to make the transition gradually. If they can find a way to become more international through a partnership arrangement, such companies would probably prefer it to the responsibilities of 100-percent ownership.

Some companies are comfortable with wholly owned acquisitions. Bridgestone's $2.6-billion acquisition of Firestone was for 100 percent of the company, although it acquired the 100-percent interest only after an earlier effort to acquire part of Firestone's tire business was interrupted by a takeover bid from Pirelli and Michelin. Sony's $2-billion acquisition of CBS Records was for 100 percent, as was Nippon Mining's $1-billion acquisition of Gould, a company with which it had had a joint venture for several years. In each case, the

327

buyer had known the seller well before the acquisition. A number of large real estate transactions have also been completed by Japanese buyers without any foreign partners. The drift is changing, perhaps, as more Japanese companies gain confidence and observe the results of actions taken by other Japanese companies.

With only a few exceptions, hostile takeover activity abroad is not pursued by the Japanese. The exceptions involve the same company, Dai Nippon Ink, whose president, Shigekuni Kawamura, received an MBA from New York University in 1958. Kawamura is the son-in-law (and legally adopted son) of the patriarch of Dai Nippon Ink, a chemicals company specializing in printing inks, which the family still dominates. In 1986, Kawamura launched a tender offer for Sun Chemical Corporation, whose chairman, Norman Alexander, had been a friend and business partner of the Kawamuras for thirty years.

In the summer of the preceding year, Kawamura had approached Alexander about purchasing Sun's graphic-arts operations. Alexander asked for $600 million, which Kawamura and his bankers, Dillon Read, thought was too high. Ultimately, Kawamura tendered for the whole company, which drove it into the arms of Chromalloy American Corporation. Kawamura then purchased the graphic-arts division from Chromalloy for $550 million.[3]

Although Kawamura had violated one Japanese tradition by launching a hostile bid (which was the only way he and Dillon Read believed the acquisition could be made), he had stuck to the principle of buying what he knew.

A year later, no doubt emboldened by his experience with Sun, Kawamura made another unfriendly offer for Reichhold Chemical Company, which resulted in Reichhold's board deciding to find another buyer for the company. This time, Dai Nippon Ink raised its bid and obtained the financing it needed from four willing Japanese banks in New York. In September, Reichhold accepted Dai Nippon Ink's offer of $540 million.[4]

In the space of a year, Kawamura had bought two U.S. companies for a total of more than a billion dollars. In March 1988, Dai Nippon's share price was trading at nearly 100 times earnings, a market capitalization of $4 billion.[5]

In 1986 a survey was conducted in Japan by the Ministry of Justice and Daiwa Securities of nearly seven thousand companies to determine their attitudes about acquisitions. More than half of these companies responded that they had used lawyers specializing in corporate acquisitions, and 12 percent reported that they had made an acquisition during the preceding five years.[6] It is likely that some of the work being done on acquisitions involved only "studying" the situation. It is also likely that a large part of the effort was aimed at foreign rather than domestic acquisitions, and many of these involved small companies. Still, the study indicates that preparations are being undertaken for greater activity.

The basic lessons that we learned in the Mitsui-Alumax case still apply, however. It is extremely difficult to persuade Japanese companies to act quickly, decisively, and opportunistically when offered something they know nothing about or are not prepared for. Many Japanese companies have been asked to become *white knights,* or friendly acquirers of companies being raided by someone else, and they have always declined. Until they have had a chance to study a matter internally, and to form a consensus, all of which takes time, they are almost impossible to move. Once they have made up their own mind, however, they can be equally difficult to stop. Yet, in today's market, if the companies are not prepared to initiate bids for companies they have selected, knowing that these bids might be opposed, then they are unlikely to succeed in acquiring the companies they have targeted.

This leaves Japanese companies with a limited field of play in a merger market that is full of bids, counterbids, tactical maneuvers of various kinds, quick actions, and high prices. Nonetheless, we can expect Japanese companies to watch and learn and to become as aggressive as they need to be in order to accomplish aims that are important to them. They will continue to be apprehensive about attracting anti-Japanese sentiments, however, which might be best assuaged by offering such high prices for the acquisitions they undertake that no one could either match the offers or complain about them. Bridgestone's final offer of $80 a share for Firestone, $22 per share or nearly $700 million higher than the price offered by the Michelin/ Pirelli syndicate, was perhaps an example. As we know, however,

Japanese companies can raise the money for such acquisitions inexpensively and therefore perhaps can afford to be generous if that is what it takes to win over a reluctant bride. The way the game is played today, it usually is.

Japanese companies have been comfortable with joint-venture and minority investments for years, and it is probably realistic to assume that more of these will occur outside of Japan. However, there are a number of pitfalls that arise from such investments.

Japanese experiences with joint ventures have mostly been in terms of small Japanese companies set up in partnership with a foreign company with know-how or technology that the Japanese lacked. There were a great many of these set up in the 1950s and 1960s. Some were quite successful, most remained dwarfs. The venture was isolated to a narrow field and, for one reason or another, was not allowed to stray. Foreign companies contributed knowledge, but Japanese managed the businesses. The foreign partners were reluctant to interfere in management or financial decisions. Insofar as these operations were located in Japan, the Japanese side dominated, sometimes to the distress and confusion of their partners.

Bill Brown is a man with deep-rooted Japanese connections. He lived there as a teenager while his father was with the U.S. Army of Occupation, grew up to become a Jesuit priest, and was assigned to Japan, where, among other things, he taught economics at a university in Tokyo. Subsequently he was sent to Harvard Business School, where he obtained an MBA. About this time he left the priesthood, married a Japanese girl studying in Boston, and returned to Japan, where he was hired by Citibank to run a joint venture it had set up with Fuji Bank in management consulting. This venture was one that George Moore thought up on the spot when visiting old friends at Fuji Bank who wanted to "do something" with Citibank to cement further their mutual relationship.

Brown was made co-managing director of the venture along with a middle-aged, middle-ranking Fuji Bank employee. Brown had had some experience as a management consultant at Peat Marwick, but no one from Fuji Bank had—indeed, they were not even aware of what management consultants did. Brown was one of a very small

330

number of foreign employees of the venture, which, before long, had a staff of nearly fifty. After a couple of years of this, Brown left the venture to join Goldman Sachs as its first Tokyo representative.

Describing the experience, Brown pointed out that one day the bank found itself in this new, odd business that really didn't exist in Japan and the next day had detailed fifty people to it. Their job was to learn all there was about management consulting and report back to Fuji Bank. They were to figure out what consulting services to offer and then offer them. Naturally, none of the staff was qualified to perform the services they were offering, but that didn't matter because Fuji Bank was reluctant to charge much, if anything, for their work because such services, at least those performed by the bank, were free in Japan. The venture of course was in the red from the beginning and unable to extract itself. This didn't bother Fuji Bank, who had to employ these people somewhere, but it was disagreeable to Citibank. The Americans wanted the venture to come up with a cogent array of services based on management and strategic-investment studies and market them for McKinsey-like fees to Fuji Bank clients. This represented a great cultural confrontation. Fuji Bank was not about to hustle their own clients aggressively. They preferred that the venture offer informational services about Japan to foreign clients of Citibank's, which would be aimed at creating new client relationships for Fuji Bank. As the Fuji Bank people did whatever the Fuji Bank co-managing director said, that is what the venture ended up doing. In the end, Citibank lost any hope of being able to influence the venture after Brown left, and the thing was ignored thereafter.

Perhaps most joint ventures are better thought-out than this one was, although, if so, that is not clear. Many joint ventures, and investments in minority ownership positions, suffer common problems that are difficult to survive. Often they are the product of two great men wanting to do something together, which is always difficult to implement. Sometimes they result from the momentary convergence of two separate strategic viewpoints; sometimes they arise because one side wants to learn something it doesn't know from the other.

The biggest difficulty with joint ventures is that as time passes the circumstances that brought the two together in the first place be-

331

gin to disappear. Strategic considerations of two dissimilar organizations, not to mention individual managers, change constantly. Unless the venture makes money on its own, or the minority investor is completely satisfied with the return he receives, these ventures are unlikely to survive or to offer much in the way of strategic satisfaction to either party.

In the late 1970s, Koichi Tsukamoto, the president and founder of Wacoal, decided he wanted to get into the lingerie business in the United States. He had identified a modest-size U.S. manufacturer of ladies' undergarments, Olga Inc., and decided to pursue it. There were several larger U.S. companies he could have chosen, many of which would have jumped at the opportunity to have Wacoal as an investor, but Tsukamoto only had eyes for Olga, a company controlled by Jan and Olga Erteszek, two Polish émigrés living in Los Angeles. The Erteszeks were not interested in selling their company, but they were intrigued by the prospect of a global business partnership with Wacoal, which had large manufacturing facilities in Taiwan and Thailand and, of course, a dominant position in the Japanese market.

I was asked by Wacoal to try to arrange for a substantial minority investment. The negotiations were extremely difficult, because Tsukamoto knew exactly what he wanted and was eager to get on with things while the Erteszeks hesitated and changed their minds. They were attracted to having a Japanese partner, but were very leery of Tsukamoto's aggressive, dominating style. Finally, a 13-percent investment was agreed, with several pages in the purchase agreement relating to how the deal should be unwound if necessary. Once the arrangement was concluded the Erteszeks tried to keep the Japanese at arm's length, but this was difficult to do. None of Tsukamoto's employees wanted to report to him that he had tried to attend a meeting but was kept out; if they wanted to succeed with the boss, the employees had to show true Japanese spirit and get in there no matter what.

Before long Olga began to feel under siege. Despite its relatively small ownership position, Wacoal was beginning to dominate the relationship. Almost as soon as it had begun, the relationship began to sour, then turn bitter. Ultimately, Olga bought Wacoal's shares back.

Wacoal had not tried to run the company; it regarded itself as a totally passive investor. However, it wanted to get on with joint marketing and manufacturing activities, about which it had strong ideas. Olga thought that these activities would soon overwhelm the company and therefore believed Wacoal's involvement with the company was anything but passive. Erteszek and Tsukamoto saw each other infrequently, and even then they needed an interpreter to converse, so they did not contribute much to smoothing out the problems that were growing among their associates. Olga employees had never seen anything like the fierce determination of the Japanese, and the Wacoal employees dealing with Olga personnel certainly regarded them as unsamurailike.

There have been, of course, a number of successful ventures, of which the 32-percent investment in Amdahl, the U.S. computer company, by Fujitsu was one of the more noteworthy. Fujitsu was able to perform some of the manufacturing required by Amdahl, and in exchange it learned much about the business of designing, manufacturing, and selling computers in the United States. Amdahl went public in the United States, making a lot of money for its Japanese parent, which increased its holdings further in 1987.

Most of these joint ventures and minority investments, like the outright mergers, involve industrial corporations. Much more frequently than in the days of Mitsui's investment in Alumax, these transactions involve advisory services that banks are increasingly capable and willing to give.

The transactions today are larger, more complex, and often involve skills in structuring and negotiation, which investment banks, commercial banks, and Japanese securities firms are all seeking to provide. Several large automotive industry ventures have occurred, such as the General Motors–Toyota joint manufacturing facility in California, and the proposed $1-billion investment by Ford in a new facility in Ohio to manufacture minivans designed and engineered by Nissan. Most of the field of activity today is outside of Japan; many firms, however, keep a sharp eye on the evolving possibilities of foreign companies acquiring Japanese businesses (two such deals involving reasonably sized companies have been done). In the longer run (perhaps the *much* longer run), there is also the prospect of merger-

and-acquisition activity between Japanese companies becoming comparable to what goes on in New York and London today.

Of at least equal importance to bankers, however, are the merger-and-acquisition activities undertaken by the Japanese in the financial services area. These have included several different types of transactions. There have been several acquisitions of U.S. banks and branches by Japanese banks, mostly in California. There have been acquisitions by banks of finance and credit-card companies. One bank, the Industrial Bank of Japan, has ventured in the United States into financial advisory services through the acquisition of J. Henry Schroder Bank and Trust in 1985 and into government securities trading through the acquisition of Aubrey G. Lanston in 1986.

Several Japanese trust banks have also entered into investment-management joint ventures with U.S. banks to capture some of the overseas-funds-management business of Japanese pension funds.

Most Japanese banks have set up merchant-banking subsidiaries in Europe to participate in Euromarket underwritings. One, Sumitomo Bank, also acquired in 1984 a controlling interest in a Swiss bank, the Lugano-based Banco del Gottardo. This acquisition was made from the liquidators of the failed Italian bank, Banco Ambrosiano, whose president, Roberto Calvi, was found hanged on scaffolding underneath Blackfriars Bridge in London in 1982. Perhaps this opportunistic and successful transaction gave Sumitomo the courage it needed to pursue aggressively the ailing Heiwa Sogo Bank two years later.

The investments by Japanese financial institutions in minority interests in U.S. investment banks have, however, attracted the most attention from competitors, regulators, and even from potential investees.

There have been three of these investments so far: 18 percent of Paine Webber by Yasuda Mutual Life; 13 percent of Shearson Lehman Hutton by Nippon Life Insurance; and 12.5 percent of Goldman Sachs by Sumitomo Bank. The Goldman Sachs deal with Sumitomo Bank in 1986 was the first of these and the only one to involve banking regulatory questions under the Glass-Steagall and Bank Holding Company acts.

Following our successful cooperation on the Mazda Motors plant

334

financing in Michigan, Tokuyuki Ono, Sumitomo Bank's senior executive in the United States, and I had a number of occasions to work together on other projects. We would meet for lunch every couple of months with our respective colleagues and suggest things we might do together. Sumitomo had already increased its lending business in the United States and was starting to branch out into other things. We suggested that Sumitomo acquire a finance company, which could be used to finance dealer inventories of their manufacturing clients, industrial and consumer receivables, and other things. It could create a company like General Electric Capital Corporation, which was then, and continues to be, the national role model for such companies.

Ono was noncommittal, but said that the head office was currently studying the advantages of a finance company, and when the study was finished we could talk again. No one was more surprised than I when I learned that the "finance company" study Ono had referred to was in fact a study that led the bank to approach Goldman Sachs to explore an investment in the firm and ways the two of us could work together in the international arena.

Two things had happened during the preceding year to propel Sumitomo in this direction. The first was that Sumitomo became concerned that its largest clients would be drawn increasingly into international capital markets for their financing requirements, and Sumitomo had little capability to serve them there. It feared losing its most important clients to either Japanese securities firms or foreign-banking and investment-banking firms. The fact that Sumitomo already had a merchant-banking subsidiary was of little weight. Their clients would seek out top players to handle their business, and Sumitomo Finance International was unlikely to be seen in that category. The second development was that Ono was singing the praises of Goldman Sachs to his colleagues in Tokyo; as our relationship improved, we offered Sumitomo, and they accepted, a large loan of subordinated capital to the firm.

Sumitomo was blocked in Japan by Article 65 regulations, which separate banking from investment banking, although most observers thought that these regulations would not last forever and that already there were signs that foreign banks were finding ways around them.

335

Sumitomo was also blocked from owning 5 percent or more of a U.S. investment bank by U.S. banking regulations. It could build a first-class investment banking operation in Europe if it spent enough money and waited long enough, but the limitations on the business by Japanese regulatory authorities were such that it had to be realistically modest in estimating the market share it would achieve. And by the time Sumitomo might achieve it, its major clients would probably have forged new banking relationships with other firms. It was a tough problem. Sumitomo consulted McKinsey & Co., which had a good reputation for performing such studies in Japan.

In the end, Sumitomo decided to form an overseas finance company, which would, like General Electric Capital Corporation, be both an active issuer of securities and a lender and investor in transactions. It would also acquire an interest in a major international investment bank, with whom it would then set up joint ventures in London and Tokyo (squeezing Article 65 as much as possible) to handle international capital-market transactions. Early in 1986, we received an approach from an intermediary, asking if we would be willing to meet with Sumitomo's president, Koh Komatsu, to discuss a proposal. Why not? we replied.

Komatsu came in to see John Weinberg in what he referred to as ''disguise'' (wearing dark glasses, I guess) so no one would know he had been there. He explained the bank's views in general terms, left a memorandum behind, and departed.

John Weinberg had told Komatsu that the firm was flattered that he had come to visit, but that it had no plans to do anything of this sort. We would study his proposal and get back to him in a while, probably not too soon, as the matter came as a surprise and would take time to consider properly. Komatsu said fine.

At the time, Goldman Sachs was considering what, if anything, it might do to raise additional equity capital for the firm, which had grown rapidly for the past several years. As a partnership, one of the few investment banks continuing to be so organized, capital had to be paid out to retiring partners. Over a period of about ten years, the entire capitalization of the firm would be paid out and had to be replaced. Incorporation and the sale of shares to the public, which

336

most firms had done, would solve the problem, but the firm liked operating as a partnership and wanted to continue that way. To do so, however, would require finding new sources of capital, either as subordinated loans (which had to be repaid after ten years or so) or as additional limited-partnership capital, which was roughly comparable to preferred stock. This had been a continuing matter of consideration for many years, since the middle 1970s, when the increasing capital requirements of the investment-banking business made the issue an important one.

On the other hand, the firm had always avoided tieups, or joint ventures, where it and another party had to be committed to one another in such a way as to preclude working with others in the banking industry. To have exclusive relationships in the investment-banking business was dangerous. The firm had turned down joint ventures with Japanese banks before for this reason. As things turned out in the other cases, due to personnel changes, changing strategic emphases, and other reasons, considered in retrospect, we were lucky to have avoided those arrangements. Had the relationships been made, it is likely that not only would the firm's other banking relationships have suffered, but the benefits we were expecting would have been very slow in coming. This was not the sort of situation that Goldman Sachs wanted to encounter with Sumitomo.

Indeed, there were a number of joint ventures in the 1970s involving British merchant banks and Japanese commercial banks. Sumitomo–White Weld (involving Crédit Suisse White Weld) was one of these. They were all heralded at the time as excellent marriages between Euromarket new-issue and corporate-finance skills and powerful Japanese banks trying to keep their clients away from Nomura and Daiwa. These ventures brought a few deals to market, but altogether they were far fewer than the deals being brought by the Japanese securities firms, who had by this time managed to persuade their clients that they should be allowed to lead manage their deals instead of foreign banks doing so. The merchant banks were uncomfortable being unable to try for business from clients outside their partner's orbit. The banks didn't find the merchant banks to be all that convincing to their clients, and they too wanted to be free to

337

work with the Americans or the Swiss if that was what their clients wanted. After a few years, all of these joint ventures were disbanded, the Sumitomo one being the first to go.

As Goldman Sachs considered the possibility of a Sumitomo tieup, we decided to turn the thing around and ask ourselves the question: "If we had to do it, what terms and conditions would we ask for to make it not just acceptable, but downright desirable?"

"How about a large investment at a great price, no interference in our affairs, no joint ventures, and full access to all their clients?" someone ventured. "I don't suppose they would go for it, but if we could get a deal like that, then we ought to be smart enough to take it."

"But what are they going to want—I mean *really want?*"

Eyes turned to me as the only expert on Japanese inscrutability present in the room.

"I think they mean what they say about not being able to put together their own first-class investment-banking capability. They have now decided to get themselves an investment in a top U.S. firm, and come hell or high water they will do so. They have picked us as their first choice, and they will be willing to concede a fair amount in order to pull this off with their first choice, but they have a list, you can be sure. We ought to tell them that the only way we could be interested would be on terms such as those just mentioned."

"Yes, but what do they *really* want?" This question came up every few minutes.

"Roughly this, in a declining order of importance: a solution to their present-day strategic problem of losing business to the capital markets; glory and prestige for having captured a winner; an opportunity to learn the business over the long term; and probably in last place, an attractive return on their investment."

"How do we present a solution to the present-day problem?"

"I don't know, and probably they don't either yet. Based on what they have said, it seems like they want to fold our respective Eurobond businesses together and be able to get credit in Japan with their clients for now having a capability in that market that is second to none. This is an unavoidably messy area."

"Why?"

338

"If our two Eurobond businesses were combined, it would restrict our freedom to work with clients of other Japanese banks, and with the other banks themselves. It might not go down well with the other major players in the market, the Japanese securities firms certainly would hate it, possibly our other non-Japanese bank clients would too. What do we do when they want to bring one of their important clients to the market at the wrong price, and we've got to take losses in the Euromarket to subsidize their need to impress their clients? I am sure there are other problems too, like the size of their Euromarket staff, employment contracts, compensation levels, all that stuff."

"You're right, messy—but not entirely untidiable, if need be."

"How do they get glory and prestige out of this? At our price, they might look pretty stupid instead."

"The focus will be within Japan, not so much outside. They like to take the initiative on things, to do the dramatic. They also like to show how competitive they can be. For these reasons Sumitomo is rightfully thought of as the Citicorp of Japan. Capturing a position in a well-respected firm like Goldman Sachs will give them an instant stamp of quality and international competence in the capital-market area. It would be seen as a challenge to Nomura and the other banks. Sumitomo will look once again like a leader and a prospectively tough competitor in an area they have been weak in. This is just a guess, however, because how glory and prestige are created in Japan is not something Westerners can relate to very easily."

"That's not much help, but if prestige is so important to them we ought to try to find out how to give it to them, particularly if it becomes a trade-off for some of the other things. It sounds, however, like this Eurobond matter is wrapped up in it. Maybe we ought to think about how we could live with something in that area if we had to." The Goldman Sachs management-committee partners knew something about negotiating acquisitions, the heart of which lies in trade-offs and concessions.

"What about their long-term interests? Aren't they really trying to learn everything about our business and our clients so they can compete with us as soon as their Glass-Steagall Act is removed?"

"Probably. They have a long way to go to learn all it takes to start knocking us out of the box, however. We could argue that we need to learn a lot about Japanese companies too and that Sumitomo could be our teacher. Still, it probably makes some sense to put an expiration date on whatever we might do.''

"All of these arguments sound pretty light to me," one of the committee said. "If we're going to do this, we have got to get a very attractive deal for ourselves. Such a deal may not be so attractive for them. The rest of what they are going to get out of the deal seems pretty intangible. Their return on investment is not. If that doesn't work out for them, I can't see why they would want to do this. And if that means there are going to be realistic limits on the price they would be willing to pay, then we ought to forget the whole thing right now, before we hurt anybody's feelings.''

"It might not be all that bad for them. Japanese banks have a comparatively low return on capital—Sumitomo's will be about 11 percent this year—but they have an incredibly low cost of capital if you look at their stock price. Sumitomo is now trading at about one hundred times this year's earnings and six times book value. If they were to invest $500 million or so in the firm, they could raise the money very cheaply. Then, if that gave them an eighth of the firm's profits, plus interest on their money, at our present levels of profitability they would be making 15 percent or so on their capital, after tax. That's not too bad for them, and they get other things too, which I believe they value more highly right now than the return on their capital.''

"In a way, it's an opportunity for us to sell an equity interest in the high-priced Japanese market at something like four times book value, without going public, giving up any control, and maybe getting something good out of their Japanese client base.''

"Will they agree to do this without any votes?''

"We will only know by asking, but they know they would not have much say in how we run the firm and probably don't care as long as we continue to run it as we have.''

A few months after Mr. Komatsu's incognito visit, we sent word back that we would be willing to talk providing they understood that we would expect a large investment at a high price, that they would have no say in management whatsoever, and that we did not want any joint ventures.

The bank said fine and sent a ten-man negotiation team to New York. The negotiations were being handled by one of the bank's planning departments, working directly for Komatsu and the chairman, Ichiro Isoda, once described as one of Japan's toughest managers.

No problem on the votes—they wouldn't require any.

No problem on the maturity of the commitment. They would be locked in for ten years, and if they or we wanted out after that, redemption would be spread over five additional years.

No problem on the size, or on the price, that is, they would invest $500 million for approximately a one-eighth interest, a price that valued the whole firm at $4 billion, or roughly four times book value. This could only be described as an exceedingly high price for an illiquid interest with no votes in a private partnership whose partners could withdraw capital at the conclusion of every two years.

No problem on setting up a general joint venture that, like Crédit Suisse First Boston, would involve our turning our international business over to a new company; they would not require it.

But they continued to want a joint venture to include the Eurobond business of the two firms, at least the Eurobond business that involved the issuance of Eurobonds for Japanese companies. Such a venture, further, was not to be some kind of wallflower, but a serious player in Japanese Euromarket activity. It would be based in London, but should have representation in Japan.

And, in order for their people to learn how our business works, so we could do more business together, they would like to send a number of their trainees to us, and we could do the same if we wanted to.

They were confident that the Japanese Ministry of Finance would approve such a transaction. They were also confident that the Federal Reserve Board, which had jurisdiction over such transactions by Sumitomo in the United States, would also.

Could we live with the Japanese Eurobond joint-venture business? The rest of the terms were pretty good.

341

We could.

Japanese Eurobond new issues were dominated by the Japanese securities firms, who did not, nor would we expect them to, leave many crumbs on the table for consumption by foreign competitors. To compete with them in Japan for Eurobond business was very difficult. Tying up with Sumitomo would upset the securities firms, especially all-powerful Nomura. These firms, however, were already our competitors around the world, not in any way our beneficiaries. Would we care if something we did competitively upset Salomon Brothers, or Warburg?

In fact, if we were ever going to compete with Nomura for Japanese business, we would need a leg up: partnership with Sumitomo with its influence with the Sumitomo Group and other Japanese companies could be an important marketing boost. We could help out on the Japanese end, but our main job would be to structure, price, and market the issues. In this respect, like the Japanese firms, we would have to expect to lose money by overeager pricing from time to time, but this could be made up from the profits of the hot equity and warrants issues that were often oversubscribed.

The other banks would not like our tying up with their most dreaded competitor. However, these banks were not influential with their clients insofar as controlling Euromarket new issues. Where they had something to offer was in the management of their own Euro issues, and in having them as swap-market counterparts. These issues, on the other hand, were almost entirely done at competitive bid, and those that weren't were not very numerous or profitable. Besides, in the last few years, we had only managed a handful of deals for Japanese banks (two of which were for Sumitomo), so we had very little to lose if the banks totally cut us off out of revenge.

The banks were not likely to curtail their credit lines and lending facilities. We had cultivated their credit assiduously and had won their high regard. We had large amounts of loans outstanding with the banks, and it was unlikely they would pull them. Japanese banks lent plenty of money to customers who were associated with their competitors in various ways. Anyway, we could replace any loan runoff that might occur.

342

We were asked to assess how much incremental revenue the association with Sumitomo could produce. I didn't know, but if all worked out well, after a few years, we might see something over $10 million annually in new revenues resulting from the relationship.

With all of this analysis in place, John Weinberg wanted to think about it for a while. He felt that there were several major points that the Goldman Sachs partnership had to understand and accept before any agreement could be made.

"First," he said, "we are not going to give up our independence just for some low-cost capital. We will continue to run the business the way we always have, for the benefit of the partners and employees. We will tell Sumitomo this and tell them that we will be telling everyone else in Japan the same thing." We were not going to come out of this looking like we had been "taken over" by Sumitomo.

"Second, if we take their capital, we are going to get some more from somewhere else. We will continue to get capital from various sources, and not become, or appear to have become, dependent on a single source. Sumitomo's got to be told that also.

"Third, if we proceed, we are going to have to do our best to give them what we agree to in terms of joint activity in the Eurobond business and in the way of trainees. We are going to have to take them seriously as a partner of the firm and recognize that they have needs, too."

One of us added that if we did the deal, Sumitomo would be an extremely powerful, influential, and competent partner. They would want us to teach them everything, bring them all our best deals, be tough as nails when negotiating deals in which their money was to be used. In general, we had to expect that they would not be pushovers at all. Tying up with them would have its moments when we would feel as if we were paddling in a canoe handcuffed to a fully armed, bad-tempered samurai warrior.

Negotiations proceeded, and finally an agreement was reached and announced. Press attention was considerable, as we expected. The reaction was favorable all around. Some were a little surprising, however.

343

I called the senior man at one of our large Japanese banking clients to tell him we wanted to continue business with the bank just as before.

"Very good," he said. "We of course will comply with great pleasure. I, however, will probably not be here to accommodate you. I am sure I will be recalled in shame for missing this wonderful opportunity that Sumitomo has taken. For you, it is a wonderful deal, for Sumitomo an even better one. For me it is a disaster. I will have to take responsibility for the loss of this opportunity. I may be invited by my bank to perform *hara-kiri*." He was joking about the latter invitation, but probably serious about the rest.

Reaction in Japan, regardless of the carefully structured description of the transaction that appeared in our Japanese press release, was that Goldman Sachs had been taken over by Sumitomo.

Our Tokyo office reported all sorts of reactions. They were immediately concerned that existing clients would not like it. Some probably didn't, but even if so, it was hard to predict what they would do about it.

Within a few days, it passed from the front pages of the financial press. The next step was to get the Federal Reserve to approve the application Sumitomo had made to it.

This took forever. This business of a foreign bank acquiring an interest in an investment bank right under the nose of the Glass-Steagall Act was controversial. Sumitomo had not done anything that a U.S. bank couldn't have done if it had wanted to make such a nonvoting investment in us. None ever had, so the issue was moot, except in the eyes of the Federal Reserve, which wanted to avoid criticism for either favoring investment banks over commercial banks, or foreign over domestic banks, or for being soft on Glass-Steagall. The Fed wished the problem would go away.

Knowing that it would not, the Fed scheduled public hearings on the matter and invited U.S. banks to testify, which some did. It read Sumitomo's submission carefully and asked a number of questions. It was concerned about de facto control of an investment bank lying in the hands of a commercial bank, a matter that the partners of Goldman Sachs had spent months being sure could not happen under any circumstances. We asked for a meeting with Paul Volcker

344

to explain how we felt about it. Not necessary, we were told. Time passed.

Finally, about six months after the announcement, the Fed called Sumitomo and said the permission would not be granted unless Sumitomo agreed to certain conditions. These were tough, and the Fed wanted an immediate answer. No joint ventures of any kind. No trainees. No joint solicitations of new business. No further extensions of credit to Goldman Sachs. And, the most arrogant of all, Sumitomo (whose parent in Japan was not regulated by the Federal Reserve) would have to raise the full value of the investment from the proceeds of the issuance of new capital.

The Federal Reserve must have thought that its regulatory pill would be too difficult for Sumitomo to swallow and that they would call it off.

Remembering our first lesson from Mitsui, that once they decided to go, nothing can stop them, I was not surprised when Sumitomo, more than a little annoyed, accepted the terms imposed upon them at the last minute in order to complete the deal that had been acclaimed all over the world. Their acceptance of the Fed's terms, however, would cause Sumitomo to lose face in Japan and, in general, would be an embarrassment and an irritant to them for some time.

The deal was closed, corks were popped, and speeches made. Komatsu and an interpreter attended the annual dinner for Goldman Sachs partners and their wives. We had meetings to discuss how we could cooperate in our businesses to our best mutual advantage.

We appointed a team to coordinate things at Goldman Sachs. They did too. We exchanged high-level visitors. We discussed specific financing proposals.

One day, we were told that the negotiation team was being reassigned. They would be replaced as principal liaison people by a couple of middle-ranking executives who had been on the negotiation team and who were now assigned to the new finance company Sumitomo was setting up, as planned. The principal liaison points now would be with the new capital-markets department, which Sumitomo had set up in Tokyo and which would be headed by people we had not yet met. The number-two man on the negotiating team, Ken Hotta,

345

was to be assigned to New York as branch manager, reporting to Ono. We held meetings to discuss various products and services, being careful to stay within the guidelines imposed by the Federal Reserve.

We showed them municipal-finance, real estate, leveraged buy-out and bridge-financing deals, capital-market transactions for Sumitomo Finance Corporation, mergers and acquisitions for themselves and their group, and many other things to do. They accepted some of these deals but turned most down, much like any other client. Our people found them very tough on rates and fees. Their people must have found us the same. Often it was difficult to tell which of us was to be "king," and which "slave," in the changing flow of transactions. The investment was proving to be quite profitable for them, and they were learning a lot, but not in the way they had wanted to before the Fed intervened.

There were rumors out of Tokyo that Sumitomo was not happy with the investment because it had lost the opportunity for the joint venture and the trainees. Some suggested that Goldman Sachs had put the Federal Reserve up to doing what they did, so we could get their money without having to work with them in joint ventures. "A strong firm," some Japanese sources were reported to have said, "would have been able to force Mr. Volcker into complying with our wishes." Anyone who has spent any time in Wall Street or in Washington knows how foolish such propositions are, but perhaps the idea influenced some Japanese who did not know any better.

The press in New York and London frequently reported comments about how much better a deal Goldman Sachs had received than Sumitomo Bank. The deal was diminished for both of us by losing the opportunity to formalize our joint efforts in the Eurobond market and to be able to point to these efforts in marketing Euromarket products in Japan, but what remains is a very good arrangement for both firms.

I am still convinced as I was at the time that the Goldman Sachs–Sumitomo association is a prototype for many similar deals in the future. Two others involving Japanese life insurance companies have occurred since, but I think many more will develop in the future.

•   •   •

346

The financial industry simply cannot escape the consequences of globalization. Most major firms do not wish to be bit players in the worldwide, deregulated, competitive, and exciting financial marketplace of the next century. We have already seen the enormous effect that Big Bang has had on the London financial community. Similarly, we are anticipating further changes in Europe as a result of the further deregulation associated with 1992. Stirrings within its heavily regulated capital market are being felt in Japan. Even this classically isolated society cannot resist being drawn into an era of restructuring, reform, and more open competition in both financial and industrial sectors than ever before. Perhaps in Japan the *pffpft* that finally became the Big Bang has not yet been heard, but most likely it soon will. And then the collapse of the last remaining walls and buttresses of financial Japan will follow.

Global bankers often depend on their home-market franchises for the strength and the resources to compete internationally. These are generally their most profitable operations, to be preserved and protected, if not by local regulation then by energetic effort. But in the United States the great era of financial restructuring will probably come to a close sometime in the next decade. Then American firms may see not only a decline in revenues from home-country operations, but also movement of the dynamic and profitable restructuring activity to Europe. Perhaps at the same time, perhaps a bit later, the restructuring wave will come also to Japan.

Few bankers are capable of assembling and financing winning teams in all of the principal overseas markets, where always they must compete with entrenched giants, many of which will have been fortified by deregulation and greater competition.

Those who succeed in the shifting global banking environment of the future will have preserved or developed the ability to attract plenty of capital and the best people—always the two most important resources in banking.

Capital is perhaps plentiful enough if one has a decent business, and one can periodically go to the public market to replenish if necessary. But capital that is available without disturbing the private ownership character of a firm, capital that protects the great entrepre-

347

neurial firms from the dim bureaucratic destiny of the commercial banks, is worth a great deal more. Japan is, for the moment, the world's most abundant source of capital, not just capital to be invested in firms like Goldman Sachs, but capital to be invested in financial transactions of all types, capital that can be accessed, managed, used as a lever. Having access to such capital is an enormous advantage, especially if a firm aims at becoming a major player in the fully globalized financial markets of the future.

The "best" people are not just clones of the folks in the head office, or at least the best are not *only* them. Where in the U.S. banking firms are the European and Japanese equivalents of the top producers of Goldman Sachs or Citibank or General Electric Capital Corporation? Where are their U.S. equivalents in Sumitomo? Or Warburg? Non-European firms often have a hard time attracting the very best people in Europe to their affiliates. Non-Japanese firms have even more difficulty attracting the best Japanese. But Sumitomo doesn't; it already has more than its share of Japan's financial best and brightest.

The best partnerships are those that work, that make much more money than the individual partners could make on their own, and in which one partner's strengths complement another's weaknesses, so that in the end, the competition sees mostly strengths, and no weaknesses. Japanese can be such partners, extremely valuable allies.

If bankers want to present a strong competitive front across the globe, then partnerships joining together players from different areas may be a road to success. But traveling this road means learning how to live with partners that have been chosen for their competence, their skills, and their competitive ferocity and power, not for their charm and willingness to please. Such partners can be difficult, living with them rarely easy, and requirements for skillful diplomacy numerous. Still, for many, such alliances may be the difference between making the list of the world's top global bankers, or not.

The best of the honorable partners are not pussycats, but tigers.

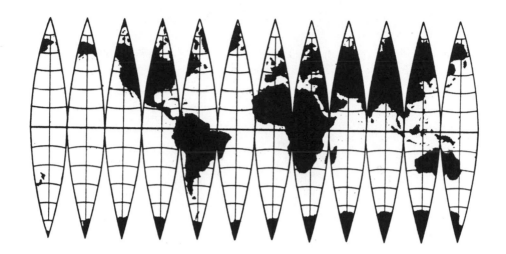

# Part IV

# Looking to the Millennium

# 11

# Megabanks 2000?

A popular parlor game among financial people during 1986 and 1987 was making lists of the top-ten global banks at the end of the century. The presumption, widely accepted, was that globalization of financial markets would draw to the center a handful of banks or financial-services firms that would be so much stronger and so much more globally integrated than the others that they would constitute a new financial "oligopoly."

Everyone who was a major player in his home market, and who aspired to global prominence in the future, had his eye on making the all-world team in the year 2000. The effort to do so would have to be great, and sustained, but the result would be worth it—a market position that could last for generations. With nearly fifteen years to go, there was still ample time to catch up to or to consolidate one's position among the world's leading fully internationalized financial institutions of the next century.

Many banks and investment banks thought like this in 1987: at least a dozen from the United States, a dozen from Japan, ten or so among the Continental giants, five or six from the United Kingdom, perhaps one or two from Hong Kong and Singapore. In addition, there were half a dozen insurance companies from the United States, Europe, and Asia who also had their eyes on becoming world-class players in a wide range of financial services.

There were others, too, who appeared interested in securing places for themselves at the top. Industrial corporations with substantial financial activities were looking at the same markets as the banks. Companies such as Ford, General Electric, American Express, Sears Roebuck, and BAT Industries already controlled substantial financial assets, provided an extensive array of services, and were looking to expand them further.

Altogether, sixty or seventy serious, apparently qualified candidates for the millennium's financial Olympic games could be identified. Of these perhaps only ten or fifteen would make it into the oligopoly. The object of the parlor game was to see whom informed observers would tap for the top spots.

There was reasonable agreement about which institutions would be included among the top twenty. After that, narrowing down to the top ten was a matter of individual taste.

In one such survey, two firms were mentioned for the top ten more often than others—Citicorp and Nomura Securities. In the next tier, those most often named included Deutsche Bank, Salomon Brothers, and Morgan Guaranty.

Then came CS First Boston and a collection of U.S. investment banks and the other Swiss banks. These were followed by British and Continental banks and more Japanese.[1]

Out of this crowd, most people thought, would come the mega-banks of the twenty-first century.

The players, however, were for the most part predicting the future based on what they knew about the present. The ones picked are basically the global finance oligopoly of today, the market leaders in banking from around the world.

Perhaps there is one thing we can be sure of regarding today's consensus oligopoly for the year 2000—it will be wrong!

Fifteen years is a very long time in finance. Over such a long period, many unexpected but dramatic things occur that change the ranking of prominent firms.

Two fifteen-year periods before, in 1957, there was no thought of globalization. The top-ten list would have comprised the list of the largest banks in the world, which was dominated at the time by American banks and did not have any Japanese on it at all. The most prominent American banks then included Bank of America, Continental Illinois, and Mellon Bank. Securities firms and merchant banks were too insignificant in size to be on the same list with the large commercial banks. If there had been a listing of top securities firms from around the world, which were then not considered important enough to rank, it would have included only a few U.S. and British houses. The U.S. firms would prove to be an impermanent lot; of the top seventeen investment banks in the United States as of the time of the ruling by Judge Harold Medina dismissing the government's antitrust suit against the industry in 1953, only five were still in business under the same name thirty years later.

If the survey had been made fifteen years later, in 1972, the American banks would still have been prominent, and the Japanese would still have been absent, but perhaps Merrill Lynch, Morgan Stanley, Warburg, Deutsche Bank, UBS, and Paribas would have been talked about. If the list had been made just a few years later, it might also have included one or more of the London-based consortium banks, such as Orion Bank, which had sprung into being for a short time as the best answer to the internationalization of banking. Orion was formed in 1970 by six prominent banks from different countries. It succeeded for a while, but coordination among shareholders became difficult, and in time the individual shareholders became competitors of the consortium. Orion was unwound in 1981 when it was purchased by one of its shareholders, Royal Bank of Canada.

The list in 1987 not only showed the displacement of American banks by European and Japanese banks, but also underscored the rapid rise through the ranks of Nomura Securities, Salomon Brothers, Credit Suisse First Boston, Goldman Sachs (with or without Sumitomo), and American Express–Shearson Lehman Hutton.

353

The list of the top ten or twenty global bankers, in other words, is quite volatile. Perhaps only four or five of the original twenty would have remained on the list throughout the full thirty-year period since 1957—banks such as Citicorp, Deutsche Bank, Morgan Guaranty, and one or more of the larger Swiss banks.

The list of top bankers in the year 2000 is bound to be as different from the list of 1987 as it and its predecessors were different from those taken fifteen years before. Already in the few years since 1987, important changes have occurred. Since the beginning of that year, we have had nine months of surging bull markets followed by a sickening crash of the stock markets and a listless recovery thereafter. The dollar has sunk so low as to render the Euromarket ineffective until recently. The United Kingdom had both Big Bang and the crash within a year with much havoc in its wake. The commercial banks have made large additions to loan-loss reserves only to encounter the Basel Committee's new rules on capital adequacy. Some firms like Nomura, Warburg, and American Express–Shearson Lehman have grown bigger and stronger, while others like Salomon Brothers have felt profit pressures and the hot breath of a predator. One new firm, CS First Boston, has been formed following the reorganization of CSFB. Ahead is Europe 1992, probable repeal of the U.S. Glass-Steagall Act (and possibly too its Japanese cousin, Article 65), further deregulation of markets and commissions in Japan, a lot more restructuring within the financial-services industry, and, no doubt, a further calamity or two somewhere in the global financial marketplace.

The principal requirements for membership in the global top tier, many thought, would be a strong base of profits and capital, a permissive home-country regulatory environment, and the ability to participate in a major capacity in each of the world's capital market centers: New York, London, and Tokyo. These features are unevenly distributed at present among the world's leading financial-services institutions.

The principal U.S. commercial banks understand the globalization trends that are affecting their business quite well. These banks have been operating large units abroad for the past twenty or thirty

354

years. They have networks of branches, contacts with clients, internationally acclimated personnel, and good entrenched positions in the key markets. What they lack so far are the skills of the securities markets, into which so much of their clients' business is being driven, and the legal authority to participate in the securities business in two of the world's three key arenas.

At home, however, U.S. commercial banks have been taking a pounding. Deregulation has allowed money-market funds, savings institutions, credit-card companies, and stockbrokers to compete with banks for retail business. Their domestic-lending business also has been under attack from foreign banks in the United States and from the securities markets.

Securitization became an important factor in the financial-services industry during the 1980s. Ever since commercial paper developed into the principal source of short-term unsecured credit for large corporations, the threat to banks of having their corporate-lending business be displaced by the securities markets has been a serious one. Commercial paper outstandings in the United States grew to exceed $400 billion at the end of 1987, as compared, for example, to $300 billion of commercial lending by major money-center banks. At the same time, banks were also losing business to the capital market as high-yield bonds issued by corporations with below-investment-grade bond ratings, now known as *junk bonds,* had developed into a business involving $35 billion of new issues annually.[2] Securities backed by home mortgages and other assets also grew into a very large business in the United States. Securitization was eroding deeply into the mainstream corporate-lending business of U.S. banks.

While others were allowed to enter their businesses, the banks were still restricted from expanding across state lines or from entering different financial and nonfinancial service areas by a web of entangling legislation, Federal Reserve regulations, and Comptroller of the Currency and Federal Deposit Insurance Corporation requirements. The domestic-banking regulatory base in the United States is confused, changeable, inconsistent, and generally hard on the banks. Nor does there appear to be much hope that the mess that represents American banking regulation will be simplified and improved in the near future.

The banks, however, have for years been free to compete in virtually any business outside the United States. Banks have had unrestricted access to the Eurobond business since the beginning. One or two have made names for themselves in this market, but for the most part U.S. commercial banks have been unimpressive in the freewheeling world of Eurobonds. Banks have suffered from the fact that Eurobonds have been a sideshow business compared to their main lending operations. Their top people in Europe are bankers, not traders. Their operations, the personnel they recruit, their compensation structures, their control systems, everything about their business has reflected its leadership by commercial bankers, with no training in or stomach for aggressive trading practices, position risks, or hedging techniques.

Also, by not being able to participate in the U.S. capital markets—which after 1984 provided for instant market access through the SEC's Rule 415—the commercial banks were not able to offer the full range of global alternatives to their clients, and therefore were distinctly uncompetitive with investment banks for the capital-market business of U.S. corporations. Not being able to service the principal companies from their home countries, where their market position and influence with clients was at its highest, was a serious disadvantage.

Unless the banks can escape the chains of Glass-Steagall, and orient themselves properly to the capital-markets environment that is sure to dominate corporate and governmental financing until the next century, their prospects for being included among the few perched on the financial Mount Olympus of the twenty-first century seem dim. Knowing this, the banks made a mighty push to encourage Congress to repeal Glass-Steagall in 1988.

It almost succeeded. The banks argued that the law blocked their necessary strategic development. They also argued that respected academic studies had shown that the entrance of banks into the securities markets would increase competition, lower fees, and provide better services for issuers, "with small- and medium-size companies being the prime beneficiaries of the heightened competition."[3]

It seemed a clear case for repeal. The Treasury was for it. The

356

Federal Reserve was also and became more vocal on the subject after Alan Greenspan replaced Paul Volcker as chairman. And the banks had good academic and free-market arguments supporting their case.

Some observers, however, noted that the banks were much weaker institutions than they had been a few years before. Most of the money-center banks had seen their bond ratings reduced, to the extent that in 1988 not one major U.S. bank-holding company retains an AAA Moody's rating, while as recently as 1980, nine had had them. In early 1988, Standard and Poor's downgraded Chase Manhattan's senior debt to A from AA −; Manufacturers Hanover Trust went to BBB from A −; and the Bank of America to BBB − from BBB (though it has since been flagged for possible upgrading). These new ratings certainly brought into question the traditional reputation of bankers as being pillars of financial stability and probity. A BBB − rating is only one notch above a junk bond.

Domestic-lending difficulties over the past several years had reduced a number of large and important American banks to even humbler circumstances. Banks such as Continental Illinois, Interfirst Republic, and the once-proud Mellon Bank, among others, had to be rescued or substantially restructured by the FDIC. If this weren't enough, Third World loans became a more serious problem. Citicorp's May 1987 addition of $3 billion to its loan-loss reserves forced most other banks with substantial exposures in Latin America to make similar charges. The capital positions of the banks did not look very strong to many in Congress, who at the time were busy working out the rescue of the savings-and-loan industry in the United States, which some expected would cost the taxpayer upward of $100 billion.

U.S. banks were probably the hardest hit among all international banks by the BIS's new capital-adequacy standards. Though the amounts to be raised are modest in most cases, the profitability of the banks has not been high enough in recent years for retained earnings fully to cover future capital requirements. Meanwhile, market conditions in the United States continue to be unfavorable for the sale of new equity securities by the banks. The BIS capital requirements could become especially burdensome for some U.S. banks, however, if further large charges to loan-loss reserves are made for Latin American loans or if a bank wants to invest in new areas, such as the securities industry, where such investments would be subtracted from the bank's core capital.

Some in Congress felt that the banks already had too much on their plates in 1988 to be able to compete in the tough business of investment banking, which, if anything, was yet another capital-intensive business. These members were concerned that however much the banks may wish to be free-market operators, they remain regulated financial institutions with the U.S. government standing behind their deposits. If things should go wrong, the stability of the U.S. banking system as well as taxpayer dollars are at stake.

Compromises were offered to keep the focus on getting rid of Glass-Steagall. "Firewalls" were proposed to keep the securities business of the banks, and its noninsured financing, separate from the banking business. Restrictions were suggested that would prevent banks with insufficient capital from participating in the securities business and that would disallow mergers between the largest banks and securities firms. The Senate passed a bill 94–2 in the spring of 1988 that substantially did away with Glass-Steagall. The House Banking Committee passed a bill through committee during the summer of 1988 that gave the banks securities powers, but with many more restrictions than were in the Senate bill. This version was quite disappointing to many commercial bankers. In September, the House Energy and Commerce Committee, whose Telecommunications and Finance Subcommittee was also given a say in the matter, proposed amendments to the Banking Committee's bill that tightened the restrictions further and confused the issue about which committee in the House had jurisdiction over the legislation.

However, it was very late in the congressional calendar of an election year. Congress recessed in October 1988 without passing the repeal of Glass-Steagall. Still, it had been close. The fifty-five-year-old legislation had only barely survived the most effective of many assaults. Surely, however, the matter will be addressed by the next Congress, where, among other new arguments, emphasis will be placed on the need to provide reciprocal banking and investment-banking privileges to banks operating in the United States in order for U.S.

banks to be granted access to the large and important post-1992 European universal banking market.

In the meantime, banks will rely upon the Federal Reserve as their source for further liberalization of the rules restricting banks in the securities area. Chase Manhattan Bank has announced that absent repeal of Glass-Steagall, it has filed for permission from the Fed to increase its investment-banking activities to 10 percent or more of the business of its securities affiliate (up from 5 percent) and to be permitted to underwrite corporate equity securities.

By the end of 1987, U.S. banks had fallen well behind the leading banks of other countries in terms of the value of their assets. Their domestic losses, their capital constraints, slower growth, and the weakness of the dollar drove even the best of the American banks well down the list ranking banks around the world by assets. Citibank fell to eighth place in the league table, Chase Manhattan to thirty-third, Morgan Guaranty to fifty-fourth, and Bankers Trust to seventy-sixth in the world standings, in which nine of the top fifteen banks were Japanese.

Ranked by an even more demanding standard, the market value of the bank's stockholder's equity at December 1987, the results were somewhat worse. Citicorp was thirty-fourth; only five U.S. banks were among the top hundred.[4]

The U.S. banks have fallen into a bind. Pressured on one side by increasing competition for their traditional business and on the other by an unsatisfactory regulatory environment, the banks have lost profitability and the ability to raise additional equity capital. Many banks have been forced into mergers; others have mapped out strategies for restructuring their businesses through combinations with banks in other regions or with other banks similar to themselves that would permit greater profitability through economies of scale. As regulatory relief is granted, albeit piecemeal as has been the practice in the United States, further combinations, perhaps even with investment banks, will no doubt be considered. Without restructuring, however, most of today's major U.S. banks will not be in strong enough positions in the future to capture any of the gobal medals.

Japanese banks, by contrast, now appear to be the ascending

stars of the global scene, certainly at least in the asset and market capitalization rankings. These banks, however, suffer from similar problems as the U.S. banks.

On the whole, Japanese banks are undercapitalized, but they are not capital-constrained; they are much more limited by regulatory factors than by anything else. The Japanese Ministry of Finance exercises exceptionally close control over the banks, with respect to both their permitted activities and new domestic business opportunities in Japan and all of their activities abroad. They are further constrained by Article 65, which prevents them from competing in Japan with securities firms. The authorities, however, have allowed foreign banks, under the constant threat of being denied reciprocal banking privileges abroad, to find back-door entrances to the securities businesses in Japan. Vickers da Costa, a U.K. brokerage firm specializing in Japanese securities that had been acquired by Citicorp, was allowed to continue doing business in Japan after the acquisition and subsequently became a member of the Tokyo Stock Exchange. Several Hong Kong companies that are joint ventures between banks and nonbanking companies have been granted securities licenses in Japan, and some of these, including six European universal banks, have joined the exchange. Japanese banks have still not been allowed such opportunities.

Japanese banks have been prevented by the Ministry of Finance from keeping pace in the securities field abroad. They have been allowed small European merchant-banking affiliates but not the freedom to manage underwritings without restriction. With little else to do, the affiliates became big players in the ill-fated floating-rate-note-and-preferred-stock market, which seized up during 1986 causing considerable losses to their parents. The banks barely keep these affiliates afloat now with small participations in Euromarket offerings of Japanese companies.

Though a few Japanese banks in the United States have invested in government-securities dealers, most have stuck to traditional commercial banking in their overseas activities. Here they have been able to acquire significant shares of U.S. and other lending markets because of aggressive pricing practices.

Japanese banks, consequently, own a large quantity of Latin American loans. Although they have not been hobbled by them so far, most of these loans have not been very fully reserved against. The new BIS standards will require most major Japanese banks to seek additional capital. This will not be difficult for them, however, because of their ability to raise large amounts of new equity through the sale of shares in the high-priced Japanese market and because of the banks' considerable holdings of appreciated real estate and equity securities in other Japanese companies.

Like the U.S. banks, the Japanese seek regulatory relief in order to enhance their ability to compete in the coming world. Financial regulation in Japan, however, is an intricate, interwoven construction that is not easily changed in substantial ways without a consensus among the principal affected parties that can take years to develop. For the time being, the Japanese banks, with access to more funds than they can use in lending and great incentives to enter the global-ized securities markets, are not permitted to do so. They are all dressed up with nowhere to go. Hence investments such as Sumitomo Bank's in Goldman Sachs appear to make sense.

European banks approach the year 2000 with a number of strengths and weaknesses in common, plus, of course, the varying individual characteristics of their national backgrounds. Among their common strengths are their secure capital bases, profitable businesses at home that underwrite their continuing stability, and, for many, a long his-tory of international activity and substantial experience in the securi-ties business.

Very few of the major European banks will require additional capital to meet the new BIS standards. Most Continental banks have written off most of their Third World exposures, and many have raised additional capital in the markets in the past few years.

Banks in Europe, as those in Japan, have been the subject of regulatory protection through preclusion of competition from sources outside the domestic-banking system. Retail money-market funds, as they are known in the United States, do not exist in Europe. In many European countries, regulators are not unwilling to see large spreads between deposit and lending rates, knowing that the differentials are used to provide profits and strength to the banking system. They know, however, that in the trade-off they have given up some of the competition, efficiency, innovation, and invigoration that might otherwise have entered the banking industry.

Many major European banks, especially British, French, and Dutch banks, have been active abroad since colonial days, and many have extensive overseas branch networks. Continental banks, operating throughout their history as universal banks, have accumulated substantial experience in underwriting and brokerage activities and investment management. In their home countries, these banks exert considerable power and influence in corporate boardrooms, in the legislature, and in the markets.

However, after years of being large, powerful, and profitable, many of these banks have grown inefficient, overly centralized, and conservative. Swiss banks, for example, make most of their interest income from lightly competitive domestic-lending businesses and most of their fee income from safekeeping and managing investments for foreigners. Many find the risk exposures and the comparatively poor profitability of business abroad to be discouraging. Yet they persevere, knowing that they have to develop their international business, but they are reluctant to delegate all of the authority that their colleagues outside of Switzerland may need to be fully competitive. The same is true for many other European commercial banks.

For the same reason, many European banks have been somewhat timid in expanding into areas permitted to them, such as the post–Big Bang market in the United Kingdom and the U.S. securities markets, where about fifteen European banks are grandfathered to conduct both banking and securities activities. Those banks with the right to conduct universal banking in the United States have not extensively exploited the opportunity, although one bank, Crédit Suisse, controls a 44.5-percent ownership of CS First Boston, into which CSFB and First Boston Corporation were merged in December 1988. The pace of change in world financial markets has been much greater so far than the pace at which European banks have reacted to it. Perhaps their unconscious strategy is to play the tortoise in the race with the hare.

The problem for the European banks is 1992. This event will

have major implications for the larger banks. They will have to both advance into other countries and defend their profitable domestic business from new competition. If domestic banking is truly deregulated in Europe, the assured domestic preeminence of the large banks may come into question. The banks will have to change, which may be an invigorating, positive experience that enhances competitiveness—and the desire of the banks to become more aggressive globally—or it may not be.

No doubt, most European banks will give first priority to getting their 1992 position straight and worry later about their competitive postures in the United States and Japan. As the steady process of globalization of markets continues, however, banks may find their strong relationships with their clients weakening. Clients all over the world, influenced by the ways of the modern chief financial officer, are already accustomed to look daily for the best deal. Unless the large European banks develop their linkages with financial markets in the United States and Japan while they are reorganizing their businesses in Europe, they may find themselves less competitive than they need to be to satisfy their traditional clients' emerging global requirements long before the year 2000 arrives.

The other type of bank in Europe, the merchant bank, or *banque d'affaires,* will be affected quite differently by the proposed 1992 reforms. These banks are oriented to the securities businesses—underwriting, brokerage and trading, and merger and other financial advisory services—that are likely to grow with further deregulation and integration within Europe. The problem for them is developing the size and capital base to perform as a global player in the face of tough competition, dangerous markets, and generally low profitability from their basic businesses.

Such a prospect has discouraged many a merchant bank in the United Kingdom from seeking to become a global investment bank. Such firms are scrambling instead to find niches in which they can hope to protect a reputation for excellence that comes from specialization. Others perhaps recognize as John Craven does "that merchant banks must become U.S.-style integrated investment banks capable of handling all types of market transactions if they are to win the big prizes."[5]

Of all the merchant banks, only a few have attempted to position themselves as U.S.-style integrated firms operating internationally from the outset. Two of these, Barclays de Zoate Wedd and County NatWest, are part of large commercial banks. Two others, Warburg and Kleinwort Benson, have a long history of prominence in investment banking in the United Kingdom. They are both seeking to expand their activities further in Europe, in the United States, and in Japan. None of the leading merchant banks has yet developed a heavyweight capability in the United States, nor, of course, in Japan. They hope to survive on an international scale as they always have, however, by "living by their wits."

U.S. investment banks were among the first to discover and exploit the effects of globalization on the markets. Unlike most U.K. merchant banks and European universal banks, the U.S. firms were in Europe to be salesmen. They had no "captive" accounts in Europe, no in-house placing power. They did not manage major amounts of money in Europe until the early 1980s, when Morgan Stanley was given several billion dollars of Kuwaiti funds to manage.

Their original aim was to help U.S. corporations raise money through the issuance of Eurobonds. They were also active in selling U.S. stocks to brokerage clients in Europe. For both purposes, they needed sales forces that were aggressive, capable, supported by research and trading capabilities, and that covered every appropriate nook and cranny in Europe, Asia, and the Middle East.

The sales forces gave the firms the contact with the market they needed to be able to come up with the best ideas, the best timing, and the courage to "buy" deals from their clients. The investment banks also were the first to integrate Euromarket and other information into their overall, global-trading and underwriting activities.

In the late 1970s, Saudi Arabia was a major influence on the markets. A high proportion of the Saudi reserves were being invested in U.S. Treasuries. Being part of the group of brokers that were servicing the Saudis at the time was an attractive source of trading profits, but more important, being in close contact with the Saudis meant that one knew enough about what he was doing in the markets to position one's trading book accordingly. This meant that a firm trading U.S. government securities in New York would be at a significant disadvantage, relative to its competitors, if it did not know what the Saudis were doing at the time.

Soon after the period of exceptional Saudi activity passed, it became obvious that foreign influences affecting the U.S. government markets would continue to be considerable. After the Saudis, the central banks of various countries became substantial open-market buyers and sellers of Treasuries and other securities. Then, of course, came the Japanese era, by which time it was totally clear to everyone that globalization of securities markets was here to stay.

Indeed, it became clear before long that globalization had affected almost all of the other businesses of U.S. investment banks. They could not represent a company that was for sale without providing a list of possible foreign buyers that the seller would want to be sure saw the offering papers. They couldn't advise a buyer as to the price he should offer for a company without knowing who the potential competitors would be in the transaction, including the likely foreign buyers. The same was increasingly becoming true in the real estate field, where Japanese investors, in particular, had been extremely active.

U.S. institutional investors, especially the pension funds, had by the mid-1980s become convinced of the wisdom of acquiring foreign equities for their portfolios. They were, in fact, selling U.S. stocks and buying foreign ones. Staying with one's clients when they moved from one market to another had always been a fundamental business practice of most brokerage firms. Now the firms would have to become competent suppliers of international securities as well as of domestic ones.

By moving quickly, something investment banks are traditionally good at doing, the U.S. firms were able to occupy some of the high ground in the early days of globalized securities markets. But this progress was probably achieved at much greater cost than any of the firms expected at the time.

Their expansion was explosive, simultaneously occurring in London, Frankfurt, Paris, Zürich, Sydney, Toronto, and Tokyo. Overheads went virtually out of control. All of the firms had hired large numbers of new employees at the same time, competing with each other in the employment market and running compensation to unprecedented levels. Many of these new people would take several years to become productive members of the teams they had joined. Others never would, either because there was no system for properly training them, or because they would leave to join a competitor before he or she had contributed very much to the business. Offices were filling with expatriates, especially experienced hands from New York, who were far more expensive to maintain abroad than locals. Also, the back offices had to expand radically; new and larger computer systems, more operations personnel, and more space in budget-busting London and Tokyo were needed. And, after 1985, the collapse of the dollar added considerable foreign-exchange cost to all the rest.

Early in 1987 it became evident that the firms could not continue to expand at the same pace of the past few years. Slowing down, and indeed some retraction, was going to be necessary. Repairs necessitated by the rapid advance would have to be made, or the firms would lose too much money. Layoffs began, efficiencies were pursued. Cost control finally had its turn.

The experience, however, made several things clear. The blind rush into new territories made much less net progress than one thought at the time. A more gradual, tortoiselike approach might have been a better alternative. Management competence and control are vital in securities operations, mainly in setting priorities and directing traffic. There are many international temptations that look good to the newly arrived New Yorker, but which in the end he learns should have been avoided.

Most U.S. firms, however, have not backed away from their commitment to globalizing their businesses. The speed of advance has been slowed across the board by tough competitive and cost conditions, but it will surely resume. The investment banks, though gradually becoming larger, more bureaucratic institutions, still manage their businesses on a much more immediate, ad hoc basis than do more traditional banking institutions. This makes the investment banks much more opportunistic and responsive to changing market and competitive conditions than others. However, they may still be too transaction-driven and narrow-focused to install the sophisticated cost-management systems that they must have if they are to rise to become part of a future oligopoly. Short of that important requirement, however, they perhaps have a head start over the other types of banks in assimilating global financial conditions and performing effectively in markets outside of their home country.

Investment banks also have become quite large on an absolute basis and now appear to possess the scale needed to be major players on a global basis. As of the end of 1987, four of the ten largest U.S. investment banks had total assets in excess of $40 billion, and one, Shearson Lehman Hutton, had total assets in excess of $70 billion. At the rate at which investment banks have been increasing their capital and their leverage—and therefore their total assets—it is likely that within a few years the top-ten U.S. investment banks will show more assets on their balance sheets than will the top-ten commercial banks. The same would also be true for assets managed by the firms. Shearson Lehman alone manages more than $100 billion of assets for others.

Japanese securities firms are different from all others. They are enormously profitable: each of the "big four" firms earned more than $1.8 billion before taxes in 1987. The most profitable U.S. firm earned less than half that amount. The firms are well capitalized: the smallest of the four had capital in excess of $5 billion at the end of 1987, more than all but two U.S. firms. They are also well valued by the market: Nomura's market capitalization alone exceeded $60 billion at year's end 1987, more than the market value of all the major U.S. securities firms put together.

Such financial power makes these Japanese firms intimidating, but they really have not used much of it so far.

Though there are some government regulations in place that might restrict their freedom of action, so far the securities firms have not acquired other international firms or top professionals from such other firms. They have bought a few deals as lead manager from non-Japanese companies in Europe and in the United States, but almost all of their market power comes from handling Japanese transactions.

Their great size and profitability comes from their domestic business. The Japanese stock exchanges still maintain fixed commission rates. Profitable retail transactions are the major part of Japanese stock-market activity, which now exceeds the trading volume of the New York Stock Exchange. The big four firms dominate the primary and secondary markets and cannot fail to earn prodigious profits.

Overseas, the Japanese firms are currently leading the league tables in the Eurobond market because of the large volume of Japanese debt-and-equity issues and because of the low volume in the non-Japanese-dollar sector of the market. The Japanese firms have developed other capabilities in the Euromarkets, and they are by no means limited to just Japanese business. However, the Japanese new-issue business that they do now puts them at the top of the market-share tables. Without it their rankings would be much less impressive.

While selling Japanese shares to foreigners, the firms also developed the capability to underwrite and distribute new issues of Japanese securities internationally. During the last ten years, more issues of securities have been made by Japanese companies in the Euromarket than by companies from any other country—by a large margin. Most of these issues were brought to Europe to avoid stringent new-issue queuing and pricing regulations in the domestic Japanese market. Today, with large financial surpluses, Japanese investors are the principal purchasers of Japanese Euro-issues. These issues are basically domestic capital-market issues that detour through Europe to avoid the technicalities of issuing at home.

Handling all of these new issues, which Japanese securities firms invariably do, has been both profitable and educational for the Japanese houses. This has prepared them to underwrite and distribute non-Japanese securities. Much, but not all, of such distributions are to investors in Japan, whom, as the Saudis of a decade earlier, all major issuers want to be able to reach. With investors and issuers from Japan both exceptionally active, the securities firms have become extremely visible.

The Japanese firms, however, suffer from some of the difficulties of the European banks: with profits so easy to make, and the Japanese-related business so abundant, why do anything else? The firms are not risk takers, except in isolated cases, nor do they have to be. They are not especially creative—most of the innovations and new securities in the Eurobond market are introduced by non-Japanese firms. Nor are they especially efficient. It is generally understood that Japanese securities firms' overseas operations lose money and always have, despite their continuing profitable business with Japanese clients. The offices are usually overstaffed and spend large amounts on entertaining a continual flow of Japanese visitors.

In the last few years, these firms have begun to hire senior personnel from among experienced financial people in the countries in which they have operations. These people, however, are never made part of the top management of the firms' Japanese parents, neither do they visit Japan often or impart their advice to colleagues in Tokyo with much effect. The firms remain totally Japanese in their global outlook. Very few decisions are made at the local level; everything has to be passed back to Tokyo, where decisions are finally taken, according to Japanese consensual practice.

In the long run, Japanese securities firms face dangers from deregulation of their domestic business, especially the end of fixed commission rates, from the exclusion of banks from the securities business, and from more competitive and efficient capital markets in Japan that will preclude the need for Euromarket round-tripping by issuers of securities.

Perhaps they will also face the more distant dangers of being required to provide the same level of performance-enhancing services for their clients that U.S. and some European bankers do. Such services as block trading, off-the-shelf underwriting, and providing first-rate research and support activities will someday be demanded by Japanese corporate and investor clients. Japanese firms do not provide such services extensively today. Their emphasis is on service and loyalty to clients, not on price and objectivity. They are used to having their clients do as they say. The swing from relationship banking to transactional relationships could be very traumatic for some of the Japanese securities firms.

Of course these firms are by nature extremely competitive and are quick learners. They are likely to adjust as the times change and to use their formidable wealth to good effect by acquiring major firms in other countries and hiring teams of people to make them run well. For the moment, however, this is not the Japanese way and therefore it won't happen.

To succeed in the coming environment of the twenty-first century, global banking firms will have to be, above all, competitive. The freshest ideas, the best rates, the quickest executions will be what matters most to the clients, who are themselves under increasing pressure to demonstrate that they are performing their jobs well in comparison with their competitors.

To be competitive, firms will have to be prepared to find the best ideas from all over the world and to be willing to risk their own capital as part of the process of serving the client. The business will be risky, and some firms will fail or be displaced in the ranks of the leaders by others. There will be a premium value placed on firms that are large and strong enough to recover from losses or mistakes and on firms that have grown wiser from their international experiences. There will also be a premium, as there usually is, on good management, solid business judgment, and sensible strategic thinking.

No one can be all things to all clients. None of the potential leaders of the year 2000 is yet so global in scope and competence that it could hope to provide all services equally well. The mighty Europeans are still ill-equipped to handle competitive capital-market services in the United States and Japan. The long-dominant American banks have been battle-weakened by poor-quality loans, domestic regulation, and competition from the securities markets. The Japanese banks are still tied too closely to the home market and too tied up by regulation to have become effective in any business but lending. U.S. investment banks are just beginning to learn what it takes to compete on a world scale, but may lack management ability and staying power when times are tough. Japanese securities firms are still too heavily dependent on Japanese business to compete on equal grounds in Europe and America. The nonbanking financial institutions are maneuvering, but haven't yet committed themselves to the battle.

The odds are there won't be a global financial services oligopoly in the year 2000, at least any more so than there is one now. The parlor game played ten years hence will no doubt result—as does the same game played today—in a selection of the leading banks and securities firms from each of the major regions. These institutions will still derive the bulk of their business, their profits, and their claims to fame from their domestic prominence. They will not likely be the same ones on today's list, however. The increasingly cruel and unforgiving requirement to perform competitively will drop some well-known names into the great realm of mediocrity or into specialized niches. Others—aggressive, determined, and probably lucky new players—will find their way into the top ranks.

It is ever thus in free markets. Now that it appears that one legacy of these turbulent times of the 1980s will be freer markets in the twenty-first century, we can expect exciting times and bountiful opportunities but also many slippery and hazardous roads ahead.

# 12

# Megatrends

"A rising tide lifts all boats," say the fishermen in Maine, "and a falling tide can put them all on the rocks." All the boats benefit or suffer equally from the ebbs and flows around them. The same is true in banking. All the players have been affected equally by the trends that have shaped the financial environment over the past twenty years or so. Some players, of course, started out in more or less favorable positions than others. And like the fishermen, some maneuvered their craft more skillfully or with better fortune than others.

The tides, however, are just the tides. They go in and they go out. They are not too difficult to identify or to track. We know they reverse periodically. What makes the difference, to fishermen and to bankers, is how one positions oneself amidst the tides to one's best advantage.

Casting a mariner's eyes astern, into our wake, we can still see the traces of past tide flows that have had profound effects on the business of banking. The roaring twenties, a crash, bank failures, and worldwide economic depression followed by necessary government intervention in private-sector financial affairs. Social policies to prevent further collapse and poverty. Violent political upheavals abroad largely caused by faulty economic policies. Chaos. Then victory. Reconstruction. Socialism as a form of government that might avert similar horrors in the future. Economic overconfidence in Western democracies. Vigorous political and economic competition from the East. Balance-of-payments deficits.

Inflation. Overmanning and inefficiency. The computer age and other technological miracles. Deregulation. Reform. Increasing competitiveness. Reinvigoration. Free-market economics on the ascendancy. The roaring eighties. Another global crash—but this one does not bring the world to its knees.

From a banker's more narrow perspective, the past fifty years would be seen as beginning with Glass-Steagall and Federal Deposit Insurance, the rebuilding of the world's banking system after the war, Bretton Woods and the creation of the IMF and the World Bank, the rise of international banking as an escape from the limitations of domestic regulation, the creation of the Eurodollar market, the forced adoption of floating exchange rates, the oil shock of 1973, petrodollar recycling to Third World borrowers, inflation, the Federal Reserve's abandonment of interest-rate regulation in favor of controlling the money supply, volatility of interest and foreign-exchange rates, pension-fund investments, deregulation, Mayday, the heyday of Eurobonds, foreign direct investment, roaring bull markets, mergers and LBOs, securitization, Japanese financial power, globalization of markets, Big Bang, privatizations, the crash, 1992, *perestroika,* and so on.

There has certainly been a lot of change. A lot of confusion too. Sometimes the trends are not all that easy to spot, much less to exploit.

Predicting what the trends are likely to be in the future can be quite difficult. Trends are not always so clear, and even when they are, they seem fickle, even whimsical. Markets go way up, then way down. The United States is the world's largest lender one minute, the world's largest borrower the next. Oil prices go from $2 a barrel in

373

1973 to $15, then to $40 in 1979, then to $10 in 1987. The dollar, once the world's mightiest currency, is now traded in the market at prices lower than at any time since World War II. Yet billions of dollars are flowing into the United States in the form of portfolio investment and direct investment in U.S. companies and real estate. Despite the collapsing dollar, the United States is where foreigners seem to want to invest. Are the trends contradictory?

Well, there are trends, and there are *megatrends*. Megatrends are like the tides—they lift all the boats. They are the major, long-term influences that shape the future more than anything else. But they are so general, and act so gradually, they are often difficult to credit in the here-and-now market mentality of bankers. Nevertheless, they are the tides in which one must position oneself for the future, where the skills of the mariner are most important. These skills are what will separate the best from all of the rest.

Two particular megatrends stand out. One has to do with technology, the other with how people choose to live.

Technology, and its pace of development, has changed the world beyond comprehension. Everyone's life has been changed by television, computers, new food production and preservation techniques, new vaccines, robotics, and telecommunications, among other things. People live longer, have more interesting lives, go more places, and have more fun than ever before. The standard of living in developed countries is high and rising, and improving in many other countries.

When I became a banker, the most exciting technological development was the plastic circular slide rule. Today the most junior employee of an investment bank has a personal computer installed in his or her workplace, access to extensive data banks inside and outside the firm, desktop publishing, a facsimile machine, instant telephonic communications with most of the world, real-time market price information for thousands of securities, worldwide news retrieval capability, and twenty-four-hour-a-day dial-a-cab and pizza-delivery services all within reach of the telephone. Traders have all this plus six or seven data screens showing instantaneous market movements, programs to compare securities, value options, and prepare strategies,

374

and a direct communications link to every salesman in the firm. Many clients, both issuers of securities and investors, have all of these things too.

Technologically, banking has jumped from a biplane crop duster to the cockpit of an F-14 Tomcat. Both fly, but the F-14 has a lot more buttons and switches to make you go higher and faster than ever before.

All of this technology (mainly in the case of bankers, telecommunications technology) has made market information more available. More opportunities are illuminated. More choices appear on the menu. The more choices and the more widespread the knowledge of the choices becomes, the more efficient the market. The more efficient it is, the better it can provide what its users want. Better trades, at less cost.

Dennis Dammerman has several options for every financial decision GE makes. Frequently, the lowest cost option is one that was not on offer a few years ago. His transactions can be executed quickly, with market-makers taking much of the risk. When pension-fund investors want quotes on large transactions involving several blocks of securities, bankers will give them instantly. The markets are efficient not only in relaying price information, but also in terms of the speed and size of trade executions.

How much better can it get? Assuredly a lot better. The age of program trading, portfolio insurance, synthetic securities, sophisticated arbitrages, and other such marvels of technology, as little appreciated as some of these may be, is only a few years old. New products and market innovations will flow even more quickly in the future. Market participants will have to have the reflexes of a *Top Gun* pilot to keep up.

This is an age of explosive technological development. Technology fuels finance as much as it fuels anything else, perhaps more so because money is so fungible. It can only continue to do so. Those who learn to use it, who train their employees to fly today's financial F-14s, will have a big comparative advantage over those who do not.

Technology has affected the way people live in many ways. It has also affected their views about how they want to live.

•     •     •

375

In the 1930s, people wanted to escape poverty and depression, and in the 1940s to escape war and destruction. Grim times ruled the Earth from 1930 until well into the 1950s in many countries. Such a long exposure to hardship made people want to live in the shelter of a beneficent government. A government that cared about them. Socialists and socialistic political parties were elected to office in the United States and all over Europe. Welfare programs, social security, national health insurance, pension entitlements, and other similar programs came into being in the Western democracies during this period. They were needed.

Socialism, however, was more efficient in sharing wealth than it was in creating it. As the smoke cleared, people started back to work creating it. When they looked up, however, they saw that a huge portion of all the world's wealth was sitting in the United States and that many countries had very little wealth. Those without became determined to reclaim their share of the world's income and surpluses. The success of these countries in recovering wealth from the United States fractured the currency system then in use. Foreign-exchange markets that repriced currencies daily replaced the old system that was designed to maintain stability. The new world would value efficiency more than stability. The free-market mentality took shape. The world came to understand the value of free trade and open competition, harsh as it may sometimes be.

Gradually people came to understand that if their economies were deprived of trade, they would shrivel, threatening another episode of the 1930s. Embracing trade, on the other hand, meant to become good at it, to be competitive. Lazy, complacent enterprises may be unable to survive in a competitive environment. Some individual companies might shrivel in such a hard environment, but the economy as a whole would prosper. New or restructured companies would replace those that dropped out. But there was no choice: Compete or Die!

The people realized this, just as they realized that their governments were inefficient stewards of industry. "If we must compete," they would say, "let's give ourselves a chance and get the government's hands off the controls."

Some people didn't agree and tried to revive the memories of the dark days. But the people had changed. They were no longer the downtrodden wrecks they had been during the Depression or after the war. They were prosperous again. They wanted to compete. They wanted a crack at a larger share of the world's wealth. "The intensive-care ward is great when you need it," they said, "but you don't want to spend the rest of your life with tubes up your nose once you're better."

Socialism waned—in Europe and in America. Jimmy Carter started the chain of deregulation in the United States with the airlines and the railroads. Margaret Thatcher repealed foreign-exchange controls and started the long march back toward free enterprise, which Ronald Reagan joined as soon as he could. A series of prime ministers in Japan led their country in stripping away protectionist postwar economic policies. Privatization came next, and was emulated by virtually everyone. Then open-market reforms in China, which had discovered state control was hopelessly inefficient. Then came the EEC's move toward total deregulation by 1992 and Mr. Gorbachev's talk of economic restructuring within the Soviet Union.

Socialists are still alive, however, and trying to get back in the game. But past governments have so mismanaged their nations' economic affairs that few have any money left for social programs, even if the socialists could get them passed. In Europe nationalization of industry did not accomplish the social objectives intended. At the same time, governments have realized the benefits of the free market in the proceeds received from privatization, the relief from subsidy payments, and the increased tax revenues. Many of these governments find that a hands-off industrial policy works best, for the people and for their reelection prospects.

Will it continue? Short of another war or Great Depression it should. Governments really can't compete with the private sector as a creator of wealth in times of relative prosperity. When the government's great protective shield is not necessary, people seem to prefer to take their chances in the free market.

However, a "compete or die" policy is tough on the losers, and no doubt there will be times when free-trade nations will fall

377

from grace into the slough of protectionism, but we can hope that these times will be few in number, that the sinners will not include any of the major trading nations, and that the lapse will be for a relatively short period. In the last U.S. elections, protectionism from either the hazards of trade or foreign investment appeared to be issues more popular with unsuccessful candidates than with the voters.

The retreat from socialism in the United States, in Europe, and in Japan has left a legacy of financial deregulation that has permitted not only the free flow of capital across borders, but also an increased freedom for most players to compete in financial services around the world. This enhanced liquidity and competitive energy have encouraged innovation and creativity in financial markets. Competition has increasingly come to turn on performance. Uncompetitive relationships do not last in this environment, in which the service user is himself under the gun of performance measurement. The search for more competitive capabilities has led bankers abroad to tap the savings of different types of investors and to bridge financial markets in different currencies. The globalization of financial markets is the result.

In the process, markets everywhere have become more familiar and more comfortable with the presence of foreigners. Foreign capital, we now know, has to flow to offset trade imbalances. Foreign portfolio investments are as important to Japanese insurance companies as they are to the U.S. government, whose securities they purchase. Some anger and anxiety concerning the level of direct investments in the United States by foreigners continue to exist in the United States, but these feelings are not xenophobic nor are they likely to result in discriminatory legislation. On the whole, Americans see the benefits of foreign investment as much as they see the disadvantages. The same is true, for the most part, in Europe, and Japan is moving in that direction.

Developing countries have little choice but to increase direct investment by foreigners, if they are able to. George Moore often said that "foreign capital flows where it is wanted and well treated," a simple lesson that many debt-laden countries would do well to relearn. It is likely that the whole world will become more accustomed to foreign capital living alongside it and come to trust it again.

All in all, the world of finance is global—it knows few borders or nationalities. Perhaps financial perspectives will light the path for others in the future.

In the meantime, global financiers have got to make their living, fishing with the tides. One such current in which they find themselves is the perilous drift of deregulation and enhanced competition in domestic markets, which threatens to wash away the secure ground of domestic profitability. This trend exists in the United States, in Europe, and in Japan—it is not a local phenomenon. The remedy is to work harder and smarter to increase competitiveness.

A second tidal flow is evident in the effects of securitization— markets are offering lower cost financial solutions than bank lending. This can only be addressed by enhancing the quality and the reach of a firm's capital-markets services.

A third current affecting our navigation is the shift from relationship-oriented to transactional business, again requiring heightened levels of performance. All of these currents are floating us out toward rougher seas of increased competition. To survive in such waters, firms will have to be competitive. Compete or Die!

U.S. banks have been subjected to these currents longer than their counterparts in Europe or Japan. They are accordingly much stronger, tougher, and smarter today than they were several years ago. They have benefited from the combat experience of the last fifteen years. But they haven't solved their regulatory problems, which prevents their becoming major factors in the securities markets. Some will deal with this difficulty patiently, conserving resources until the day comes when they will be allowed by Congress to cross over into investment banking. Others, impatient for this to happen, may sell off their banking charters and reconstitute themselves as investment banks, a painful metamorphosis, but one that may be necessary if the bank is truly serious about investment banking.

If all of these trends continue, the global financial landscape will be greatly changed from what we have known it to be. The ongoing events surrounding the Europe of 1992, the final days of Glass-

Steagall and Article 65 promise us that there are still many more changes ahead, notwithstanding the fact that global banking institutions and their global bankers have already endured more change in the past fifteen years than at any other time in memory.

The frontier days are not yet over.

# Notes

### 1: BORN-AGAIN BANKING

1. Vincent Carosso, *Investment Banking in America* (Cambridge, Mass.: Harvard University Press, 1970), pp. 3–15.

2. Ibid., pp. 80–82.

3. Kuhn, Loeb & Co., *Investment Banking Through Four Generations* (New York, 1955).

4. Carosso, *Investment Banking,* p. 80.

5. *The New York Times,* April 1, 1905.

6. E. B. Potter, ed., *The U.S. and World Sea Power* (New York: Prentice-Hall, 1955), pp. 431–50.

7. Cyrus Adler, *Jacob Henry Schiff* (New York: American Jewish Committee, 1925).

8. Carosso, *Investment Banking,* pp. 192–241.

9. Paul Kennedy, *The Rise and Fall of Great Powers* (New York: Random House, 1987), p. 274.

10. Carosso, *Investment Banking,* pp. 197–98.

11. Ibid., p. 212.

12. The 1987 *Information Please Almanac*.

13. Kennedy, *Rise and Fall,* p. 330.

14. Allin Dakin, "Foreign Securities in the American Money Market, 1914–1930," *Harvard Business Review* (1932).

15. Adler, *Jacob Henry Schiff*.

16. Barrie Wigmore, *The Crash and Its Aftermath* (Westport, Conn.: Greenwood Press, 1985).

17. Dakin, "Foreign Securities."

18. Carosso, *Investment Banking,* p. 307.

19. Edward Schrader, interview with the author, 1972.

20. George S. Moore, *The Banker's Life* (New York: W. W. Norton, 1987), p. 203.

21. Harold Cleveland and Thomas Huertas, *Citibank, 1812–1970* (Cambridge, Mass.: Harvard University Press, 1985), pp. 258–65; and Moore, *Banker's Life*.

22. Fran Schumer, "Banking on the Future," *Barron's,* April 12, 1982.

## 2: GLOBAL BANKING TODAY

1. Sources: Morgan Guaranty Trust, *World Financial Markets* (January 1988); and *Federal Reserve Bulletin* (March 1988).

2. Christopher Lorenz, "Real Bankers Don't Manage," *Financial Times,* November 6, 1987.

## 3: PAUL REVERE'S RETURN?

1. Yoshi Tsumuri, 'Made in America, Managed by Japan," *The New York Times,* November 16, 1986.

2. Source: Bank of Japan, *Financial Statistics* (1987); U.S. Department of Commerce, *Survey of Current Business* (June 1987).

3. John Burgess, "After Years of Courting Foreign Investors, Americans Reconsider," *International Herald Ttribune,* February 23, 1988.

4. John Plender, "No Foreign Capital, Please," *Financial Times,* May 16, 1988.

5. Raymond Vernon, "Foreign-Owned Enterprise in the US: Threat or Opportunity?" Research in progress at Harvard University, June 20, 1988.

6. Source: U.S. Department of Commerce, *Survey of Current Business* (March 1988).

7. "Proposed Semi-Conductor Merger Creates US-Japan Standoff," *The Wall Street Journal,* March 13, 1987.

8. Plender, "No Foreign Capital."

9. Peter F. Drucker, "Low Wages No Longer Give Competitive Edge," *The Wall Street Journal,* March 16, 1988.

10. Frederick A. Moses and Thomas A. Pugel, "Foreign Direct Investment in the United States—The Electronics Industry," *Research in International Business and Finance,* vol. 5 (New York: JAI Press, 1986).

11. Source: *Federal Reserve Bulletin* (July 1988).

12. Source: U.S. Department of Commerce, *Survey of Current Business* (March 1988).

13. Moody's *Bond Record,* November 1978.

14. DeAnne Julius and Stephen Thompson, *Capital Flows and International Economic Relations: The Explosion of Foreign Direct Investment Among the G-5* (London: The Royal Institute of International Affairs, 1988).

## 4: HOSTAGE TO THE BANDIDOS

1. Morgan Guaranty Trust Company, *World Financial Markets* (December 1988).

2. World Bank Debt Tables, 1988, and William R. Cline, *Mobilizing Bank Lending to Debtor Countries* (Institute for International Economics, June 1987), pp. 4–5.

3. World Bank Debt Tables, December 1988.

4. Robert Graham, "Tough Guy Plays It Close to the Edge," *Financial Times,* July 28, 1987.

5. Veronica Bauffati, "Queues Become the Hallmark of Peruvian Austerity," *Financial Times,* November 16, 1988.

6. Barry Eichengreen and Richard Portes, *Settling Defaults in the Era of Bond Finance* (London: Centre for Economic Policy Research, 1988).

7. Alan Riding, "Debt Pact Attacked in Brazil," *The New York Times,* November 9, 1987.

8. William R. Rhodes, "An Insider's Reflection on the Brazil Debt Package," *The Wall Street Journal,* October 14, 1988.

9. *The Economist* (October 29, 1988).

10. Robert Graham, "An Old Hand Gets Off on the Wrong Foot," *Financial Times,* March 6, 1989.

11. Morgan Guaranty Trust Company, *World Financial Markets* (December 1988).

## 5: EUROMONEY

1. Crown Agents discontinued managing these funds in 1983 and subsequently went out of business. It is referred to in the narrative to illustrate the role played by a prominent institutional investor at the time.

2. Bondware Database, Euromoney Publications PLC, 1988.

3. Morgan Guaranty Trust Company, *World Financial Markets* (January 1965–1987).

4. Ibid.

5. Cary Reich, "The Three Faces of Credit Suisse White Weld," *Institutional Investor* (November 1976).

6. "Credit Suisse White Weld, Boston Tea Party," *The Economist* (July 22, 1978).

7. "Michael von Clemm, End of a Legend," *Euromoney* (March 1986).

8. Matthew Winkler, "Kingpin in London Eurobond Market Thrives on Moving Swiftly and Being Unpredictable," *The Wall Street Journal,* December 14, 1984.

9. "Texaco Completes Sale of $1 Billion of Debt in Europe," *The Wall Street Journal,* March 26, 1984.

## 6: THE BIG BANG

1. "London Stock Exchange, Give an Inch, Save a Mile," *The Economist* (July 30, 1983).

2. Andreas Whittam Smith, "Up the Revolution," *Barron's,* April 22, 1985.

3. David Lascelles, "The Parent Pulls in the Leash," *Financial Times,* February 24, 1988.

4. Craig Forman and David Manasian, "Morgan Grenfell Group Cuts 450 Jobs and Halts Its Market-Making Activity," *The Wall Street Journal,* December 7, 1988.

5. Stephen Moore, "Institutional Investors Move Could Cut Equity Sales Overseas by British Firms," *The Wall Street Journal,* April 30, 1987.

## 7: BREAKING WITH SOCIALISM

1. Peter Jenkins, *Mrs. Thatcher's Revolution* (Cambridge, Mass.: Harvard University Press, 1988), pp. 9–16.

2. Ibid., p. 16.

3. Gary Humphries, "French Without Fears," *Euromoney* (December 1987).

4. Guy de Jonquieres, "1992 and All That," *Financial Times,* April 18, 1988.

5. European Communities, "The Economics of 1992," *European Economy,* no. 35 (Cècchini Report), Office des Publications des Communautés Européennes, Luxembourg, 1988.

## 8: PACIFIC GRIDLOCK

1. Karel G. van Wolferen, "The Japan Problem," *Foreign Affairs* (Winter 1986/87).

2. Source: Goldman, Sachs & Company, December 1988.

3. Rimmer de Vries, "The U.S. Budget, Savings, and Investment," *World Financial Markets* (August 1988).

## 9: SAMURAI FINANCE

1. Paul Aron, *Japanese Price-Earnings Multiples Revisited* (New York: Daiwa Securities America, Inc., October 19, 1984).

2. Aron Viner, *Inside Japanese Financial Markets* (New York: Dow Jones–Irwin, 1988), p. 17.

3. Charles R. Elliott, *Japan Investment Strategy Highlights* (New York: Goldman, Sachs & Company, June/July 1988).

4. Ibid.


## 10: HONORABLE PARTNERS

1. Eamonn Fingleton, "Tokyo Takeovers Are for Japanese Only," *Euromoney* (February 1986).

2. Stefan Wagstyl, "Boone Pickens Goes East," *Financial Times,* April 17, 1989.

3. Andrew Tanzer, "With Friends Like These," *Forbes* (June 30, 1986).

4. *The Wall Street Journal,* September 30, 1987.

5. *Japan Company Handbook* (Tokyo: Toyo Keizai Shinposha, Summer 1988).

6. Aron Viner, *Inside Japanese Financial Markets* (New York: Dow Jones–Irwin, 1988), p. 88.


## 11: MEGABANKS 2000?

1. John Carson-Parker, "Tomorrow's Superbanks," *Global Finance* (November 1987).

2. Sources: *Federal Reserve Bulletin* (January 1988); Goldman, Sachs & Company, January 1988.

3. Thomas A. Pugel and Lawrence J. White, "An Analysis of the Competitive Effects of Allowing Commercial Bank Affiliates to Underwrite Corporate Securities," in Ingo Walter, ed., *Deregulating Wall Street* (New York: John Wiley, 1985).

4. "Market Capitalization of Banks as of December 31, 1987," *Euromoney* (February 1988).

5. David Lascelles, "Specialize If You're Not a Global Player," *Financial Times,* September 26, 1988.

# Glossary

**ADRs**   American depositary receipts, a form in which U.S. investors can hold and trade foreign shares in the United States.

**Arbitrage**   The buying of a security in one market and simultaneous selling of the same security in another market at a slightly higher price.

**Asset-backed securities**   Securities issued in the markets that depend upon the assets by which they are secured for the generation of all interest and principal payments. These securities are the product of "securitization."

**Bargains**   Volume discounts of brokerage commissions offered to institutional customers of U.K. stockbrokers.

**Belgian dentists**   Market appellation for "typical" Eurobond retail investor.

**Big Bang**   October 27, 1986, the day when the London Stock Exchange adopted several important changes to its rule book that resulted in negotiated commissions, foreign entry into the exchange, and dual-capacity operations for members.

**Book runner** *The* lead manager of an underwritten issue, who is responsible for pricing, syndication, and stabilization.

**Bought deal** A transaction, usually a Eurobond new issue, that is purchased by an underwriter(s) for resale to the market without a preceding marketing effort to affirm the rate level. Such transactions differ from negotiated transactions in which the underwriters defer purchasing the issue until they have first attempted to market the issue.

**The City** The City of London, a square-mile municipal subdivision that contains the Tower of London, the original castle of William the Conqueror. For many years the City has been the financial center of the United Kingdom.

**Clearing banks** Large commercial banks in the United Kingdom.

**Commercial paper** Short-term promissory notes of corporations sold in the money markets. In the United States, the maturities cannot exceed 270 days without requiring registration of the notes with the SEC, unless exempted.

**Currency swap** The exchange with another party of future debt-payment obligations of interest and principal in one currency for that of another.

**Debt-for-debt swaps** Exchanges of outstanding foreign currency bank loans for newly created debt instruments with substantially different terms and investment characteristics.

**Debt-for-equity swaps** Exchanges of outstanding foreign-currency bank loans for local currency to be invested in equity ownership of local assets.

**Deregulation** The freeing-up and adding to the liquidity of markets as a result of the removal or reduction of regulations affecting financial transactions.

**ECUs** European currency units, a financial unit of account used by members of the EEC that consists of each country's currency indexed on a trade-weighted basis.

**Eurobonds** Bonds issued in the international-capital market denominated in one of several Eurocurrencies, or currency holdings outside the country of origin.

**Euro-commercial paper** Commercial paper originated and sold in the Eurocurrency markets.

**Euro-equities** Equity securities issued through the Eurobond distribution network and process.

**Foreign bonds** Bonds issued by foreigners in the national capital market of another country in accordance with rules and practices of that market.

**Foreign direct investment** Investments by foreigners in plants and equipment, real estate, substantial stakes in corporations, and joint ventures.

**Gilt-edged securities, or gilts**   United Kingdom government securities.

**Globalization**   The process of interlinking financial markets in different countries into a common, worldwide pool of funds to be accessed by borrowers and lenders alike.

**Gray market**   An electronic over-the-counter market in Eurosecurities made by intramarket brokers on a "when-issued" basis.

**Gross spread**   The net underwriting discount paid by the issuer. It consists of a management fee, an underwriting fee, and a sales commission. The gross spread on Eurobond offerings is substantially larger than that for U.S. offerings because of the need to offer two-tiered pricing.

**Innovation**   New ideas for new products or procedures in financial markets.

**Institutionalization**   The increasing domination of securities markets by financial institutions managed by professional money managers.

**Interest-rate swap**   The exchange with another party of future debt-interest-payment obligations in the same currency.

**Internationalization**   The process of adapting firms and institutions to global financial markets.

**Iron Lady**   Margaret Thatcher.

**Jobbers**   U.K. dealers in securities that functioned before Big Bang similarly to specialists on the NYSE.

**Junk bonds**   Bonds rated below BBB−, or investment grade, by rating agencies.

**Kamikaze offer**   An offer by a prospective underwriter for a bought deal at such a low yield to investors that other underwriters will decline to participate in it. An effort to buy market share through a loss leader.

**LBOs**   Leveraged buyouts, the acquisition of a company through the use of large amounts of debt, which is tied to the assets of the target company for repayment.

**Leveraged lease**   An arrangement through which a corporation can sell a property to a financial intermediary and lease it back. The transaction is financed by selling equity interests to investors who wish to own the tax benefits of the lease and by borrowing the remainder of the funds required from banks or insurance companies.

**LIBOR**   London interbank rate (offered side), the base lending rate between banks in the London Eurocurrency market.

**Loan-loss reserves**   Reserves maintained by commercial banks for possible future losses on loans in their portfolio.

**Mayday**   The day when the NYSE adopted negotiated commissions, May 1, 1975.

**Merchant banking**   (1) The activities of a U.K. merchant bank, including lending, securities activities, and money management; (2) the term applied to the investment-banking activities of certain U.S. commercial banks; and (3) large-scale corporate lending and investments in special situations generated by U.S. investment banks.

**Merchant banks**   Banks in the United Kingdom that descend from banks specializing in trade finance, and as such are known as ''accepting houses'' (for accepting and discounting trade bills for their customers). Today merchant banks provide corporate finance and various advisory services similarly to U.S. investment banks.

**NIFs**   Note issuance facilities provided by banks that guarantee a customer the ability to sell short-term notes in the Euromarkets at a minimum interest rate.

**Nonperforming loans**   Loans held by banks that regulators require to be so classified if interest payments are more than ninety days in arrears. Interest income may not be accrued by the banks on nonperforming loans.

**The Old Lady of Threadneedle Street**   The Bank of England.

**Perpetual FRNs**   Floating-rate securities with no fixed maturity date. Investors must rely on market liquidity for repayment of principal. Used by banks as a form of capital.

**Plain vanilla issue**   A bond issue with standard terms, no extra frills.

**Program trading**   Trading of securities that is governed by computer programs, for example, trading to provide customers with indexed investments.

**Recapitalization**   The process of replacing a company's capital structure with new debt-and-equity securities, usually involving a much greater concentration of debt.

**Reregulation**   The adoption of new regulations to govern financial markets or environments that have been stripped of old regulations.

**Rescheduling**   An agreement between banks and borrowers (often Third World countries) for the lengthening of the maturity of the repayment schedule of an outstanding loan.

**Rights offering**   The issuance of shares to existing shareholders at a discount to the market as a means to raise new equity capital.

**Rule 415**   A rule promulgated by the SEC in 1984 (test-marketed after 1982) that allowed companies to issue securities in the United States without delay caused by registration with the SEC; the rule permitted the bought deal to be adopted in the United States.

**SAMA**   The Saudi Arabian Monetary Authority, which functions as the central bank for the kingdom.

**Securitization**   The process of transferring assets from banks and other lending institutions to financial markets where the assets become freely traded.

**Single capacity**   The London Stock Exchange rule requiring members before Big Bang to act either as a broker or as a jobber (dealer), that is, in a single capacity, but not as both (dual capacity), which only became possible after Big Bang.

**Spread over Treasuries**  The way bond issuers and traders discuss the market in corporate bonds; the market is measured by the number of basis points over a comparable maturity U.S. Treasury benchmark security.

**Stripped**  A procedure in which the annual interest-payment obligations are separated from the repayment of principal on a bond and sold to different investors.

**Synthetic securities**  A package of securities, such as a Eurobond and a currency-swap agreement, that converts an original security into one with different currency or other characteristics.

*Tokkin*  A tax-advantaged investment fund in Japan for corporations to invest excess cash in marketable securities.

**Two-tiered pricing**  Eurobond pricing system in which a large gross spread is affixed to issues so they may be sold either to retail investors at a yield to investor that reflects the official offering price, or at a higher yield to institutional investors reflecting a discount from the official offering price. This practice is not allowed in the United States.

*Zaitech*  The practice in Japan through which corporations make investments of their excess cash, or funds raised through financings, in marketable securities of other companies strictly for the purpose of making a profit on the transaction, that is, without regard to traditional purposes of financing or for intercorporate relationships.

**Zero-coupon bond**  A highly volatile security that pays no interest until maturity, at which time all interest due over the life of the bond is payable along with the return of the original principal.

# Acknowledgments

Many people have helped me in the preparation, the writing, and the re-writing of this book. I am indebted to them all and, of course, completely exonerate them from any responsibility for errors of omission or commission that remain within the book.

I am especially grateful to those who helped me decide to undertake the project, an unusual one for a retiring investment banker newly taking up a demanding and unfamiliar career as an academic. Among those boosters of courage are numbered my wife, Marianne, my friend Ed Novotny, and my agent, Mike Cohn, to whom Ed introduced me.

Chief among those who helped shape the book, and give it what body and soul it has, was my publisher Truman Talley, who stepped down from the heights of famous and surefire authors to take a chance on me.

Among those who helped me with research, useful ideas, and helpful comments were Katherine Cray, Mary Elizabeth Poje, and Mary Anne Reilly of the Goldman Sachs research library, and Jill North of the firm's London Equity Capital Markets Group. Never did they tire of my unending requests for obscure information about times, places, events, and other matters that were not part of their regular world. Also I am sure that without the help of Carol Amsterdam and her several colleagues in the Goldman Sachs computer end-user department and my uncomplaining secretary, Mary Allen, I would have never gotten the thing done.

Finally, I want to express my thanks to those of my friends and colleagues who painstakingly reviewed the text and made valuable comments and suggestions. These include William H. Brown, Michael H. Coles, Charles Elliott, Henry H. Fowler, William C. Landreth, Peter J. R. Spira, David J. Watkins, and Professor Ingo Walter of New York University.

Others who helped without knowing it, particularly by sharing their stories and insights into what was going on during the last twenty years of international finance, were James Abegglen, Ernest Bloch, John Browne, Christopher Castleman, John Chiene, John Craven, Dennis Dammerman, David Davies, the late Otmar Emminger, Pieter Fisher, Robin Fox, Brian Garraway, Tatsuro Goto, Robert Hormats, Ken Kawashima, Henry Kissinger, Andrew Large, the late Stanley Miller, Yuko Oana, Shijuro Ogata, Tokuyuki Ono, "Dusty" Rhodes, Peter Sachs, the late Edward Schrader, Yuji Shirakawa, Alberto Togni, and John Whitehead.

Roy C. Smith

New York, March 1989

# Index

398